Holistic Analysis of Household Decision-Making

Adoption of Agricultural Technologies and Development in Ethiopia

Yohannes Kebede Gebre-Mariam

University Press of America,® Inc.
Lanham · Boulder · New York · Toronto · Plymouth, UK

4501 Forbes Boulevard
Suite 200
Lanham, Maryland 20706
UPA Acquisitions Department (301) 459-3366

Estover Road
Plymouth PL6 7PY
United Kingdom

Printed in the United States of America
British Library Cataloging in Publication Information Available

Library of Congress Control Number: 2011931311
ISBN: 978-0-7618-5641-2 (paperback : alk. paper)
eISBN: 978-0-7618-5642-9

™ The paper used in this publication meets the minimum requirements of American National Standard for Information Sciences—Permanence of Paper for Printed Library Materials, ANSI Z39.48-1992

Dedication

This book is dedicated to my late father, Ato Kebede Gebre-Mariam and late brother, Engineer Begashaw Kebede Gebre-Mariam.

Table of Contents

LIST OF FIGURES

LIST OF TABLES

LIST OF APPENDICES

Preface

Yohannes K.G. Mariam, Ph.D.
Falls Church, Virginia
August 9, 2010

Technology adoption has been seen as key to the development of more productive agriculture in lesser developed countries. However, the adoption and adaptation of new agricultural technology occurs at the level of farm families where decisions are made based on perceived risks and benefits of the new technology, and its fit within the knowledge and practices of existing agricultural system.

The processes and consequences of household decision-making are investigated using uni-disciplinary methods of decision analysis involving statistical, mathematical, psychological, anthropological and econometric techniques brought together in a holistic manner. The study is focused in the Ada and Selale regions of the Central Ethiopian Highlands. The technologies studied are fertilizer, improved crop varieties, pesticides and cross-bred cows.

Differences in goals and strategies of households are caused by inequalities in ownership of physical resources, and access to local institutions and indigenous knowledge. Village and regional institutions greatly influence access to physical (e.g. land) and non-physical (knowledge and information) resources. Successful intervention strategies are those that recognize region- and experience-specific potentials.

When compared with physical resources, non-physical resources exert greater influence on decisions to adopt technologies and on the efficiency with which inputs are used in the production of grain and milk outputs. Producers are willing to take risks in enterprises in which they have the advantages of favorable location or experience. The degree of risk-averse behavior of households reduces the probability of adopting technologies in both study sites. Natural factors (e.g. rainfall) and policy variables (e.g., land tenure and market) increase production risk. Indigenous production knowledge and schooling consistently reduce variability in production.

The research findings reported in this book contribute to the literature on agricultural households in two ways. Firstly, as far as I am aware, this study is the first of its kind in adopting a multidisciplinary quantitative and qualitative approach to investigate the processes and consequences of household decision-making. Furthermore, it is the first attempt to make use of the linear structural relations model (LISREL), previously only used in the fields of education, sociology and psychology, to examine agricultural decision-making.

Secondly, several studies have argued that adoption of technologies, regardless of their number, is essential to increase food production. This study has consistently found that, given the existing socio-economic environment, producers can attain greater increases in food production and the efficiency with which it is produced if they adopt at least two innovations. Moreover, previous household studies have emphasized the importance of physical resources to attain food self-sufficiency. In this study, detailed investigation of the processes and consequences of household decision-making has revealed that non-physical resources such as indigenous knowledge and institutions play a fundamental role on the ability of producers to achieve increases in food production. For development to be sustained, therefore, investments in physical resources should be matched with those in non-physical resources.

Acknowledgement

I would like to express my heartfelt thanks and appreciation to the late Professor Thomas Owen Eisemon, Director of the Centre for Applied Cognitive studies of McGill University, Professor John Galaty (chairman of my doctoral thesis committee), and Professor Harold Garth Coffin (co-supervisor of the doctoral thesis committee) without whom I would not have started the joint doctoral program with the departments of Anthropology and Agricultural Economics, McGill University. I am indebted to them for their contribution in designing the course work, preparing the field questionnaire, analyzing the data and writing the dissertation that led to the production of this book. I am very grateful to Professors Dan Aronson and Don Attwood for their critical comments and suggestions in writing the thesis.

Several institutions supported my course and research work. The Department of Agricultural Economics and the Faculty of Graduate Studies and Research of McGill University, the Max-Bell Open Fellowship (Canada), the International Development Research Centre (IDRC, Canada), and the Rockefeller Foundation (U.S.A.). The International Livestock Centre for Africa (ILCA, Ethiopia), The Finnish International Development Agency (Finland), and the Ministry of Agriculture in Addis Ababa and Selale (Ethiopia) cooperated in supplying information and providing technical support.

I would like to thank the following staff of ILCA: Dr. Ray Brokken, Michael Smalley, Mohammed Sallah and his staff, Awoke Negewo and his staff, Gashaw Geda, Getachew Baye and Shiferaw Hailu. Staff of the Selale Ministry of Agriculture staff and Regional office of All Ethiopian Farmers' Association cooperated in the design and implementation of the field research.

I would like to thank staff of the Department of Agricultural Economics who have bore with me since 1985. My appreciation also goes to fellow graduate students in the Departments of Agricultural Economics and Anthropology of McGill University. I thank my Ethiopian friends at McGill University for their companionship. I am very grateful to the editorial assistance of Ms. Janet Forrester and Sheenagh Pietrobruno, and Mrs. Janet Broxup.

This research would not have been possible without the cooperation and enthusiasm of the Ada and Selale farmers who provided me with the opportunity to complete my research with no direct benefit for themselves. I would also like to thank my research assistants, especially Mr. Belay Kebede, whose courage, determination and friendship was invaluable to the completion of the field research.

Last but not least, the deepest gratitude goes to my parents, brothers and sisters for their unreserved support and encouragement since 1968. In particular, I would like to thank Demere Hailemariam and my late brother Engineer Begashaw Kebede. My sincere appreciation and gratitude also goes to my wife, Mrs. Beza Belayineh and my two princesses, Phoebe and Priscilla Mariam. Most of all, I wish to offer my heartfelt thanks and respect to my mother, Weizero Atsede Belayineh, who has been the only source of inspiration throughout my academic career.

CHAPTER I

PRELUDE TO TECHNOLOGY ADOPTION AND DECISION-MAKING

1.1. Introduction

The purpose of this study is to examine decision-making processes of, and to evaluate the consequences of these decisions for smallholding peasant farmers of the Ethiopian highlands.[1] Crop and livestock production innovations are investigated as a case study. In this study, the analysis will concentrate mainly at the micro or individual farmer level. Nevertheless, macro issues that directly impinge on micro-economic processes of decision-making are briefly discussed. Before examining decision-making among households of the Ethiopian highlands, three major points ought to be considered. These relate to the concepts of decision, household and technology.

There is no such thing as a strictly individual decision. There is no decision which does not depend on information, knowledge, goals, norms, etc. derived from other individuals, enterprises, villages, or the society at large. Decisions are seldom made or acted on by isolated individuals, because any decision affects the morphology and activities of the household as a whole (Rubenstein, 1975; Wilk, et al., 1984; Arnould, et al., 1984).

Secondly, there are several definitions of what a household is (see Wilk, 1991). These definitions range from those pertaining to a farming family (e.g., Netting, et al., 1984; Wilk, et al., 1984, 1989; Schultz, 1964; and Ellis; 1988) to those that consider a country or the world as a household (see Hopkins, et al., 1982; Smith, 1983; 1984; Wallerstein, 1974, 1979). Common features to most definitions of a household are that its members cooperate for its survival, it functions under a single roof or compound, and it may expand and contract in size overtime.[2] Given the level of this study, therefore, it is sufficient to provide a working definition of what a household represents. This study adopts the following working definition of a farm household:

> "the food-producing, consuming and reproducing unit, including any member of the family living in the same residential or sleeping unit, and participating, to a varying degree, in various activities directed towards securing subsistence and more" (Kusterer, 1989).[3]

The third issue relates to the concept of technology and innovation. Technology can be seen as an idea or a form of organization and an object to be used for productive purposes (Sperling, 1987). It can be a means to an end or an end in itself. Once it is accepted by a society, it becomes part of the totality of culture. Culture can be understood as a way of thinking and believing, an abstraction from behaviour, a store-house of pooled learning, a set of standardized orientations to recurrent problems and a set of techniques for adjusting both to the external environment and to other men (Geertz, 1973:5-6). Thus, all methods of production can be referred to as technologies. The extent to which a farmer or decision-maker has experimented with a method of production distinguishes new and old technology. This study argues for culture-specific definition of technology, because it is given meaning by culture and its adoption depends on its ability to produce changes for those who gave it a meaning.

The culture of technology can be seen as a framework through which new ideas, materials and information interact, are given meaning and are acted upon. Producers decide to adopt new technologies to the extent that: i) their economic advantages out-weigh their expenditures, ii) they do not as a result acquire lower social status, and more importantly, iii) they can comprehend and give meaning to new technologies within the totality of their social, economic and cultural setting.

Securing food is the primary goal of village households. Food production is made possible through methods of production or technologies. Technology is part of everyday life. If technologies are thought to be means to solve food production problems, and if they are part of everyday life, then examination of decision-making with respect to technologies should help reveal the processes that go into production. Accordingly, it should help obtain information for the design of new policies.

This introductory section argues for the relativity of definitions of the three key concepts. Specifically, decisions made by individuals are influenced by several factors outside their control. Since it is impossible to account for the influence of all factors, we try to provide an approximation. Secondly, it is impossible to have a universal definition of household for comparative studies. Therefore, a household as a unit of analysis should reflect the societal definition. Thirdly, not every new method or idea is technology. It is only when an idea or object, given meaning by those who make use of it, results in changes in their form of production and standard of living that it will be considered as a technology. Before stating the problem that instigated this study, a definition of decision-making and its variants is provided in the next section.

1.2. What is Decision-making?

Decision-making is a ubiquitous activity inherent in the behaviour of individuals, organizations and society. It has been described as a characteristic of "purposeful" systems (Dunn, 1984). A corollary to this statement is that the extent and importance of decision-making activities within a system is dependent upon the motivation with which it seeks to attain objectives or goals.

Decisions can be categorized as intuitive, programmed and analytical (Dunn, 1984). Intuitive decisions are those choices which individuals make without conscious thought as to the alternatives and their relative evaluations, either innately or through habit and experience. Under some situations we know what to do and in other situations we may not know what to do, but have a set of strict guidelines to follow. This would be a programmed decision, which, in principle, should be capable of being automated. Finally, there are those important decisions about which we have to think carefully. We must analyze the possible outcomes and their consequences. A "decision problem" exists when an individual has alternative choices, each with significant consequences, and he or she is unsure about which choice is the best. If the consequences of a decision are uncertain, then we have a risky choice (Dunn, 1984).

A decision problem consists of i) actions available to the decision-maker, ii) states of "nature" (e.g., rainfall during rainy season, prices after harvest, social networks in peak labour demand seasons), iii) probabilities attached to the states of nature affecting the decision problem, iv) consequences of an action, v) the criteria for choosing between the acts (e.g., maximizing income), vi) the process of conducting experiments to obtain additional information about the likelihood of outcomes of an event occurring, given a state of nature, and vii) the strategies for action which are conditional on the experimental outcome observed (Dunn, 1984; Anderson, et al., 1977).

The distinction between producing crops or livestock either for reproduction of the farm and household or for the market is key to the study of peasant households. Often these decisions are not in the hands of a single decision-maker or household. Households make various kinds of production, consumption and reproduction decisions. The livelihood of a household is dependent on actions carried out by its members at a given time and place. It is useful to investigate how decisions made by a particular household are influenced by factors in the decision-making environment, such as inter-household decision-making, markets, patterns of social ties and government policies.

A major cause of failure in development strategy has been the inability of planners to obtain evidence about how decisions are made and what kinds of decisions are important to households under different circumstances or in different socio-economic environments. This study attempts to fill part of this gap through analysis of information processed, the decision-making environment, production knowledge possessed and the goals of decision-making. Moreover, it will try to quantify consequences of these decisions.

1.3. Problem Statement

Households make decisions to achieve various goals including production, consumption, and reproduction, among others.[4] Various micro and macro variables influence household decision-making and the consequences thereof (Figure 1a). The figure assumes that the macro-structures that reflect aggregate

Figure 1a. Micro-Macro Structures & Household Decision-Making: A General Framework

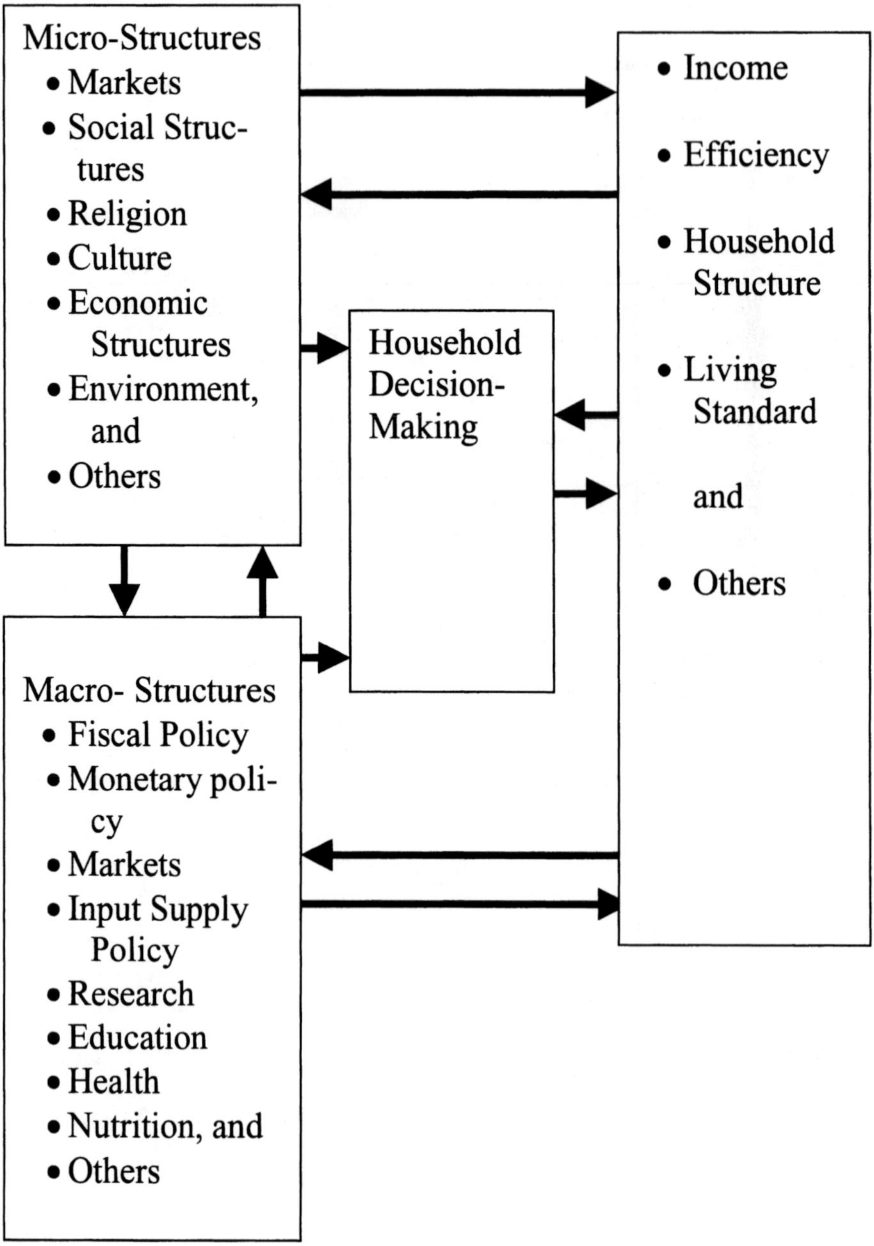

measures can not exist without the base or micro-structures. For most variables, it is difficult to conclude whether their impact is significantly larger at the micro or macro levels. Furthermore, it requires substantial empirical study to justify whether or not some factors or variables originate at the micro or macro level.

Notwithstanding these difficulties, the criteria used to classify a variable should be derived from where each aspect originates or where its greatest impact would be. For instance, in a country such as Ethiopia there are several ethnic groups. Each ethnic group has its own distinct culture. But Ethiopia also offers a larger world of culture. Nevertheless, it is plausible to argue that the impact of a specific culture is greater on the function and structure of households at the village or regional level than at the national level. Access to secular education and health care depends on government expenditure and its effect is often felt or evaluated at the national level. These micro and macro structures influence household decision-making which in turn influences the design of policies and the operation of micro-structure. The outcome of household decision-making can be seen on household income, production efficiency, household structures, living standards and so on. Examination of household decision-making based on this complex model is difficult.

A simple yet realistic framework for the conceptual definition of the problem investigated in the succeeding chapters is given in Figure 1b. The conceptual framework is broadly divided into two parts. The first part deals with variables shaping the processes that lead to actual decision-making. The second part examines the quantitative outcome of decisions. Previous studies of decision-making at the household level have concentrated either on the first (e.g., Webb, 1988), or the second part (e.g., Ellis, 1988). The factors involved in the linkages between the two parts have not been examined in detail.

The conceptual model indicates that there are three types of variables. The first set includes exogenous variables such as crop and grazing area, education, experience, labour, oxen, feed, number of cows, seed and wealth. The second set includes endogenous variables such as the number of technologies adopted and production knowledge.[5] The third type of variable, an outcome of the first and the second sets of variables, measures production efficiency. Models that incorporate these three types of variables to study household decision-making in agriculture are few. This is an area in which this study will potentially contribute to the literature.

Food production is possible through access to and appropriate combinations of physical and non-physical resources. Several studies have argued that decision-making and consequences of decisions are influenced by physical resources and prices (Ellis, 1988; Lipton, 1968; Stevens, et al., 1988; Hayami, et al., 1985). These studies have not given due consideration to indigenous knowledge and institutions as resources affecting the processes and consequences of production decisions (e.g., Schultz, 1975; Jamison and Lau, 1982). The result of the present study indicates that, when land and labour markets are fixed, indigenous knowledge plays a significant role in shaping the process of decision-making and the resulting efficiency of production.[6] This finding is examined throughout

Figure 1b. Socio-economic Variables and Production Efficiency: A Testable Model

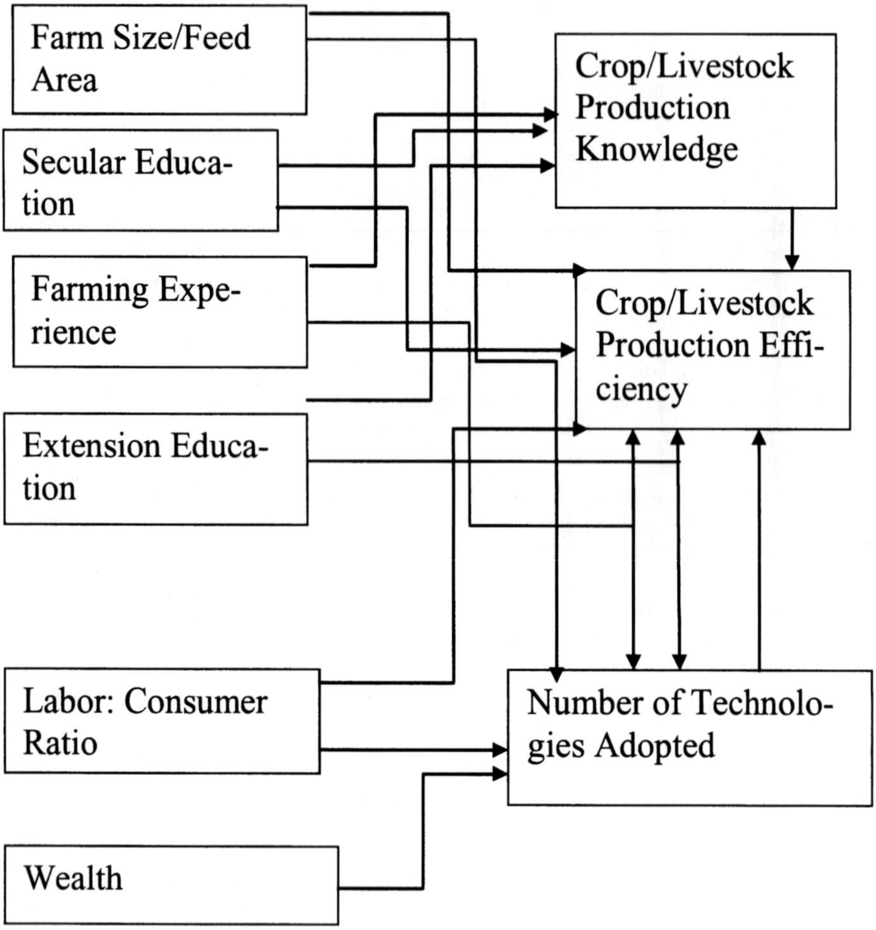

LDCs have not given due consideration to indigenous knowledge as a resource. Similarly, decision-making processes have not been investigated in-depth. However, variables that influence indigenous knowledge, such as schooling, have been the focus of several studies (e.g., Schultz, 1975; Jamison and Lau, 1982).

There are several realms in which households make decisions. These include crop choice, labour allocation, marketing, consumption, education, reproduction of producers, child care, and health. The most important requirement is that people have to produce to survive before making other kinds of decisions.[7] Thus, we would expect that food production decisions would take precedence over other kinds of decisions. This study investigates production decisions of agricultural households in Ethiopia.

To enhance the overall development of a country, agriculture should ideally provide enough food for the population. Inputs necessary for increased food production are scarce. In answer to the question of why Ethiopia is poor, Ricket (1991) argued that:

> "The tempting answer is the dominance of traditional agriculture in total economic activity. To accept this, however, is to confuse effect with cause. Ethiopia is not poor because of its traditional agriculture; but its agriculture is traditional because the country is poor" (Ricket, 1991:13).

Agriculture is the sector where development efforts should concentrate on ensuring basic needs, i.e., food. The issue that needs to be addressed is what methods of increasing production can cope with increases in population. Strategies to increase agricultural production include expansion of arable land, increasing productivity of resources, such as land or labour, or both (Saith, 1985; Ghose, 1985; Teka, et al., 1984). Technological innovation may offer a method of increasing production per unit of scarce resource, as a viable agricultural development strategy. To make technology useful for the people who are using it, it is essential to know what constitutes technology and how it should be generated and diffused. Furthermore, the manner in which a technology is designed and provided to smallholders, and the socio-cultural nexus within which technology is introduced, should be examined.

Several strategies have been introduced in peasant agriculture to enable increases in food production. The technology transfer paradigm was dominant until the 70's, while the induced innovations and farming systems approaches were used in the 80's and 90's. Recently, development strategies have changed their focus to the paradigm of sustainability. However, these strategies have not considered the ability of producers to comprehend new ideas, given their experience, the information they have gathered, their socio-economic resources and the structure of their decision-making, prior to the introduction of new technologies.

Not only are key elements of decision-making processes ignored, but also what producers should do is often pre-defined by policy makers. Also, intervention strategies often involve more than one innovation. Studies of the adoption of new agricultural technologies, however, have tended to concentrate on single technologies. But in reality, households are confronted with several technological options, both new and traditional. They may adopt one or more technologies simultaneously, a single component of a technology, or none of them (Feder, et al., 1982). The key to understanding household decision-making and to design policy is to examine how producers arrive at the choice of technologies and how the consequences of the adoption of these technologies compare with existing practices.

The introduction of new production technologies must be evaluated with respect to: i) their acceptability to smallholders (i.e., compatibility with existing social, cultural and economic characteristics of households); ii) their consequences for production risk and household risk-taking behaviour; and iii) the kind of support services needed to accompany the new technologies. The crucial roles of indigenous knowledge and risk management strategies in technology adoption have been given little emphasis in studies of small farming systems (Warren, et al., 1988). It is, therefore, important to find out what kinds of knowledge and information are used in making decisions regarding production and technology adoption.

Producers in the Ethiopian highlands are not isolated from the rest of society. As in many other developing countries, they are affected by government policies, international markets, social networks and other households. Crop and livestock enterprises are highly inter-dependent. Intervention in one enterprise will inevitably affect another. An in-depth study of all production technologies would require more time and resources than were available. Thus, four key production technologies were studied in this research.

The experience of many LDCs with respect to introduction of technologies is dismal. Some have enjoyed success and others have not. Even success has not been without cost. A typical example of a "success" story in Africa is that of the Swedish International Development Agency's (SIDA) integrated rural development project in the Arssi region of Ethiopia. However, this project resulted in eviction of smallholders and greater income disparity (Cohen, et al., 1988; CADU reports, 1971, 1972).

In Ethiopia, crop and livestock production technologies have been introduced to smallholder mixed farmers by the Ministry of Agriculture, external donor agencies and international research centres such as ILCA.[8] The most important technologies include improved crop varieties, fertilizers, pesticides, cross-bred cows and various kinds of soil and water conservation techniques. These technologies compete for the limited physical and financial resources of smallholders.

A number of studies have investigated the importance of pesticides, fertilizer and improved crop cultivars (e.g., Ayana, 1985; Waktola, 1980; Feder, et al., 1985). Very few studies have been made regarding adoption of cross-bred cows (ILCA, 1985).

Studies of technology adoption in Ethiopia and many developing countries have treated crop or livestock farming separately even in situations where one can not exist without the other. They have not examined the processes that lead to decisions regarding adoption of a combination of innovations, and have treated decisions as if they were made by a single member of the household. They lacked detailed investigation about the impact of inter- and intra-household decision-making, and the impact that markets and government policies may have on patterns of adoption. Moreover, these studies have not related the adoption of technologies to the impacts they exert on the quantity and efficiency of production. In order for a "compromise policy" to be able to facilitate sustainability, these and related issues should be examined.

1.4. Objectives and Hypothesis

1.4.1. Introduction

Any technology introduced on one plot or to a single household may influence farming practices or allocation patterns on other plots and households. Moreover, the management strategies of smallholders may vary from farmer to farmer. To gain an understanding of the pattern of selection of technologies and their corresponding impact, farms managed by different groups of farmers should be compared rather than farms under similar management. That is, farmers who have used new technologies should be compared with farmers who have not. In the research sites, the existence of other production technologies, in addition to those introduced by the International Livestock Centre for Africa (ILCA) and the Finnish International Development Agency (Finnida), implies that farmers may have adopted one, two, three, four or none of these innovations. However, little is known about how smallholders have arrived at these decisions and about their consequences. This is the underlying issue investigated in this study.

1.4.2. Objectives of the Study

The objectives of this study are: 1) To identify the goals/objectives of households and to examine the structural relationship between household strategies and goals, with the aim of identifying which strategies play a major role in the attainment of which goals; 2) To examine the kinds and sources of information used by farmers, and to quantify indigenous production and marketing knowledge; 3) To analyze decision-making regarding household activities and to uncover factors influencing the nature of household decision-making; 4) To examine adoption of cross-bred cows, fertilizer, pesticides and improved seed technologies; 5) To assess the risk-taking behaviour of households in terms of the outcomes of decisions already made rather than by using idealized utility functions; 6) To evaluate the production efficiency and risk of adopting a single technology (e.g., cross-bred friesian/boran dairy cows) as opposed to a combination of crop and livestock production technologies; and 7) to evaluate the validity or adequacy of the conceptual model previously presented (Fig. 1b).

1.4.3. Hypotheses

The analysis of the data collected from this research will be informed by both testable and non-testable hypotheses. In addition to assessing statistically testable hypotheses, the logical explanation for patterns or behaviour of households will be pursued. The testable assertions include the following:

1. Strategies that recognize the comparative advantage of each study site will explain most of the variation in household goals. For instance, livestock feed or purchases of cross-bred cows are anticipated to be of greater significance in sites more involved in animal husbandry.
2. Farmers' resource allocation strategies and adjustment mechanisms are hypothesized to follow either: i) safety-first/ disaster avoidance, or ii) risk minimization strategies as distinguished by techniques of crop and livestock production that are designed to minimize the adverse effects of inadequate rainfall, incidence of pests and lack of traction animals.
3. a) It is hypothesized that households with a high worker:consumer ratio will be willing to devote a larger proportion of their labour resources to relatively labour intensive innovations (e.g., dairy cows) than households with a low worker:consumer ratio. b) Households with relatively more experience will devote their resources to innovations with more certain outcomes than to those with high returns.
4. Indigenous production knowledge gained by farmers through many years of experience will significantly influence production and adoption decisions relative to knowledge gained from contacts with extension agents, development agencies or other sources.
5. a) Not only economic but also social, cultural and political factors will significantly influence decisions regarding adoption or non-adoption of a single or a combination of technologies. b) The existence of modernization conditions such as access to technologies and markets will play a significant role in adoption decisions.
6. Whether or not households possess significant wealth or resources and gain adequate off-farm income, they are hypothesized to be moderately to highly risk-averse. Farmers who adopt one or more technologies are hypothesized to be less risk-averse than those who do not. This behaviour of households will exert a negative and significant influence on the adoption of innovations. Moreover, the degree of risk-averseness is hypothesized to decline with increases in the number of technologies adopted.
7. Farmers who adopt improved production technologies (e.g., cross-bred cows, fertilizer) are hypothesized to exhibit a higher level of technical efficiency compared to those who do not adopt these technologies. Smallholders who selectively adopt combinations of innovations may show superior efficiency compared to those who adopt a single technology.

8. New production technologies are hypothesized to increase production risk or yield variability compared to locally available or "traditional" production techniques or inputs.

Finally,

9. The conceptual model presented is hypothesized to provide an accurate representation of household behaviour.

To reject or support these hypotheses, various statistical models are used. Accomplishing the objective of this study also requires qualitative analysis of household objectives, goals, household decision-making and the environment in which it occurs.

1.5. Significance of the Research

The goals of this study are to examine the processes, patterns and consequences of household decision-making through analysis of information, knowledge, and the decision-making environment. The study evaluates the consequences of household decision-making for adoption of innovations and for resulting levels of income, production efficiency and risk. To achieve these goals, the study focuses on decisions regarding adoption of cross-bred dairy cows, fertilizer, pesticides and improved crop varieties in two sites in the Central Ethiopian highlands. The study is based on data collected from interviews.

Results of this research should provide valuable guidance to policy makers in designing methods for the generation and introduction of new production technologies for smallholder mixed farmers under dryland agriculture. The research provides evidence on what constitutes household decision-making and its environment, which should be considered before designing development strategies. Policy makers can design ways of mitigating the adverse effects of social, cultural, technical and economic factors on production decisions and on the efficiency with which resources are used. Intervention strategies that rely on these kinds of information from the target population can produce sustainable growth better than those dependent on top-down prescriptions.

1.6. Organization of the Book

The thesis is divided into ten chapters. The next chapter investigates strategies of agricultural development followed by the government of Ethiopia and how these strategies relate to core paradigms of development. It also examines the role that technology played in these strategies. Chapter three presents the nature of households, in general and with specific reference to the peasants in the study sites. Chapter three examines the socio-economic nexus within which technologies are introduced and provides the framework for explanation of behaviour or outcomes in the study. Chapter four discusses problems associated with the use of uni-disciplinary methods of decision analysis and proposes an interdisciplinary approach in the choice of methods to evaluate the decision-making processes and their consequences.

Chapter five examines the relationship between goals and strategies, access to information, and differences in production knowledge among farmers in and between regions. Chapter six analyses the environment within which household decision are made and what effects they may have on decision-making are investigated using selected parameters. Factors influencing adoption of technologies and risk-taking behaviour are discussed in Chapter seven. Chapter eight presents the outcomes of traditional inputs and selective adoption of combinations of innovations on production efficiency. The conceptual model will be tested in Chapter nine. Chapter ten presents conclusions and recommendations.

1.7. Notes

1. The highlands are defined as areas higher than 1500 meters above sea level and with a mean daily temperature of less than 20^0C during the growing period (Getahun, 1978).
2. In fact, so-called modernization is not a transition from one type or form of household to another, but is basically a change in the activity spheres that underlie the household's form (for further explanation on this issue see Wilk, et al., 1984).
3. In the Central Ethiopian Highlands, *a compound* defines a household. A compound may include more than one houses. Additional houses are built by parents of the husband or wife, or adult male members of the family.
4. Caution should be exercised against taking a dogmatic definition of goals because, for some, goals can represent means or strategies, and for others they can be ends or aims.
5. In this context, exogenous variables refer to factors that, given the structure of the model, determine the value or magnitude of another (output, decision or endogenous) variable. It must be noted, however, that in this simple conceptual framework, structural relationships between exogenous variables, and bi-directional relationships between exogenous and endogenous variables are excluded. This is primarily because of the requirements of the statistical analysis and to ease the interpretation of relationships between variables. An elaborate framework is provided in Appendix 1.
6. This condition also holds true if the effect of other variables, such as that of policy, is held constant.
7. Even in the U.S. agriculture, Harper and Eastman (1980) found that operators of production units view agricultural activities as meeting personal, non-monetary needs first and income or profit second.
8. ILCA stands for "International Livestock Research Center". It is an International Agricultural Research Centre, under the Consultative Group on International Agricultural Research (CGIAR), located in Ethiopia.

CHAPTER II

TECHNOLOGY AND AGRARIAN DEVELOPMENT IN ETHIOPIA: THE HISTORICAL ORIGIN OF THE PEASANT PREDICAMENT

2.1. Introduction

Over the past five decades, several paradigms of development were implemented by developing countries and Ethiopia is no exception. Common to all paradigms is that unbalanced emphasis is placed on either agriculture or industry. Few studies proposed the need to correct structural deficiencies before embarking on a development strategy (Streeten, 1981; Meier, et al., 1984; Adleman, et al., 1973). Some studies have argued that regardless of the strategy of development followed, a country, which cannot feed its people, will pay the costs of underdevelopment (Gibbon, et al., 1992). That is, poverty, malnutrition and disease will affect the majority of the population. Thus, non-agricultural policies should be designed in such a way that peasants would not be discriminated against.

In examining the place of agriculture in the development process, it is necessary to incorporate the temporal and spatial dimensions through which a country evolved. The political history of a country can reveal problems which are bypassed by strategies emphasizing spatial features. For instance,

> "... The essential fact is that Ethiopia has come late to modern economic growth. History and geography combined to keep the country largely isolated until relatively late. Cut off from the mainstream of technical progress and lacking the stimulation that comes from international commerce, the economy could not progress more quickly than simple techniques and nature allows...." (Ricket, 1991, pp. 13).

Ethiopian agriculture and the peasantry who till the land can be seen in the context of two successive political systems: until 1974 a "feudal" system; between 1974 and 1991 a socialist one.[1] Since May 1991, however, Ethiopia has adopted a mixed economy with varying degree of changes in agriculture, industry and the service sectors.

This chapter discusses the state of development theory, the importance of technology as a means to increase food production, and strategies followed in

Ethiopia. It is hoped that this chapter will identify major obstacles to development and will highlight the relevance of this research to the design of future development strategies in Ethiopia. Furthermore, this chapter will demonstrate that food production problems of the Ethiopian highlands can only be tackled through the introduction of *technologies that are meaningful* to peasants.

2.2. Theories of Development

Development strategies followed in Ethiopia can be categorized as capitalist, socialist, or both. These strategies include new technologies as their major components. This section reviews selected paradigms of development followed by analysis of strategies implemented in Ethiopia.

Theories of development can broadly be divided into Marxist and non-Marxist. The major non-Marxist theories of development include: Laissez-faire or free market (Smith, 1776; Solow, 1957; Todaro, 1984; Stevens, et al., 1988; Meier, 1976); the population trap theory (Malthus, 1933); the demographic transition model (Schultz, 1981; Boserup, 1965, 1981); the conservation model of agricultural development (see in Stevens, et al. 1988); the theory of comparative advantage (Schultz, 1964, 1981); agri-led strategy (Myint, 1975; Lewis, 1984; Fei and Ranis, 1964); import-substitution industrialization (Nixson, 1982; Hirschman, 1968, 1984); location theories (Schultz, 1953); stage theories of growth (Rostow, 1960); the high pay-off input model (technology embodied in capital) (Schultz, 1953, 1964, 1981); and export of manufactured goods (Wilber, 1984; Prebish, 1964; Streeten, 1982).

Marxists theories variously stress the inevitable transformation of socio-economic structures, the role of control over the reproduction of producers in shaping paths of development, and the need to examine the world as a single entity (world systems theory) (Marx, 1970; A.G. Frank, 1967; Laclau, 1971; Wilber, 1984; Amin 1974, 1976; Meillassoux, 1972, 1981; E. Wallerstein, 1974, 1979). Both Marxist and non-Marxist approaches are characterized by economic determinism, give limited attention to changes that evolve from within the system (i.e., induced changes) and ignore the role of socio-cultural components of a society in the design of development strategies. At large, considerations of households enter in a passive way and are assumed to operate in a "moral economy" or "economy of affection" (Scott, 1975; Hyden, 1980, 1983, 1986).

Modernization theorists, on the other hand, stress the crucial role of non-economic elements of a system as causes of development or underdevelopment. These theories focus on: the dominance of beliefs and values (e.g., Foster, 1962; Long, 1977); psychological and motivational variables (Weber, 1956; McClelland, 1961; Hagen, 1962; Inkeles and Smith, 1974); political variables (Uphoff and Ilchman, 1972); social structure variables (Kunkel, 1970; Parsons and Shils, 1951) and social learning (Kunkel, 1970; Dunn, 1971). Moreover, these theories identify values, institutions and *technology* as the fundamental variables affecting development (Elder, 1968; Myrdal, 1968; Hoselitz, 1960;

Nair, 1962).

Paradigms that combine some aspects of the above approaches have not gained a lot of success. For instance, Polanyi (1954, 1958), Parkin (1972) and Geertz (1963) held that the development of any society cannot occur without simultaneous changes in economic, social, cultural and institutional components. However, they over-valued the role of resources at a micro level and gave less emphasis to macro-structures - that is to government policies which may influence micro-level development efforts.

All theories of development may be right in their own terms but are not complete. Morphologically, Calkins (1989) summarized the [past] status of development theory as:

> "Indeed, it is as if the theorists and practitioners of economic development were actors in the fable of the three blind men and the elephant. The first, who touched the trunk, thought the elephant was like a tree. The second, who touched the flank, thought it was like a wall. And the third, who touched the tail, thought it like a rope. Each was correct in describing one part of the anatomy, but failed in capturing the organic composition of the whole. The men would have been better off if they had projected what an elephant might look like from such essential postulates as its size, its strength, and its ability to uproot trees" (Calkins, 1989:4).

Marxist, non-Marxist, modernization, or other development theories envisaged to date understand technologies or techniques as the principal methods of stimulating accelerated growth or development. Most of them have failed to live up to their promises. What went wrong is an issue that cannot be addressed fully in this thesis. Of particular relevance are the questions: Is new technology required by those for whom development is intended? Are the conditions necessary for the generation and utilization of innovations "right"? The next section provides a synopsis of this issue.

2.3. Technology and Agricultural Development

2.3.1. Definition of Technology

Sperling (1987) describes two aspects of technology. One aspect is the knowledge and skill required to construct a material world. This aspect of technology relates to applied science where know-how and information flow are essential to progress. The other aspect connects the skills and knowledge that comprise technology with the material aids they produce (Sperling, pp.173).[2]

Progress is equated with technical change and strategies to technological intervention. Technology is not a concept reflecting the material aid only, but most importantly, it is composed of ideas about how the state(s) of the world can be transformed and the practices that do so. Thus, technology is a social and cultural concept.

The concept of technology, both as applied knowledge and its material aids, may be less revealing when applied to less developed nations. Because of the absence of changes in the physical or material world among peasants of less developed countries (LDCs), variations and complexities in know-how and in how it is used are by-passed. In fact, the complex nature of the environment in which technologies are applied requires realistic methods of analysis to reveal what goes on in these societies. Technology not only is the product of social experience but carries with it such aspects as identity, symbolic representation, aesthetic values and division of labour. Whether labelled as traditional or static, it is through indigenous techniques or technologies that households and societies at large perpetuate themselves and the system. Therefore, it is possible to study societies through investigation of technologies used in the system and to analyze their interactions, functions and contributions to the eco-system in time and space (for further discussions see Firth, 1939; Larick, 1986; Lechtman, 1984).

2.3.2. Technology and Development

Several agricultural development strategies were designed because of the belief that traditional methods of production could not stimulate growth. Changes were expected via the introduction of new technologies. Governments of LDCs were urged to invest more in research to generate new technologies. However, patterns of research and extension expenditures in developed and developing countries were moving in opposite directions.

Agricultural research expenditure as a proportion of the value of agricultural production in the developed world between the 1950s and the 1980s was more than 1.2 percent. The intensity of extension (scientists' man-days per U.S. $ 10 million agricultural product) for the same period was less than 0.5. The respective figures for research expenditure and intensity of extension in LDCs were less than one percent and more than 0.6 percent (Ahmed and Ruttan, 1988).[3] The premise behind this pattern of expenditure was that technology generated in the developed countries could be transferred to LDCs. In other words, the LDCs should simply strengthen the extension effort to accommodate the increased inflow of research results from the developed nations.

As a result of the assumption of transferability of technologies, government and development experts concentrated on the mechanisms through which technologies could be transferred or diffused. Rogers (1971) proposed the diffusion model. This model assumes that reasonable increases in agricultural production may be obtained by devoting considerable resources to: a) increasing the flow of information to farmers about new agricultural technology and new institutional arrangements such as credit, and b) teaching tradition-bound farmers how to make more economically rational management decisions about the use of resources they possess (Molnar, et al., 1983).

Despite numerous studies dealing with technology and development, projects that include technology as their major component have not helped to change the material world of LDCs. It is inappropriate to argue that the know-how assocated with new technology cannot be understood by smallholders of the LDCs, for the simple reason that this knowledge may be devoid of social meaning. That is, they may understand why a technology is introduced but they also realize its consequences. If in their opinion the economic and social consequences of adopting a technology do not warrant its use, they will not adopt it. The know-how and material aid that constitutes a technology will not be integrated into the body of social experience (Norman, et al., 1981; Warren, et al., 1988).

Alternatives to transfer paradigms were those concerning innovations induced from within the system. This approach also focused on factor endowments (land, labour, capital) and factor prices. However, the social aspect of induced innovations was not treated as equally important with the economic aspect (Ruttan, 1978; Binswanger, et al., 1978, 1984).

Past efforts at technological intervention have not provided universally positive evidence about their effect on development. Techniques or technologies which can expand the body of knowledge and thereby change the material world are required to minimize problems associated with the scarcity of resources. It is, nevertheless, important to understand technology in terms of who, why, how and what, both at micro and macro level.

We have seen that most development strategies have used technology as a driving force, yet there is no universally accepted definition of development or technology. It is only if development strategies involve the people for whom they are intended in the process of formulation, planning and implementation of projects that sustainable growth can be achieved.

The following section discusses issues related to Ethiopia's resource endowment, potentials, constraints and strategies of development followed in the pre-and post-1975 period. Detailed examination of their consequences will reveal factors that should be addressed in designing development strategies. It is in this way that this study might be useful for policy makers.

2.4. Pre and Post-1975 Agrarian Development in Ethiopia

2.4.1. Background

Ethiopia is located between 3^0 to 18^0 latitude north and from 33^0 to 48^0 longitudes east.[4] It is bordered by the Sudan in the west, north and southwest, by Kenya and Somalia in the south and southeast, and by Somalia, Djibouti and the Red Sea in the east.

The country is inhabited by 50.8 million people, of which 87.63% are rural and 12.47% are urban.[5] Of the total population, 50.26% is male and 49.74% female. The annual growth rate of the population is 3% (CSO, 1989). The land area covers 1.22 million square kilometres, of which 40% is categorized as highland (i.e., lies between 1500-4000 metres above sea level). The average population density ranges between 8 inhabitants/km^2 in the lowlands to 74 inhabitants/km^2 in the highlands (Tagoe, 1983).

Agriculture accounts for nearly 50% of the gross domestic product (GDP), 90% of the export earnings and 80% of rural and urban employment. Sixty-five percent of the total land surface is agricultural, of which nearly 5.7 million hectares is under crop production. Crops account for 80% of the gross value of agricultural production, livestock and livestock products for 20%.

With respect to land use, perennial and annual crops, forests and pasture account for 15, 4 and 51 percent respectively. The cropping pattern is 87% cereals, 11% pulses and 2% other crops (MOA, 1990). The average farm size is less than 2 hectares (ha) per household. Less than 0.5% of the total agricultural land is irrigated, with potentially irrigable land of about 2.2 million hectares (Teka, et al., 1984; FAO, 1984). The annual rainfall varies from 400 mm per annum in the semi-arid areas to 2700 mm per annum in the southwest.

2.4.2. Agricultural Systems in Ethiopia

Despite its location in the tropics, the Ethiopian climate is conditioned more by altitude than by latitude. This accounts for variation in the type of crops grown, population distribution and the nature of economic activities. Therefore, it is important to know the kinds of agricultural systems as classified by altitude.

The two dominant agricultural systems in Ethiopia are the mixed farming of the highlands and pastoralism in the lowlands. The mixed agriculture exhibits several sub-systems (see Getahun, 1978; Sisay, 1980). Mixed farming involves 88% of the total population and 95% of cropped lands.

Traditionally there are three altitude-physiographic zones: the "Kolla" (tropical) below 1500 meters above sea level (mts a.s.l.), the "woina-dega" (subtropical highland) between 1,500 and 2,500 mts a.s.l. and the "dega" (highlands) above 2500 mts a.s.l. If an area is located at an altitude of less than 800 mts a.s.l., it is called "bereha" (desert). The combination of altitude, temperature and rainfall determines the range limits of vegetation.

Getahun (1978) classified the Ethiopian highlands into three major agroecological zones according to their resource base and development potential (Table 2.1.). These include the high-potential cereal/livestock zone (HPCL), the low-potential cereal/livestock zone (LPCL) and the high-potential horticulture/livestock zone (HPHL). For detailed discussions about the features of these agricultural systems see Getahun (1978, 1980) and Sisay (1980).

Table 2.1. Major Highland Agricultural Zones and their Characteristics

Categories	LPCL[1]	HPCL[2]	HPHL[3]
Area(km^2)	134,000	149,000	139500
Population density (persons/km^2)	72	73	74
Proportion of total human highland population %	35	32	33
Livestock density (TLU/km^2)	30	55	27
Proportion of total highlands livestock population (%)	26	41	33
Ratio of livestock/human population	0.42	0.75	0.36
Climate	sub-humid/ semi-arid	sub-humid/ humid	humid
Topography	mountainous/ escarpment	rolling pla-teau	dissected plateau
Growing period(days	90 to 150	150 to 240	>240

Source: Gryseels, et al., 1988.

LPCL[1]- refers to Low potential cereal livestock zone; HPCL[2]-refers High potential cereal livestock zone; HPHL[3]- high potential horticulture-livestock zone; TLU stands for tropical livestock unit. 1 TLU= 250 kg live weight.

2.4.3. Agrarian Development Strategies in Pre-1974 Period

Ethiopia was one of the first countries in Africa to formulate a national economic plan. The country launched three five-year plans during the period 1957-1973. In the first two five-year plans, investments to increase production in the peasant sector were less than five percent of the total government budget (Habtu, 1988; Sisay, et al., 1978). Import substitution industrialization or industry-led growth, export of primary commodities and large scale commercial farming were given special treatment during this period (Habtu, 1988).

In 1968, a major policy change gave priority to the small-scale farmers (Getahun, 1978; Sisay, et al., 1978). The percentage share of productive investment in the peasant sector increased from 5% to 11%. Subsequently, a number of agricultural projects such as Chilalo Agricultural Development Unit (CADU), Wolaita Agricultural Development Unit (WADU), Extension and Project Implementation Department (EPID) and others were implemented in the peasant sector. Their objectives were to: i) supply inputs, mainly fertilizer and improved seed, ii) provide short-term credit to smallholders for purchases of inputs, and iii) provide agricultural extension services to smallholders. Nevertheless, these agricultural development projects resulted in eviction of small farmers, high degree of income polarity and unemployment (Hoben, 1973; Sisay, et al., 1978; Griffin, et al., 1985).

Obstacles to agricultural development in the pre-1974 period included: i) the structural defects of the land tenure system, ii) the relatively high taxation imposed on the farmer, iii) poor marketing infrastructure and services, iv) limited access to innovations, and iii) services required by feudal landlords, state and church. These and related factors have kept the peasant sector only a little above subsistence (for details see Rhamato, 1984; Wolde-Mariam, 1984; Wubineh, et al., 1988; Pankhurst, 1961, 1966). Agricultural development projects in pre-1974 period acknowledged the severity of these problems. Their intention, however, was to modernize agriculture, which it was thought would improve the welfare of the masses.

Increases in human and livestock population in the highlands, limited use of soil conservation and soil fertility increasing techniques contributed to cultivation of marginal lands, destruction of vegetation cover and changes in ecological balance of the environment. These changes reinforced by continental and global climatic changes contributed to inadequate rainfall. In fact, the years of low rainfall correspond to the years in which Ethiopia was affected by drought (Appendices 2 and 3; Dejene, 1990; Wolde-Mariam, 1984).

In 1974, the combination of drought, poverty and lack of democracy resulted in a series of student and peasant revolts. The alliance of these groups with the armed forces brought about the downfall of the feudal regime of Emperor Haile Sellassie after four decades of aristocratic rule. Capitalist or feudalist states were, for those who were politically active, thought to bring misery to the rural poor. The military regime, assisted by the socialist bloc countries, adopted a socialist pattern of development.

2.4.4. Agrarian Development Strategies in the Post-1974 Period

In 1974, the feudal regime was abolished and replaced by a pro-socialist military regime (1974-1991). The primary goal of the regime was to destroy the feudal relations of production and improve the means of production. Consequently, a number of strategies were implemented. Their objectives could be summarized as the redistribution of wealth or assets (e.g., land, buildings and factories) to the oppressed, or to the state as a proprietor. The ultimate objective was to create "a state of the working class".

According to a 1975 proclamation, the government's guiding principles and framework for development were: i) equitable distribution of benefits of growth; ii) eradication of poverty, hunger, disease and illiteracy; iii) expansion of the production basis of the economy to meet basic needs and the building of a strong and self-reliant economy; and iv) creation of an institutional framework for people's participation in economic and social development (PMAC, 1975).

The measures taken to attain these principles include: i) redistribution of property rights (esp. land) and giving primacy to agriculture as an engine of growth; ii) establishment of people's institutions to mobilise and organize the rural population to participate in the development process;[6] iii) nationalization of key industries to serve the needs of the population; iv) creation of a central planning and resource allocation organization (PMAC, 1976).

The land reform of 1975 abolished private ownership and made all rural land, including forest and mining land, the collective property of the people. Land was prohibited from being sold, exchanged, mortgaged, rented or inherited. Hiring of labour was prohibited. Tenants were freed from all kinds of debt to their landlords and allowed to retain agricultural implements and a pair of oxen that were owned by the landlords. Each household could cultivate a maximum of 10 hectares, but the average farm size could vary by family size. Large scale commercial farms were nationalized and given to evicted tenants and landless agricultural labourers. Moreover, a portion of the confiscated large scale farms were converted into state owned farms. Settlement farms were established for peasants affected by drought or famine.

Strangely, there was no drastic change in the industrial or agricultural development strategies of the pre- and post-1974 government policies. Input substitution industrialization, export of primary commodities and establishment of large scale agricultural farms were also the major strategies pursued by the socialist regime. The only difference between the feudal and socialist regime was that while the former adopted private ownership, the latter opted for state or public ownership. The agricultural policy of the socialist regime was to devote physical and financial resources to the creation and expansion of state farms and cooperatives. In both types of regimes, smallholders produced at least ninety percent of total grain and livestock products and cultivated approximately the same percentage of the arable land. Furthermore, the enabling economic environment, including provision of credit, fertilizer and pesticides, was biased against smallholders (Table 2.2.). Yet, peasants were forced to follow strategies designed by policy makers and international agencies which dictated the fate of their very existence (Belete, 1989; Constable, 1985).[7]

The pre- and post-1974 policy makers, instead of investigating the roots for decline in the growth of agricultural production, focused on imported technologies as a means to boost production. Smallholders of the Ethiopian highlands, amidst the environmental disaster, were presented with innovations without appropriate pricing, marketing or credit policies.[8] Growth in agricultural production was minimal because enabling conditions for increases in agricultural production through technological interventions were not adequate (Belete, et al., 1991; Saith, 1985; Ghose, 1985; Teka, et al., 1984; Ahmed, et al., 1984a and b). Thus, it is essential to understand who participates in the generation and diffusion of technologies and the socio-cultural nexus within which innovation occurs during the next period of development.

The radical transformation of the economy from a classical feudal regime to a socialist state in 1974 brought significant changes in several parameters of

Table 2.2: Distribution of Loans to Agriculture by Farm Type, 1981-1987 (in million birr)[1/]

Year	State Farms	Cooperatives	Private Farmers	Total Loan
1981	295.9	48.4	7.6	351.9
1982	294.2	50.8	0.7	336.7
1983	202.4	51.5	1.7	255.6
1984	219.9	54.4	1	275.3
1985	267	37.1	2.5	306.6
1986	306.1	65.9	2.4	374.4
1987	250.2	46.7	4.4	301.3
Average	262.24	50.69	2.9	314.54
Total	1835.7	354.8	20.3	2201.8

Source: MOA, 1989.
1/ until 1992, 2.07 Ethiopian birr= 1 U.S.$.

growth. A summary of these consequences as reflected by selected indicators is presented in the following sections.

2.4.4.1. Consequences of the 1975 Land Reform Policy

Farmers were organized into 20,000 peasant associations (PAs) and 4000 service cooperatives (SCs) with a membership of about 6 million subsistence farm households. The formation of PAs was considered a breakthrough in terms of the possibility they offered for introducing innovations among the peasants (Cohen, et al., 1975, 1988). The primary reason for establishing PAs was to disseminate the socialist ideology and create the structure for the transformation of private farmers into cooperatives. However, the absence of sufficient conditions for increased food production in peasant agriculture, such as fertilizer, credit, appropriate pricing and marketing policies, undermined possible gains from institutional development.

The objectives of cooperativization, as a method to promote large scale production with modern technology, were not fully achieved. Despite the provision of incentives such as lower taxes, access to credit, seeds, fertilizer, oxen, tractors, consumer goods, building materials and extension field staff, the formation of PCs was slow. The anticipated role of PCs as a method of establishing large-scale farms which would be transformed into state farms could not be achieved.[9]

There were several reasons for slow growth of PCs. Most private farmers who joined PCs against their will put less effort into the activities of PCs, resulting in lower per capita production and less incentive for farmers to join PCs. Also, cooperativization policy prohibited family members below age 18 from joining PCs. But most importantly, members of PCs were required to sell all their produce to the agricultural marketing corporation (AMC) at low fixed prices; thus non-members become less interested in joining PCs. Furthermore, the basis for remuneration of labour was physical presence and did not take into account the amount and quality of work accomplished by a member (Rhamato, 1984; Ghose, 1985). As a result of these factors and the agricultural liberalization policy of the socialist government announced in February 1990, 95% of the 3,316 PCs with a total membership of 290,009 households in 1991 were dissolved.

The reasons for the failure of cooperativization are not only economic but also socio-cultural. Producers were expected to transform their values, aspirations and norms from an individualized to a group pattern. With the development of property, individual property rights, group or communal ways of life tend to be transformed into individualized patterns (Engles, 1972). The changes that took place in Ethiopia were intended to reverse these patterns but were inconsistent with people's values (see Cohen, et al., 1988; Bauer, 1985).

Until 1982/83, about 88 settlement farms had been established consisting of 38,813 households and many settlement farms assumed the form of PCs. The cultivated land of state farms grew from 273,550 to 1,936,640 hectares between 1975 to 1986 (Belete, 1989).

2.4.4.2. Consequences of the Socialist Development Strategies

Despite efforts to expand state farms and cooperatives, individual farms were found to perform better in terms of their capacity to absorb labour, their efficiency of resource utilization, their rate of profit on capital, and their level of crop yields compared to other types of farms (Constable, 1985). Yields of most crops in the peasant sector showed a positive growth while PCs and state farms experienced negative growth.

State farms did not achieve financial viability. The successful operation of state farms was hindered by the lack of qualified technicians, incompetent managers, over-staffing and a variety of other problems (Belete, 1989).

The socialist government of Ethiopia (1974-91) performed better in providing services which are essential for an improved standard of living (health, water and roads); but the performance of the productive sector was disappointing (Table 2.3.). This is typical of countries which have followed socialist development strategy compared with countries that implement a capitalist development strategy (see Leys, 1975; Hyden, 1980; Lele, 1976). What could we learn from these strategies? Economic, technical and social problems that may have contributed

Table 2.3.: Percentage Change in Selected Parameters of Growth between 1974 and 1991, (Using 1974=100 as base year)

Categories		% change
Gross Domestic product (GDP)		359.32
Distribution of GDP (%):	Agriculture	-27.59
	Industry	14.29
Energy Imports as % of merchandise export		212.5
Percentage share of merchandise import:	Food	142.86
	Fuels	66.67
Total Long term debt service as % of:	GNP	258.33
	Export of goods and services	239.47
Crude birth rate per thousand		20.93
Crude death rate per thousand		-28.57
Infant mortality rate per thousand		-19.39
Life expectancy		25
% age group enrolled in education :	Primary	227.27
	Secondary	650
	Tertiary	100
General public service and defence		410.38
Education		235.08
Health		275.83
Agriculture		666.1
Roads		114.66
Population density (/square kilometre of agricultural land)		76.74
Index of food production per capita		-16.51
Population per physician per '000 pop		-10.00
Population per nurse		-80.87
Access to safe water (% of total population)		87.5
Adult literacy rate		37.6
Population (million)		53.2[1/]
Density (persons/sq km)		44[1/]
Rural (%)		89[2/]
Urban (%)		11[2/]
Annual Growth Rate (%)		3.5[2/]
Literacy (%)		60[3/]
Distribution of Labour (%): Agriculture		90[3/]
	Industry	3[3/]
	Service	7[3/]

1/,2/ and 3/ refer to figures in 1991, 1990 and 1988 respectively.
Source: Compiled from FAO Production & Trade Year Book, African Statistical Year Book, & CSO (Ethiopia), 1974-91.

to negative growth in the agricultural sector are examined below.

2.4.5. Problems of Agricultural Development

2.4.5.1. Farming Practices and Resource Base

Ethiopia has the largest population of smallholder mixed farmers in Africa. A slow-down in growth rate of agricultural production both in the shorter and longer run cannot be attributed to any one specific factor but rather to several unfavourable natural, socio-economic and political conditions affecting the agricultural sector. Some of the factors related to crop and livestock production are summarized below.

Smallholders face a multitude of problems. Most farms are located in hilly, rugged and mountainous areas where optimal soil and water management techniques have to be applied to maintain soil fertility and produce enough for the subsistence requirement of smallholders. About 4 to 5 million ha of land are left fallow because of a shortage of oxen (MOA, 1982, 1984; Constable, 1985; CADU, 1968-74, 1984). In the Central Ethiopian highlands, soil or stone bunds constructed to reduce the severity of erosion tend to be destroyed by heavy rain and livestock.[10] Farmers face shortage of farming implements and oxen necessary to prepare crop fields on time. The distribution of production inputs is inefficient and their prices are very high. New agricultural technologies identified by planners are inappropriate for the system they are intended to serve and research is unrelated to the needs of smallholders. These factors have contributed to little use of techniques that are intended to increase crop production in the Ethiopian highlands (Wolde-Mariam, 1986; Constable, 1985; Belete, 1989; Saith, 1985; Ghose, 1985).

Most livestock are concentrated in the highlands, though some are in the hands of lowland pastoralists. But livestock productivity, measured in terms of milk or meat per livestock unit, is very low (Gryseels, et al., 1983; ILCA, 1981). The increase in livestock numbers has caused over-grazing of pastures and fallow land (Constable, 1985; Tagoe, 1983; Jahnke, 1982). The need to feed the large number of livestock as well as secure sufficient grain for subsistence has, among other things, led to fragmentation of land holdings, cultivation of marginally arable lands, expansion of farming to lands with slopes of 30 percent or more, and destruction of existing vegetation cover (Constable, 1985; Kebede, 1988). Prevalence of livestock diseases, poor livestock marketing systems and poor methods of animal husbandry have contributed to declining livestock productivity (Jahnke, 1982; Gryseels, et al., 1983).

2.4.5.2. Marketing and Pricing Policy

Appropriate producers' prices together with an efficient marketing system offer the most effective incentives to farmers to increase agricultural production (Abbot, 1987). In Ethiopia, the lack of an efficient grain marketing system, marketing infrastructure, marketing facilities and services, and accurate and timely market information, led to state intervention in marketing. The Ethiopian Grain Corporation (EGC) was set up in 1960 (Tecle, 1975). But the corporation was weak due to lack of working capital, application of poor price policies and lack of market information.

The agricultural marketing corporation (AMC) was established in 1978, after the socialist revolution. The objectives of AMC were to eliminate private traders so that farmers would receive a "fair" price and to secure food grains for an increasing urban population. Moreover, AMC was expected to match centralized control of production with marketing. Fixed low prices were implemented and a grain marketing quota was imposed, regardless of the amount of grain produced by peasants (see Lirenso,1986; Saith, 1985; Griffin, et al, 1985; Belete, et al., 1991).

State, settlement and cooperative farms were allowed to sell all their marketable grain to AMC at a price which was 20% higher than the farm gate price for private farmers. The response of Ethiopian farmers to the fixed and low price, and to the restricted trade and quota system, was reflected in the steady decline of total and marketable production (Belete, 1989; Lirenso, 1986; Figure, 2). Ten to twenty percent of grain production of major crops had to be sold to AMC at a very low fixed price.[11] The result was a declining share of marketable grain sold by peasants to AMC, which aggravated shortages of food supply to urban and grain deficit areas (see Figure 2; Sen, 1984).

2.4.5.3. Shortcomings of the Land reform

Production for subsistence or reproduction of the farm is the aim of most households in LDCs. This subsistence requirement is fulfilled mainly by agricultural production. Policies aimed at changing the relationship between land and the people who till it play a significant role in the development of agriculture (Rhamato, 1984; Putterman, 1985; Hyden, 1983).

The most important agricultural change in the post-1974 period, affecting the livelihood of 90% of the population, was land reform. Mistakes in the design and implementation of reform can bring severe consequences for agricultural development. Land reform was the basis on which institutions such as PAs and PCs were established. The failure of smallholders to increase food production was partly attributed to mistakes in the design of land reform. Theoretically, land reform should include not only methods of changing relations of peasants with land, but also provisions for optimal use of land and production augmenting inputs.

Figure 2. Cereal Production and Sales by Private and State Farms ('000 tons)

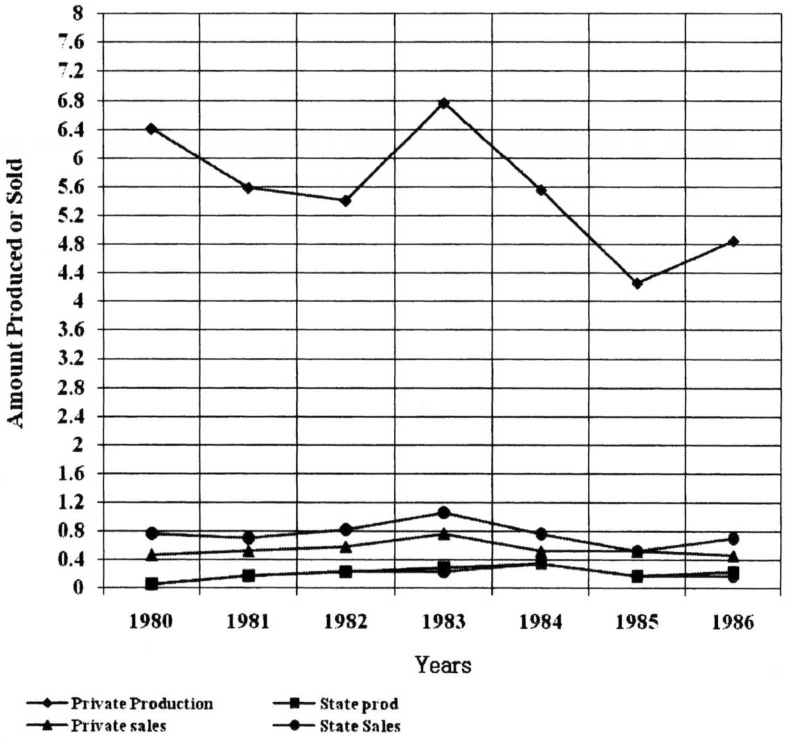

The major objective of distributing land equally was no more than marginally met. According to Belete (1989), the land reform has several shortcomings. Differences in man/land ratio has resulted uneven distribution of land between PAs.[12] The land reform policy encouraged reallocation of land between PAs. The sizes of holdings for individual farmers were reduced to small and uneconomical levels in terms of available family labour and animal power. Moreover, continuous redistribution among peasants resulted in fragmented agricultural lands and a feeling of insecure tenure (Belete, 1989; Kiros, 1976, 1980; FAO, 1982).[13] State farms penetrated into areas traditionally belonging to pastoralists, thus creating conflict between pastoralists and state farms. The land use policy did not give clear guidelines regarding soil and water conservation, which was an essential part of the optimal resource management strategy under Ethiopian conditions. Furthermore, Ghose (1985) estimated rates of underemployment in rural areas to be between 25-40%. Despite this anomaly, labour use policy within the framework of the land reform proclamation prohibited hiring of labour, and peasants were exclusively limited to family labour use. This aggravated rural underemployment.

A land use fee or tax was the same for every private farm family irrespective of farm size, fertility of land, and the amount or type of income obtained. Implicitly, farmers were taxed via a low procurement price of cereals paid by the Agricultural Marketing Corporation (AMC). Thus, taxation as a mechanism of incentive for better utilization of land was ineffective.

The economic implication of insecure tenure was to reduce incentives for increasing yields through fixed capital investments such as improvement of land. When the tenant-landlord relationship was abolished through the land reform, the mechanism of surplus extraction was eliminated. Peasants with no obligation to pay rent and tributes tended to increase home consumption (Gill, 1974; Saith, 1985; Ghose, 1985). Agricultural output either remained static or declined quite remarkably after the reform. Consequently, marketable output declined sharply. To increase marketable output of food, the government expanded the area cultivated by state farms and producers' cooperatives. In spite of these efforts, food production was not increasing at a rate sufficient to feed the rapidly growing population (Ricket, 1991; Figure 2).

In general, the consequences of land reform were loss of security; increased unemployment and underemployment of labour, land and capital; fragmentation of crop lands; reduced rural-urban migration; increased home consumption by peasants; and reduced production.

2.4.5.4. Training and Research

The development and expansion of training institutions serving agriculture was slow (Ahmed, et al., 1984). Research approaches were characterized by lack of focus on integrated farming systems (e.g., crop-livestock farming). There was no periodical adjustment of research to match farmers' needs. Little emphasis was given to the importance of the socio-economic characteristics of farmers in farm management, policy and marketing research. The nature of research reflected personal interest or training and professional recognition rather than the needs of smallholders. There was lack of coordination between research and extension. Extension agents were few relative to the number of beneficiaries. The field extension staff:peasant ratio was about 1:6000. Extension agents were pre-occupied with PCs and received limited transport services. Furthermore, there was an overlap and duplication of activities of different agencies involved in extension (Tagoe, 1983; Belete, 1989; Belete, et al., 1991; Cohen, et al., 1988; Teka, et al., 1984).

Sedentary farming spanning over 3,000 years has given the Ethiopian farmers a rich experience (Pankhurst, 1961, 1966; Levine, 1974). The ability of the indigenous Ethiopian farmer to make the best use of available resources is a rich heritage upon which agricultural research and development must be based if real progress is to be attained (IAR report, 1985, 1988; Ricket, 1991; Dejene, 1990). Peasants have been given the opportunity neither to participate actively in the conceptual identification of their basic agricultural problems nor to contribute to the generation of appropriate technologies to meet their fundamental needs. In this respect, peasants have been relegated to the role of passive objects and spectators of the very process that should be of vital interest to them. This state of affairs has contributed greatly to the inapplicability and irrelevance of most research findings (Getahun, 1978, 1980; Belete, et al., 1991; Ahmed, et al., 1984; Belay, 1977).

If the research and development system is not an organic part of the rural milieu, then it is, ipso facto, a foreign body and consequently incapable of exerting as beneficial an influence as its potential would allow. On the other hand, lack of enabling conditions may not allow testing of new technologies although they may be capable of producing changes. A sound approach might be that the researcher, the extension agents and the peasant work as a team in the identification of the basic problems and the generation of appropriate technologies (Belay, 1977; Getahun, 1978).

2.4.5.5. Institutions Dealing With Agriculture

There are a number of institutions dealing with agricultural development. Until 1991, the major government institutions involved in agricultural development are the Ministry of Agriculture (MOA), the Ministry of State Farm Development (MSFD), the Agricultural and Industrial Development Bank (AIDB), the Ministry of Coffee and Tea Development (MCTD), and commissions with overlapping duties. Most agricultural development efforts, however, are in the hands of MOA.

The Agricultural Input Supply Corporation is involved in the production and distribution of improved seed and fertilizer. Despite investments in facilities and manpower, only 7-8% of farmers in the country used fertilizer annually between 1977-1986 (MOA, 1989). Of the farmers who used fertilizer, only 16% purchased improved seed. Discouraging marketing policies and high fertilizer and improved seed prices have contributed to slow growth in consumption of these inputs. Not only is national fertilizer use very low, but it is restricted to only a few regions of the country (Belete, 1989; World Bank, 1983; MOA, 1989; Table, 2.4.).

Until 1986, AIDB was the only agency providing rural development credit. Since then, the Commercial Bank of Ethiopia (CBE) has been included as a potential credit institution. Production and marketing credits were extended to finance: (a) inputs such as fertilizer, seed, chemicals; (b) processing and marketing of agricultural products; (c) improved livestock production methods; (d) tools and implements; and (e) raw materials for small scale industries. AIDB requires a 15% down-payment and at least 75% repayment of previous credit to

Table 2.4.: Distribution of Marketed Grain Outputs and Fertilizer Use

	Share of Marketed Output (%)		Share of (%)	
Regions	Total1/	AMC Share1/	purchases2/	use2/
Shoa	17	41	29.68	44.37
Gojam	14	72	31.76	9.01
Gondar	19	17	7.25	1.57
Keffa	24	4	1.1	4.95
Wellega	16	19	4.36	4.66
Illubabor	15	2	0.22	0.39
Arssi	29	40	18.22	20.48
Bale	15	64	2.63	1.51
Sidamo	52	1	0.34	5.69
Wollo	12	8	3.55	0.66
Hararghe	15	5	0.86	2.94
Others	n.a	n.a	n.a	3.77

1/ The figures refer to 1985/86. 2/ average values for 1979-87.

get another loan. These are stringent conditions which may reduce incentives for peasants to purchase production inputs. Although local lending institutions, friends, merchants or relatives are available, merchants charged an interest that was as high as 40% per year (Lirenso, 1986; Rhamato, 1984). Therefore, rural credit as a means to stimulate increases in production was limited. The small amount of credit available to agriculture was disproportionately allocated between peasants and state farms (see Table 2.2.).

2.4.5.6. Transport and Communication

Three-quarters of the highland farmers live more than 5 km from the nearest road (Belete, 1989). About 50% of farmers in the highland areas use donkeys, mules or horses for transportation (MOA, 1984).[14] Due to the nature of the topography, transport services are limited. A lack of adequate communication and transportation and rugged terrain have kept rural Ethiopia isolated. This situation has traditionally depressed farm-gate prices and caused the prices of industrial consumer goods and services to be high for remote and relatively inaccessible agricultural areas. Furthermore, only peasants living close to major roads were reached by development projects.

2.5. Development Strategy in Retrospect

The socialist government brought radical transformation to the traditional land tenure system. It affected the lives of more than 85 percent of the country's population. It allowed peasant to retain their output. Nevertheless, growth in output did not increase. For an agrarian reform to produce increase in output, it should be accompanied by other complementary changes such as availability of credit, appropriate pricing policy and availability of production increasing inputs. War ravaged the country. The performance of the economy was deteriorating (Ricket, 1991). Ethiopia experienced forced villagization, quota allocation to grain, misappropriation of public funds by politicians, fierce opposition from rebels, inappropriate marketing policies, and biases in allocation of government budget to state and cooperative farms as opposed to private small farmers. Global changes in Eastern Europe and the above factors contributed to the downfall of the socialist regime.

The grassroots institutions, which were created to serve the socialist regime, are still functional. However, political leaders of PAs are expected to support the current regime, which follows a blend of socialist and capitalist development strategies. Specifically, it has retained public ownership of land, but allowed a free market and the free operation of private traders.

The new transitional government of Ethiopia outlined its economic policy in 1992. According to the manifesto, although the past decline in agricultural output is in part attributed to outmoded technology and vulnerability to climatic conditions, inappropriate policies are largely seen as responsible for the failure of agriculture. The new economic policy emphasizes that the role of the state should be limited to areas where the private sector cannot perform well (e.g., investment in research and schools), in order to create an enabling environment for private investment. Promotion of private investment, mobilization of external resources, involvement of national and regional administrative organs in development, priority to peasant agriculture, maintenance of state ownership of land, expansion of large-scale private farms, minimizing government intervention in marketing, conservation and development of natural resources and a viable industrial policy are a few of the policies adopted by the transitional regime (Kokeb, International Ethiopian Newspaper, February, 1992).

2.6. Summary

Within the context of the theories of development surveyed in this chapter, strategies implemented by the governments of Ethiopia have swung from one paradigm to another. Furthermore, the strategies followed make use of new technologies or techniques as a prime ingredient to initiate development. These strategies have downplayed the role of involving people in the development process.

The pre-1974 feudal regime of Ethiopia placed the peasantry under tight control through various systems of surplus extraction (Hoben, 1973; Pankhurst, 1961, 1966). The two prime causes of rural poverty and inequality in the pre-1974 period were limited access to land use and ownership, and dependence on weather (Saith, 1985). Development projects aimed at helping peasants, despite recognizing the effects of land tenure, resulted in eviction, cultivation of marginal lands, and losses of soil and forest resources. Elites and peasants in coalition with the armed forces overthrew the feudal regime in 1974.

After 1974, land ownership was vested in the state, peasants were given usufructuary rights and land allocation was based on family size (Saith, 1985). The land reform abolished the modes of surplus extraction and freed peasants from the obligation to pay rents and tribute, resulting in increased home consumption. The socialist development strategy required transformation of the small-holding system into one based on large-scale farms in order to generate surplus to stimulate industrialization, to create a state of the working class. These factors led to the establishment of state farms, cooperatives, and settlement farms. The land reform, by changing the relations of production, created necessary conditions for the redistribution of the benefits of growth. However, it lacked the sufficient condition of a buoyant agricultural sector (Griffin, et al., 1985). Peasant's poverty was not solved because of discouraging land and income tax policies, regional concentration of development efforts, low procurement prices, and market fragmentation as a result of restriction of inter-regional grain movement (Griffin, et al., 1985; Ricket, 1991). The system of tenancy, while keeping the peasantry impoverished, had preserved traditional cultivation practices.

During the feudal regime, the small, fragmented, and politically unorganized peasants posed no threat. Therefore, a top-down prescription of development strategy was implemented. In the post-1974 period, peasants were politically organized. Nevertheless, the political representatives were used as a means to bring peasants under the effective control of the regime. This is similar to the functions of the "chika shums" in the pre-1974 period (see Hoben, 1973; Pankhurst, 1966, 1967).[15]

The failure of agricultural development strategies in Ethiopia have resulted in a decline of food production, increased home consumption thus causing shortages of marketable surplus, rising food prices in major cities, increased food imports to fill the gap between domestic food demand and supply, lower adoption rate of technologies, less participation in projects, involvement of peasants in illegal markets, and resource degradation (World development, 1988; Lirenso, 1986; Wolde-Mariam, 1986; Gryseels, et al., 1983; Habtu, 1988).[16] These costs of under-development combined with climatic variation have made Ethiopia synonymous with a place of drought and famine.

In spite of problems facing smallholders, mixed farming offers possibilities for improvement because of the diversity of traditional cropping practices and livestock management techniques (Dommen, 1988). However, opportunities offered by mixed farming can be realized only if new production technologies respond to the needs of smallholder mixed farmers, and take into account their production objectives, knowledge and patterns of decision-making (Hayami and Ruttan, 1985; Binswanger and McIntire, 1987). The study of household decision-making processes will make it possible for problems to be identified, and economically, culturally and socially acceptable solutions to be found.

Food production in the Central Ethiopian highlands can be increased through the expansion of cultivated lands, increasing the productivity of resources, or both. The first alternative cannot provide a long-lasting solution because the highlands are overpopulated with humans and livestock. Furthermore, the fertility status of most soils in the highlands is poor. The lowlands are grazing areas for the livestock of pastoralists. Therefore, the only possible avenue to increase food production is to enhance the productivity of inputs used in the production process from using combinations of fertilizer, improved crop varieties, irrigation and better management (Griffin, et al., 1985).

Given appropriate conditions to induce development, new technologies could help increase food production and generate surplus which could be used to finance developments in other sectors of the economy. Conditions necessary for successful development could be identified through a detailed examination of past and on-going intervention strategies. In the past, because of the nature of the projects, the type of policies implemented and the patterns by which technologies were delivered, increases in food production were of a short term nature. This study examined the socio-cultural nexus and the economic modalities within which technologies were provided in two regions of the central Ethiopian highlands.

Comparative analysis of the regions was necessary because of differences in socio-economic characteristics of the two regions and approaches to technological introduction (see Chapter III). Examining differences in intervention strategies within and between regions could provide an understanding of peasants' decision-making processes, variations in their adoption decisions, consequences for farm income, risk and efficiency, as well as to assess prospects for a better standard of living among peasants of Ethiopia.

2.6. Notes

1. The feudal political system refers to a type of governance where most of the land was owned by the state and landlords. Peasants who tilled the land had only user rights and paid most of what they produced to the state and landlords as a service for the use of the land.
2. A similar definition is given by Rymon and Or (1991). According to these authors, for development to be self-sustaining all the necessary <u>hardware</u> (physical components of a technology) and <u>software</u> (know-how, organization, and institution) should be in place before implementation.
3. The percentage difference in the pattern of expenditure is small. However, the value of agricultural production of the developed countries is significantly larger than that of less developed countries (Hayami and Ruttan, 1985). Thus, the monetary expenditure for research and extension between developed and developing countries may show greater difference.
4. With one of the provinces, Eritrea, independent by May 1993, the geographic location of Ethiopia in terms of latitude would be between 3° to approximately 15° North.
5. This figure excludes the population of Eritrea (about 3 million).
6. People's institutions include peasant associations (PAs), producers' cooperatives (PCs) and service cooperatives (SCs).
7. The controversy over whether to follow smallholder peasant production or large scale farming (state farms & cooperatives) in Ethiopia was discussed in Cohen, et

al.(1988).

8. Most farmlands have lost their soil fertility because of erosion. Grazing or pasture lands are diminishing because of erosion, and increases in livestock and human population. There is very little vegetation cover and rainfall is often erratic. This is what is meant by environmental disaster.

9. There were three stage of cooperativization: Malba, Wolba and Woland. These stages differed by the percentage share of land, labour, and capital that were brought under communal ownership. Under Malba, only land was a common property. In Woland, all inputs of farming were communally owned except a small garden plot of not more than 2000 m^2. Once a cooperative reached the stage of Woland, the next step was to convert peasants to paid workers so the farm would operate in a manner similar to those owned directly by the state.

10. More than 50% of the land area suffers from serious erosion, resulting in a loss of about 20 tonnes of soil per hectare per year (Constable, 1985). Total loss is estimated at about 1 billion tonnes of soil per year which is equivalent to about 60,000 ha of arable land with a depth of 1 meter (Tagoe, 1983).

11. This was the case until the 1989/90 cropping season.

12. Lack of uniformity in the fertility of soils also resulted in some producers harvesting more grain than others, thus creating inequality among peasants within a PA.

13. Redistribution of land was common in many PAs because of new farmers joining PAs, complaints of members to gain equal shares of fertile soils, and allocation of PAs' land to its members who formed PCs.

14. The remaining 50% of farmers in the highland use vehicles or head loading.

15. Chika Shum is a district headman. The office carries with it duties of tax collection, adjudication, promulgation of official decrees, supervision of occasional work projects, levying men for military service, among other duties. The district over which a chika shum has authority is the closest thing to a village unit (see Levine, 1965; Hoben, 1973).

16. According to Wolde-Mariam (1984), there is no universal definition of drought. For the purpose of this study, drought is defined as rainfall induced shortage of material needs for the reproduction of the farming unit and producers. Famine, on the other hand, is the state of not having food for consumption, which in extreme cases lead people to-literally consume whatever is stored in their body.

CHAPTER III

FROM HOUSEHOLDS TO REGION: BACKGROUND AND DESIGN OF THE STUDY

3.1. The Analytical Unit: Without Definition and Boundary

3.1.1. Introduction

Factors that contribute to lack of agreement over the definition of household are examined in this Chapter. The study sites and design of the study are described. Furthermore, the socio-economic organization of households and the farming system are discussed. This chapter is intended to provide an understanding of the socio-economic nexus in which agricultural technologies are introduced and to identify parameters that should be emphasized in assessing household decision-making and its consequences.

Pooling and sharing of resources, food processing, cooking, eating and sheltering tend to take place in the household, which has therefore become a standard unit of analysis for social, ecological and economic purposes. Governments continue to collect data at the household level, and models of development are aimed at households. Any primary category that has become so ubiquitous in social surveys must necessarily be used in further research if we have any hope that our interdisciplinary knowledge can become cumulative. Furthermore, the household neither disintegrates nor completely transforms itself. The question is: What is a household?

The household as a unit of analysis has posed various difficulties. For some it is an artificial census unit, for others the essential building block of all social systems, the key to historical transformation of the world system, a product of social rules and kinship system, the place where gender roles are negotiated and maintained, a basis for demographic analysis, the key cultural symbol, or at worst a piece of ethnocentric reification that has been imposed on more complex reality (Wilk, 1991; Netting, et al., 1984; Wilk, et al., 1989; Guyer, 1986; Berry, 1984, 1986; Jones, et al., 1986; Moock, 1986). Moreover, the household has been labelled as a domestic group (Goody, 1972), production-consumption unit (Cleveland, 1980), or budget unit (Seddon, 1976).

Households perform various activities within a certain framework of organization. Thus, it is possible to make use of the structure and activities they perform to define a household. A household's function and structure, whether in sedentary or pastoral economies, is determined by an established social rule of organization. Dalton (1967) commented on the role of social organization in African economies:

> Neighbouring societies sharing the same physical environment often produce markedly different ranges of output, with different technologies used within differently organized production groups. Such economic and technical differences are largely attributable to differences in social organization (Dalton, 1967:15).

Hence,

The question, what forces, institutions, or rules direct labour, land and other resources to specific lines of production, can be answered only with reference to community social organization (Dalton, 1967:68).

According to Dalton, therefore, it is social organization that brings about changes in economic organization. Nevertheless, it is difficult to empirically establish the direction of causality, because not all parameters of social organization are quantifiable. The issue of causality will not be addressed in this study. What is going to be argued is that existing forms of social organization within which peasants operate play a fundamental role both in intra- and inter-household decision-making regarding production, marketing and consumption.

3.1.2. Structure and Function of Households

Defining households requires knowledge of their structure and functions. Household structure in its broadest interpretation can be summarized by four categories of information: personnel and household composition, production activities and division of labour, consumption activities and inter- and intra-household exchange, and patterns of power and authority (Wilk, et al., 1989). The quality of interaction among households affects the composition and structure of the household.

The number of persons living under the same roof influences household decisions because members form a common resource pool, claim income generated by the household, share poverty, and manage resources in common. Structural relationships within households take the form of specialization of tasks in the management of resources. Certain household members are allocated the management of particular resources. The degree of specialization may influence particular kinds of decisions. Membership in political or religious groups may influence the number of people living in the same house and the functions they accomplish. There is no consensus on whether the structure of a household should be defined by the people living under one roof, by the specialization of functions performed by members of a household, or by the patterns of exchange or authority within or between households (Wilk, 1991; Wilk, et al., 1989).

We see people act and can define households with respect to activity groups. The various activities performed by members of households, among others, can be classified into general spheres corresponding to what has been termed the domestic functions of production, reproduction and inheritance, among others. Accordingly, we can identify five categories of activities per formed by households: production, distribution, transmission, reproduction and co-residence (Wilk, et al., 1984).

The functions or activities of households described in the literature overlap with the structure of households (see for example Wilk, et al., 1984). There is no clear evidence on whether the structure of a household determines the functions a household performs or vice versa. Furthermore, there are no exhaustive lists of activities performed by household members. Therefore, studies and development strategies pertaining to households in different countries must be specific.

Two theoretical issues block an analytical approach to intra-and inter-household decision-making processes. First, does the household function as a corporate social unit? This depends on whether individual actions are directed towards maximizing the welfare of the household or the self. The second obstacle is the problem of setting boundaries around the household unit. Laslett (1984) and Sahlins (1972, 1976) pointed out that there is no society in which households are totally isolated and self-sufficient. Households are always connected to each other, and penetrated by other affiliations through age, kinship, gender and class. Therefore, households do not always have clear boundaries.

The discussion given above suggests that differences in and overlaps between the structure and function of household systems make it difficult to posit a cross-culturally accepted definition of household (Wilk, et al., 1989). If similarity in the structure and function of households can not be established across cultures, then development strategies designed for one culture can not be meaningfully implemented in another culture. The choice of an appropriate unit of analysis should, therefore, consider that understanding the socio-economic organization of a society is an essential step in household studies.

3.1.3. The Decision-Making Environment

The household is the one unit through which the whole economy can be studied as a connected and functioning whole (Firth, 1966). It is central to any understanding of the organizational principles of the economy (Bohannan and Bohannan, 1968:14). This implies that households are not isolated, and that they are integrated within a more complex society. Thus, their decision-making strategies and consequences of these decisions reflect influences from outside the household or the farm.

Productive resources and profitable opportunities in Africa are often mediated through established social relations and regulatory procedures. People invest in social relations in order to improve their access to labour, credit and various forms of property (see also Parkin, 1972; Berry, 1984). The importance of social networks, markets and other institutions within which individuals operate is documented elsewhere (Polanyi, 1944; Polanyi, et al., 1957; Barlett, 1980; see also Chapter VI of this thesis). Intervention strategies should design ways of using elements of the decision-making environment such that households can make "rational" decisions. Investments in social relations, therefore, may have long-term effects on food production by influencing the organization or management of productive activity (Berry, 1984, 1986).

Section 3.1 indicated that a viable development strategy has to rely on information gathered from households. It outlined reasons for the lack of a universal definition of household. The section also emphasized the need to examine elements of the decision-making environment and the modalities through which they influence household decisions. Before examining the socio-economic organization of households, important features of the study sites in relation to the Ethiopian highlands and technological options available to households in the study sites are discussed below.

3.2. The Ethiopian Highland and the Study Sites

The highlands of Ethiopia cover 490,000 km^2 or around 40% of the country's surface area. This is equivalent to four percent of the land area of Sub-Saharan Africa (Jahnke, 1982). The highlands of Ethiopia are inhabited by 88% of the country's total human population.[1] Ninety five percent of crop land, 70% of livestock production and 90% of overall economic activity are concentrated in the highlands (Constable, 1983; Gryseels, 1988; Jahnke, et al., 1984). Several studies have argued that if Ethiopia is to use its agricultural potential for development, the focal geographic or altitude zones should be the highlands (Getahun, 1978, 1980; Sisay, 1980; Belay, 1977). The highlands offer diverse production techniques and opportunities for development (Getahun, 1978; Sisay, 1980). However, they are also the reasons for sluggish growth in food production (see Chapter II; Constable, 1983).

The research sites are situated in Selale and Ada districts of the central Ethiopian highlands.[2] The two study sites have similar farming systems and both belong to the high potential cereal livestock zone. That is, they are dominated by smallholders producing cereals, pulses and livestock.

Selale is representative of the high altitude zone (more than 2000 meters a.s.l.). It is characterized by cool weather and rugged topography. The major crops grown in the area include oat, teff, barley, wheat, horse beans and field peas.[3] The average farm size is 2.9 hectares. Of this area about 3.5% is kept as fallow land. The average grazing land (permanent pasture) per household is 0.8 hectares, i.e., about 30% of total holdings. Seventy percent of the average holdings is cultivated. Ten percent of the farmers do not own livestock. The average livestock holding is 1.3 cows, 1.5 oxen, 0.55 bulls, 0.8 young animals, and 0.76 calves (Finnida, 1989). The demand for livestock products in this area is relatively small. Most farmers belong to the Oromo ethnic group.

Ada is characterized by mild climate and represents the country's large middle-altitude cropping zone (1500 to 2000 meters). The major types of crops grown include teff, wheat, barley, horse beans, chickpeas and field peas. The average farm size is 2.1 hectares. The average grazing land per household is 0.4 hectare. The average livestock holding is .93 cows, 1.86 oxen, 0.48 bulls, 0.33 young animals and 0.64 calves (Gryseels, et al., 1983). Twenty one percent of the farmers in the region have less than two, and seventy nine percent two or more, oxen. Ada is located closer to big cities with high demand for livestock products. The farmers derive from both the Amhara and Oromo ethnic groups, and exhibit relatively liberal attitude towards socio-cultural changes (Gryseels, et al., 1983; Belay, 1977). That is, interventions affecting intra- and inter-household and village production, consumption and reproduction are not new. The social rules governing the response of households to external stimuli are more accommodating than Selale.

Considering general economic trends, the state of the Ethiopian highlands and the study sites, the following propositions can be drawn (see also Finnida, 1989; Belete, 1989; MOA, 1990): i) It is unlikely that the non-agricultural sector will grow faster than the population for the next decade. Thus, out-migration from agriculture is not expected to grow unless land use policies initiate such movements. ii) It is unlikely that population growth rates and the number of livestock will decline. iii) It is impossible to introduce large scale mechanized farms capable of producing sufficient food for the whole population in most areas of the Ethiopian highlands because of human and livestock density, topography and lack of capital. iv) Creation of side activities such as small scale cottage industries will likely not expand fast enough to reduce stress on arable land. v) The introduction of imported high yielding crop varieties may be doomed to failure because of lack of long term resistance to disease, biases in allocation of credit between different kinds of farm enterprises, high purchase prices, and inappropriate marketing or pricing polices. vi) The introduction of pure exotic livestock breeds is not feasible because of their lack of resistance to disease, high purchase price of the cows, the low prices of milk and low butter production per litre of milk, and other marketing constraints (Beyene, et al., 1985; Wubineh, et al., 1988; Dejene, 1990; Rhamato, 1984; Belete, et al., 1991). The following section reviews technological options available to peasants of the Ethiopian highlands.

3.3. Technologies studied

What feasible development alternatives remain, and how can we identify the development needs of producers in the highlands? Before suggesting alternative ways of increasing food production, the importance of crop and livestock enterprises to households in the highlands study sites are outlined.

The highlands are characterized by mixed farming systems involving crop and livestock enterprises. These enterprises compete for limited financial and physical resources. They are complementary because income generated from one enterprise may help to expand the others (Gryseels, et al., 1985). Crop production is practised with the help of traction animals. Shortage of traction animals results in declining crop production (Belete, 1989; Kebede, 1988).

Livestock accounts for 35% of the total agricultural output of Ethiopia. Cattle provide traction power as well as other benefits to smallholders.[4] They are a source of organic fertilizers (manure) and dung cake for sale or for home use as fuel. Furthermore, livestock represent a medium for storing wealth. To farmers, cattle and small ruminants such as sheep and goats play a vital role in stabilizing household income. For instance, the contribution of crop and livestock enterprises to the gross farm margin was found to be 47% and 53% respectively (Gryseels, et al., 1985). Under situations where crop marketing decisions are influenced by uncertain prices or government intervention, farmers tend to shift resources to cattle and small ruminants as the most secure form of saving and/or investment (Kebede, 1988; Lirenso, 1986).

Despite the contribution of cattle to the income of smallholders, technical and non-technical constraints on livestock production also hinder increases in the production of crop and livestock products. One of the most important constraints is feed supply. Increases in human population, declining quantities of grazing land and crop residues, diminishing carrying capacity of existing pasture are the primary causes for shortages of animal feed in the Ethiopian highlands. To attain adequate food production, it is essential to increase the productivity of both crop and livestock enterprises.

Smallholder mixed farmers of the Ethiopian highlands could increase their income through innovations in crop and livestock production, off-farm employment or combinations of these activities. Most success stories in the LDCs vis-a-vis increased food production are linked to the use of improved seed, fertilizer, pesticides or combinations of these technologies (Ruttan, 1978; Hayami, et al., 1985; Molnar, et al., 1983; Tecle, 1975). Similarly, successful results in livestock production are credited to improved feeding, disease control, management practices and breeding (ILCA reports, 1981, 1983, 1985). Similar conclusions can be made for the Ethiopian highlands (Cohen, et al., 1988; Tecle, 1975; Sisaye, et al., 1978; Sisay, 1980; IAR, 1989; Wubineh, et al., 1988).

There are a number of studies on crop production increasing technologies in Ethiopia (see for example, Gill, 1974; Waktola, 1975, 1980; Tecle, 1975a and b; CADU reports 1977, 1984). What is not yet clear, at least for many parts of Ethiopia, is the feasibility of cross-bred cows and other livestock-related technologies, not as a single intervention but rather as competing with and complementary to other innovations, such as fertilizer, improved crop varieties, pesticides and indigenous crop-livestock technologies.

Agencies such as the Chilalo Agricultural Development Unit (CADU) and the Extension and Project Implementation Department (EPID) of the Ministry of Agriculture have introduced cross-bred cows in Arssi and Shoa administrative regions since the 60's. More recently, the International Livestock Centre for Africa (ILCA) has introduced cross-bred cows and other technologies around Ada and Baso districts in northern parts of the country since 1981. The smallholders of Selale, however, gained access to cross-bred cows in the 1987/88 season. The performance of cross-bred cows under on-farm trials has shown encouraging results (Gryseels, et al., 1985, 1986). Farmers in Ada and Baso regions of the central Ethiopian highlands were given a five year credit to purchase cows and construct sheds (Gryseels and de Boodt, 1986). About 50 households voluntarily adopted cross-bred cow technology.[5] The milk yield of cross-bred cows was more than seven times and its effect on cash income was more than three times that obtained from local boran breeds (Gryseels, et al., 1983, 1985). The introduction of genes of Holstein-friesian into the tropical zebu breeds resulted not only in increased milk production but has also in an increased period of lactation (Mekonnen, 1983).

Besides their milk production potential, cross-bred cows have been used for traction purposes (Gryseels, et al., 1986). The choices among different classes of livestock for the purpose of traction vary depending on the cultural values of smallholders, their economic feasibility and the availability of feed resources. Experiments were conducted using cross-bred cows for traction and milk production in the Baso area of northern Ethiopia (Agyemang et al., 1985). The result suggested that the number of hours worked by each pair of cross-bred cows was sufficient for cultivation of 2.5 hectares. This study, however, did not evaluate the cultural acceptability of using cows for traction.

Loss of soil fertility and the incidence of pests and disease have contributed to declining food production in the Ethiopian highlands. Consequently, pesticides and inorganic fertilizer are used by many peasant communities living less than 20 km from input distribution centres (CADU, 1977; MOA, 1989).

The spread of improved crop varieties is generally linked to varietal developments at the international agricultural research centres such as CIMMYT.[6] Crops tested for Adaptability have been released in Ethiopia since the early 70s. Their resistance to new strains of disease, however, is low. This has caused fear and lack of confidence among smallholders, and has led them to continue using traditional varieties which are relatively resistant to disease.

Ethiopia is one of the historical centres of crop domestication. There are a variety of local crop cultivars. Some cultivars are multipurpose. For instance, oats produce a large volume of straw with high degree of palatability for livestock. Oats are also used to produce local beer and livestock feed. Most importantly, they resist frost. However, this crop also harbours certain kinds of disease hazardous to humans. Thus, the government of Ethiopia wants to enforce abandoning cultivation of this crop. But, instead of reducing the area under this crop, producers are expanding its cultivation. This phenomenon suggests that improved crop varieties must perform extremely well in order to be widely adopted in the Ethiopian highlands.

The common feature of crop production innovations is that their effect on the level of production, moderated through the soil and influenced by rainfall and sunlight, is uncertain. This makes adoption a risky venture. Smallholders, on the other hand, can anticipate potential milk yield increases from the use of cross-bred cows. However, management requirements related to feed, disease control and in-breeding problems may affect the performance of cross-bred cows.

New and traditional production technologies compete for the limited resources of smallholders. Given their resource endowments and constraints, it is unlikely that smallholders will adopt all technologies at the same time. Instead they may adopt technologies sequentially and/or in selected combinations. Responses to technological innovations in livestock production need to be studied not alone but as competing with and complementary to crop related technologies.

Given the choices of innovations, how do farmers decide to adopt different combinations of technologies or not to adopt any one of them? What is the impact of a single or selected combination of technologies on farm income and other household characteristics? How does the socio-economic organization of households affect adoption of technologies? "Appropriate" agricultural policies and research programs in crop and livestock production can be designed and implemented *if* these and related questions regarding fertilizer, pesticides, improved seed and cross-bred cows are addressed in a *holistic* manner.

3.4. Design of the Study and Selection of Sample Size

The research was carried out over a period of 17 months in 1990-1991. The field work was carried out in two sites separated by about 180 km. Administrators, staff of the Ministry of Agriculture and officials of peasant associations at district, sub-district and village levels were briefed on the objectives of the study. The study involved interviews, observations and participatory methods of data collection. High school graduates from the study sites were given tests on their skills at translating English into Amharic and Oromigna. Their approaches to interview and their views of peasants were evaluated through interview.

Ten enumerators for Selale and two for Ada were selected. They were given training on how questions should be asked, the duration of each interview, the appropriate demeanour towards farmers, and their understanding of the questions (see Appendices 5a and 5b). For the research to produce reliable results, it was stipulated that at least one cropping season should be completed. Furthermore, sufficient time was to be allowed to observe other activities within the farming communities. The survey was carried out between January 1990 and May 1991. After selecting assistants and completing pre-interview arrangements, the next step was determination of the sample size.

The households who received cross-bred cows and were selected for this study in the Ada and Selale areas numbered 26 and 89 respectively.[7] A confidence level of 95%, coefficient of variation of crop yields of 96 percent and precision level of ± 20% resulted in a sample size of 89 farmers.

In Ada region, test farmers were located in four peasant associations that were 15 kms apart. Furthermore, the number of farmers who were willing to participate in the study was less than 35. Given the financial limitation, twenty-six farmers were chosen for the study in the Ada region.

After determining the sample size, the need to use farmers who joined ILCA and Finnida/MOA programs as test groups necessitated the use of systematic selection of the control group. The methods of technology assessment employed by Finnida and ILCA were not feasible.

The selection of farmers who had received cross-bred cows in the Selale region was biased towards those who were well-to do.[8] Households who were given the opportunity to purchase cross-bred cows must have the resource to purchase or grow locally available feed and to build shade for cows. Furthermore, they must have acquired some level of training in management of cross-bred dairy cows. Of course, those who are close associates of extension agents and politicians have received cross-bred cows. A preliminary study by the Finnish International Development Agency (Finnida) project, using randomly chosen farmers who own cross-bred cows (test) and those who do not (control), showed that test farmers were better-off with respect to all economic and social variables considered. Despite the success claimed by this study, I and my colleagues from ILCA felt that the design of the study implemented by Finnida was not appropriate because it compared farmers who were better-off before the project and who received cross-bred cows with farmers who were poor. Thus, it may not be feasible to use results from this kind of study to design projects of similar nature in other regions.

ILCA is expected to make use of the farming system research in the generation and testing of new technologies. It is after on-farm verification trials that the results are recommended to MOA for dissemination. However, ILCA provided a relatively risk free environment to farmers by supplying inputs at low cost and offering them attractive fringe benefits for participating in the testing program. Data collected from this kind of research can not reflect the situation where producers manage their milking cows.

The present study was designed in such a way that all test farmers were compared with farmers who exhibit similar socio-economic characteristics (control farmers) but were different in ownership of cows. On average, the control farmers were to have a comparable number of oxen, cows, sheep/goat, family size, age (farming experience), education, annual farm income and farm size (crop and grazing) with the test farmers. Moreover, the two groups had to exhibit similar ethnic, climatic and geographical characteristics. To accomplish this task, the following two-step procedures were followed. A group of farmers involving political leaders and elders in each peasant association were asked questions such as, "With whom do you think farmer "A" compares with respect to income, livestock holdings, living standard, etc., except that he does not own cross-bred cows?".

Secondly, each test farmer was asked questions such as, "To whom do you think you are comparable regarding income, livestock holding, family size, etc., except that you own cross-bred cows and the other farmer does not?". This method of identifying the control group was difficult and socially controversial.[9]

Nevertheless, there was no other alternative way of selecting the control group which would reduce the bias inherent in comparing randomly selected farmers with systematically chosen beneficiaries.

Several studies have pointed out that households who hold important positions in the political or religious institutions of a village or society gain greater access to resources. Furthermore, they may use their power to influence decision-making of households and the degree to which households are integrated into a more complex system (see Netting, et al., 1984; Wilk, 1989, 1991; Guyer, 1986; Moock, 1986). Thirty nine farmers who are executives of political institutions were included in the study to examine their influence on household decision-making and its environment in the study sites.

The study sites are divided by PAs. The choice of PAs from which study farmers are selected is based on distance from main road, existence of farmers who have received cross-bred cows in the PA and willingness of farmers to participate in the study. [10] A total of 16 PAs were selected from the Selale region. In the Ada region, however, transportation problem and availability of few farmers who received cross-bred cows in some PAs necessitated the selection of only three PAs. The PAs selected for the study, distance from the main road, and the number of test and control farmers are presented in Table 3.1.

3.4.1. Representativeness of Sample Farmers

A sample of entity or item is considered to be representative of the population from which it is obtained if the sample selection procedure satisfied appropriate statistical criteria (ref). However, methods that may not use statistical criteria may be employed to ensure representativeness of a sample. Producers who received cross-bred cows, thus categorized as test farmers, were chosen by local officials and staff of the Ministry of Agriculture on the basis of selected criteria (see p. 86).

Farmers who received cross-bred cows were economically better-off compared to other peasants in the region. Thus, stratification and systematic sample selection procedure was employed to ensure that control and test farmers were comparable with respect to selected parameters except that control farmers did not own cross-bred cows. Because of their socio-economic status, therefore, the results obtained from test and control farmers may not be the same as results of a study that would include poor farmers. However, variations in income and other parameters of standard of living among farmers in each region were minimal because land was state owned, crop and grazing land was fixed, market for labour was abolished, inputs and outputs markets were monopolized (see Finnida, 1988). [11] Despite the method of sample selection, therefore, the results of this study could be extrapolated to other farmers in each study region with out significant changes.

Table 3.1.: Peasant Associations, Distance from Main Road and Number of Study Farmers

District	Peasant Association	Distance from Main road	Number of Farmers	
		(km)	Test	Control
Selale	Gulele Biro	3	10	10
	Bosecha	3	5	5
	Sego Kara	3	13	13
	Arero Tosgne	3	6	8
	Anokere	2	7	9
	Elamo Efreso	12	7	7
	Jibara	5	4	7
	Jate	7	3	5
	Ado	3	3	6
	Mechela Arkiso	2	5	7
	Illukura	13	6	9
	Wayu	3	8	11
	Kasim	5	7	10
	Tumano	4	4	7
	Mechela Endode	3	4	7
	Archo	7	7	7
Sub-Total			89	128
Ada	Udie/Denkaka	6	4	4
	Babogaya	4	6	6
	Tedecha	4	16	17
Sub-Total			26	27
Grand-Total			115	155

Sources: Assembled from base-line data, 1990-1991.

The choice of control farmers involve controlling for differences with respect to selected parameters (see p. 86). In the Ethiopian highlands these factors are principal determinants of household stratification. Factors that were not included in the selection of control farmers (e.g., indigenous production knowledge, number of relatives, whether or not a household's ancestors were immigrant, etc.) may cause differences in household income and behaviour, and may compound the effect of the new production technologies. Including many stratifying variables would result in few test/control farmers per PA. This means that to study 270 farmers 270 (number of subjects covered by this study), the geographic coverage of the research site should be increased. This was not possible because of financial limitation. In lieu of obtaining large sample size while at the same time stratify households with respect to important parameters was the only feasible method of sample selection.

3.4.2. Why Inter-Regional Comparison?

Selale and Ada districts exhibit similar production systems because they are located in the high potential cereal-livestock zone. However, there are also sharply contrasting differences. Comparative analysis of the two sites is hypothesized to reveal significant differences in patterns of adoption of technologies, the efficiency with which producers utilize new technologies and their effect on household income and other characteristics.

Selale peasants have only been exposed to innovations relatively recently.[12] The Selale Dairy Development Project that provided cross-bred cows to farmers of this region was funded by the Finnish International Development Agency (Finnida) and the Ethiopian Ministry of Agriculture (MOA). MOA has had vast experience regarding approaches to technological introduction since the 1960's. Successful adoptions were achieved in areas where farmers were responsible for the management of new technologies. Farmers could not purchase non-farm produced livestock feed (e.g., noug cake) because of higher prices. Thus, the project was designed in such a way that farmers would not face difficulty in the management of cross-bred cows after the project is completed. Selale farmers were instructed that inputs that were necessary for the management of cross-bred cows were available in their locality, and that they should take full responsibility for the management of cross-bred cows. This kind of approach to technological introduction is similar to farmer-managed trials of innovations within the framework of the farming systems research guidelines (Gilbert, et al., 1980; Shaner, et al., 1982).

International agricultural research centres are expected to collaborate with National Agricultural Research Centres (NARCs) and implement principles of the farming systems research. However, ILCA's approach to technological introduction was different. ILCA provided free advice, subsidized veterinary services, concentrate feed and insemination services to farmers who were willing to purchase cross-bred cows. Compared to farmers in the Selale region, therefore, farmers in the Ada region enjoyed a relatively risk-free environment. Thus, the approach to diffusion of technologies in the Ada region was different from that implemented in Selale area. Furthermore, Selale and Ada regions also exhibit relatively different socio-historical development. The extent and speed with which the Oromo and Amhara ethnic group assimilate in Ada and Selale regions differ (see section 3.5. below). Therefore, comparison of the communities in the two sites should be useful to assess the consequences of differential approaches to diffusion of technologies. The findings from this research should provide evidence to policy makers regarding alternative approaches to technological introduction.

3.4.3. Types of Information Collected

The research involved interview, observatory and participatory methods. Discussions were conducted with farmers of different ages and sexes to obtain answers to questions which require consensus or inter-household comparisons. Data collected from interviews include the following: goals; strategies; types, sources and kinds of information; production knowledge; elements of household decision-making; socio-economic characteristics such as age, schooling, relatives, wealth, income, expenses, area, crops planted, number of livestock, milk yield, technologies adopted, and so on (see Appendix 5b). Furthermore, two households in two PAs were chosen for intensive participatory study in the Selale region. Observations were routinely made on selected households.

3.5. Socio-Economic Organization of Ada and Selale Households

3.5.1. Introduction

The social and economic organization of households in the Ada and Selale regions exhibit several similarities. For instance, most farmers in both regions belong to the Oromo ethnic group. Both regions belong to the high potential cereal-livestock zone (HPCL) and practice mixed farming. Nevertheless, there are few major differences between the study sites. Social, institutional, geographic, market structure and economic features of the study sites are described below. It is hoped that the background information provided in Chapters II and III will help to illuminate explanation of the results of empirical analysis presented in Chapters V to IX of this thesis.

3.5.2. Historical Background to the Study Sites

As early as the 14th century, regions bordering the core Amhara homeland were controlled by the Amhara. Amhara culture, religion, political system and language became dominant within the Amhara sphere of influence. Selale and Ada regions belong to these Amhara controlled regions (Levine, 1974). Next to Amharic, the language of the Amhara ethnic group, Oromigna, the language of the Oromos, became the medium of communication throughout Ethiopia as early as 1620 (Levine, 1974; Pankhurst, 1961).

The Oromos invaded the core land of the Amhara and regions controlled by other ethnic groups in the 15th century from their homeland located between the rift valley and the Bale Mountain. Prior to this expansion, the Oromos were nomadic pastoralists (Pankhurst, 1961; Levine, 1974).

A typical feature of the Oromos is that wherever they settle, they adopted the culture of their new neighbours rather than spreading theirs. They adopted plow cultivation of grains from the Amhara. Nevertheless, they maintained a distinct identity, political autonomy and Oromigna as their mother tongue. Some maintained more elements of their pastoral way of life, such as raising more cattle. By the 18th and early 19th century, nearly all the Oromos of the Shoa province had became Christians, and intermarriage between the Oromos and the Amharas became relatively frequent.

The Amhara and the Oromo show contrasting differences with respect to social life. The Amhara system is hierarchical and individualistic. The Oromo system is egalitarian and solidaristic. Religious and political functions are segregated in the Amhara institutions, while fused among the Oromos. Furthermore, the adaptation of the Oromos is contrastingly different than any ethnic group. They are the only ethnic group that pursued four types of livelihood, pastoral nomadism, hoe cultivation, plow cultivation of grains and trade. The greater degree of acculturation of the Selale and Ada Oromos to the neighbouring Amhara since the 16th century is shown in that the Oromos exhibit several features of the Amhara culture and life style.

By the 18th century, Ankober (a city located about 200 km from Selale and Ada) became the capital city of Ethiopia (the reign of King Sahle Sellassie). In the early 19th century, Emperor Menelik II moved the capital city to Addis Ababa (the current capital city). Since the establishment of Addis Ababa as a capital city, availability of large tracts of grazing lands because of low population density and proximity to the capital city attracted Amharas and the Oromos to engage in modern livestock husbandry in northern, western and southern parts of Shoa province. The eastern part of Shoa (e.g., Ada), however, was considered appropriate for crop farming. The Amhara culture of neighbouring regions such as Bulga, Shenkora and Ankober, proximity to the capital city and other emerging new cities (e.g., Nazareth), and the fertile nature of the soils of Ada were major factors in the development of crop farming in the Ada region. These historical developments resulted in greater acculturation to or assimilation of the Shoa Oromos by the Amhara. Consequently, the traditional village and household organization of the study sites is very similar to the Amhara social organization of Menz, Ankober, Bulga or other northern parts of Ethiopia.

3.5.3. Village Organization in Ada and Selale Regions

Traditionally, peasants of the Ethiopian highland live in scattered villages. Most of these villages are located more than five kilometres away from major highways and are not accessible to development projects. Private traders have to travel long distances to purchase grain for urban markets. Governments use local chiefs and political leaders as a means to implement policies.

In the pre-1975 period, a village was composed of 4 to 10 households. These households could be close relatives, members of an extended family, or non-kin. Since 1975, the village configuration has changed because most landlords left the area. Farms owned by small farmers and landlords were shared between two peasant associations, such that households were forced to live in different PAs.

A major transformation in the pattern of settlement and organization of village life occurred in 1986, when the government implemented the villagization program. Villages dispersed through out the highlands were converted into "pseudo-city" type residential areas. The major objectives of this program were to facilitate provision of social services such as health care, water supply and education. Furthermore, villagization was expected to provide the infrastructure to increase access to production inputs such as fertilizer, and pesticides. Implicitly, however, it was a strategy to strengthen control over peasants, facilitate the formation of collective farms, and expand the socialist ideology.

The consequences of this type of villagization program for household organization and economy were significant. We will start with a commentary from Ato Kebede (a study farmer from Selale): From Village to Cities: From a Citizen to a Refugee: The Parable of an Endangered Species:[13]

> "My father, grandfather and great-grand fathers taught me the value of living together. They taught me how to treat guests, neighbours, friends, relatives. If I grew up with a colleague in a village, I introduce him to strangers as a brother. We shout "Ejolie Degem!" (children of a place called Degem) at any place and time. We were respected simply because we have identified ourselves by the name of the place inhabited by our ancestors. Now, what do we say? "ye mengist lijoch?!" (children of the state or government!). We are like a refugee in our home, in our own land and in our own territory. We have no identity, except that we are "tokuma" (oneness or togetherness)".

Policies that changed the traditional pattern of village life to a "pseudo city" type of residence made peasants accessible to government and development agencies. The Government of Ethiopia could more easily enforce laws, could provide incentives to form cooperatives, and political cadres could live in these villages to teach socialist ideology.

A typical design of a village in the post-villagization scheme looks like the following: Well-to-do farmers, those with galvanized aluminium roof houses, took the front block of the villages. The next row was occupied by big houses with thatched roofs and at the rear end were small houses with thatched roofs. These villages were located close to the main road and cities. Peasants living over an area of 800 hectares were forced to live in one, two or three villages.

The findings of this research may have to be modified if producers decide to move out of the pseudo-cities. Study farmers were asked about their intention to move out of or to stay in the pseudo-cities. There were several problems of moving back to their original villages or hamlets. The building materials have lost their strength because their house was demolished more than once. The galvanized tin has been punched twice or more, thus was porous and could not protect the house against seepage of water in the rainy season. New galvanized tin costs about 20 birr per piece.[14] Obtaining grass to cover the roof instead of tin and wood to support the roof is difficult. Therefore, several farmers planned to return to their hamlets or villages over a period of three to four years.[15]

There are various factors determining the composition and number of houses in a village. The pattern and number of houses forming a village in pre- and post-villagization period were influenced by: i) the availability of arable and grazing land; ii) its closeness to a city and/or main road; iii) its ethnic composition; iv) the degree of out-migration; and v) the existence of off-farm activities. In chapter two, it was noted that the man/land ratio under a PAs control varies. The larger the land area under the control of a peasant association, the less the number of households living in a village. That is, households tend to live close to their crop and grazing lands. The closer the arable and grazing lands to the main roads or cities, the more the number of households per village located close to cities or main roads.

The Oromos show a greater tendency to live together than the Amharas (Levine, 1974; Pankhurst, 1961, 1966). The strength of relationship among extended families is relatively high among the Oromos. Thus, one would expect more households per village in Selale than in the Ada region. Proximity in residential areas, however, is not the only factor determining the strength of relationship. In the Ethiopian highlands, factors such as closeness of friendship, relationship by marriage, religion and blood, membership in mutual self-help groups and having common ancestral place of origin (being children of immigrants from northern Ethiopia) influence the size of a village and the strength of relationship. In addition to these factors, the farther a village is from the main road and the slower the out-migration is, the larger will be the number of households per village. Government intervention through opening schools helps consolidate larger villages by creating a common interest among households.

Before villagization, the number of households per village was relatively higher in the Ada compared to Selale region. Farmers in the Ada region own relatively small farm size compared to those in Selale. Ada is located closer to the major cities. The Oromos in Ada show a high degree of assimilation with the Amhara compared to the Oromos of the Selale region (Levine, 1974; Pankhurst, 1961, 1966; Belay, 1977). Peasants who reside close to cities tend to imitate patterns of city life (Korten, 1972). That is, their interpersonal relationships exhibit limited dependency on neighbours, friends, relatives and more on market (impersonal) relations compared to those in remote rural areas. This is typical of the Ada area.

The pattern of settlement in villages closely follows the area of concentration in agriculture, crop or livestock husbandry. Selale has more grazing area. The Oromos traditionally value animal husbandry as their area of expertise or culture. Their life is intertwined with livestock. However, the Oromos have migrated to almost all parts of the country (Levine, 1974). They have been intermarried and acculturated with various ethnic groups. In the study sites acculturation took place with the Amhara ethnic group. Under this situation, the Amhara culture of plow farming competes for land and labour with livestock husbandry. Peasants have to reduce grazing areas and labour allocated to livestock with the expansion of crop farming. Thus, relative to Oromos who maintained their "ethnic identity" (in southern and western Ethiopia), the Oromos of the study sites practice crop farming with a greater degree of intensity (as in Selale).[16] Where acculturation is high and residential areas are close to big cities, households concentrate even more on farming (as in Ada).

3.5.4. The Structure of Authority in Village-Level Institutions

Traditional institutions are greatly affected by changes in the political system which reflected neither the Oromo nor the Amhara political structure.[17] Rather, an exotic political system that intended to eliminate traditional institutions in favour of "modern" was established. Changes in the structure of village-level institutions occurred through the establishment of peasant associations and cooperatives. These institutions gained greater power over traditional forms of political institution or social networks because of their close working relationship with politicians and administrators.

Institutions in the pre-1974 period were established by households who belong to the same parish or whose ancestors migrated from the same place. Traditional authority structure was observed in culturally defined norms of respect and obedience, or by-laws, which reflect the consensus of members. Representatives of a village are chosen on the basis of traditional customs and norms such as being religious leader, age, wealth, education, and so on.

A peasant association is an institution defined by a specific land area (800 hectare), representing individual farmers at sub-district, district, regional and national levels. Officials of PAs are chosen based on their mastery of the socialist ideology. Election is conducted in the presence of government cadres to ensure that elected officials are politically able and willing to commit themselves to socialism.

Authority structures in PAs reflect a blend of traditional norms and modern bureaucratic operations. More fertile land may be allocated to members of households who have gained higher social status in a village compared to ordinary farmers. However, allocation of farm size should conform to the guidelines of the land reform proclamation (i.e., land should be allocated by family size). Peasant associations are still active grassroots organizations. Their elected officials, however, have now been removed from power because they supported the socialist government.

This section described a few of the changes that have taken place in village organization and structure of authority. These changes influenced household organization. Household organization and changes that have occurred since 1974 are described below.

3.5.5. The Physical, Functional and Symbolic Elements of the Household

3.5.5.1. The Physical Element

A household can be identified as comprising physical, functional and symbolic elements (Webb, 1988). The physical elements differentiating one household from another are a house (with a kitchen, living and bed rooms), Goteras (stores), a barn, and a small plot for growing vegetables and preparing dung cake, fenced-off from pathways on the outskirts. Moreover, the physical element of a household includes a plot for piling and threshing the harvest, and feeding cattle in the immediate neighbourhood of the fence. The size, divisions and construction materials (thatched roof or galvanized aluminium sheet roofs) identify whether or not household members: i) are well to do, ii) have more children, iii) have attended secular schools and iv) have a good social reputation.

Household membership is often synonymous with co-residence. Physical confines and availability of dwelling space condition the size and composition of the household unit (Wilk, et al., 1984). Each household unit in the Ada and Selale areas is identified by a physical structure or compound delineated from the neighbourhood by a fence. Before 1974, the size of a compound greatly varied depending on the farm size. Since then and especially after villagization, households were entitled to have a maximum of 2000 m^2 for gardening, building barns, build stores and construct feeding places for livestock.

When farmers were forced to live in pseudo-city patterns, most of the elements defining a household were lost. Irrespective of ethnicity, children interact more with neighbours. The household farm yard had been a sacred place when farmers lived in villages or hamlets. Peasants preferred not to expose how many storage units they owned by fencing-off pathways using shrubs and bushes. For most households, objects inside a house could only be observed or used by close relatives, friends or neighbours. Villagization has eliminated these value systems. If cities are simply a collection of houses with relationships exhibiting more of economic than social and moral values, then that is what villagization has done.

3.5.5.2. The Functional Element

Defined by the physical element, members of the household share various rights, duties and material possessions, and respect the authority of a single male head, unless the wife is widowed. Each member contributes directly or indirectly to the survival, reproduction, and advancement of all elements of the household. Moreover, relatives and friends will respect and support each other, and cooperate for mutual advancement. The household in the pre-villagization period would have at least one close relative as a neighbour, but only by chance were neighbours related after villagization.

In the Amhara families relations between spouses, and between parents and children are relatively tense and oriented to the expression of deference, whereas relations in Oromo families are relatively autonomous, emotionally expressive, and good humoured. Observatory-participatory study in the Selale and Ada regions show that the behaviour of households in these regions is slightly different from what is indicated above. Acculturation has moderated the specific life styles and produced a favourable relationship between parents and children.

All activities within and outside the house are primarily carried out by household members. A typical household consists of two adults, two to three children under fifteen years of age, a child or two above fifteen years of age, and one dependent (greater than 60 yrs old- parents of husband or wife).[18]

Members of households are given functional responsibilities to execute specific activities. Decision-making follows after members are given functional responsibilities. Often, the head of the household assigns functions to its members. Consequently, household decision-making among peasants of the Ethiopian highlands tends to be hierarchical (Korten, 1972; Pankhurst, 1961, 1966; Levine, 1974; Bauer, 1985). This hierarchical decision-making is relatively dominant in the Ada region. In the Selale region, hierarchical decision making is more common in PAs neighbouring the core Amhara territory. For instance farmers in Wayu peasant association are closer to the celebrated orthodox mon astery (Debrelibanos) and where several Amharas live in a city called Debre-Tsigie which is located a kilometre from Wayu.

Decision-making often involves an exercise of authority. In general, the head of the household represents the members in many social and economic activities within a village. Nevertheless, each member of a household is given authority or functional responsibility to carry out certain duties both on a daily basis and as long as he/she remains within the unit. This focus on functions performed by members of a household permits comparisons of households across societies even though the structure and organization of the household itself may vary among those societies.

Fundamental to the study of social behaviour is an understanding of the goal orientation provided for the individual by his culture. The goals of Ada and Selale farmers can be divided into: i) securing subsistence, ii) acquiring wealth by investing in cattle or small ruminants and iii) gaining respect or prestige through building a big house and being a good farmer (see also Pankhurst, 1961, 1966; Levine, 1974; Bauer, 1985).

Household survival in the study sites in particular, and in the Ethiopian highlands in general, is dependent on the accuracy of timing and efficiency of farming operations. In this regard, each member of the household has to be assigned mutually supporting activities. There are highly standardized role-rules that members of the household are expected observe.

The role-rules of decision-making are different between the Selale and Ada areas. In Selale, most members of a household seem to have more responsibility than do those of Ada households.[19] Furthermore, within each region, the Amhara households exhibit greater dominance of the husband over most aspects of decisions than the Oromo households (Bauer, 1985; Hoben, 1973).

Men are expected to do all field work such as plowing, planting, weeding, harvesting, threshing and putting grain in store. The husband is expected to guarantee the family food, shelter and security. It is at his discretion that strangers are welcomed, non-blood related children are adopted and the family is represented at village or mutual help groups (see also Bauer, 1985). The husband is also engaged in part-time herding or feeding of livestock, and selling of animals and crops.

Women are primarily engaged in food preparation; caring for children and members of the household who stay at home for health or other reasons; weeding, harvesting and threshing of pulse crops; collecting fuel wood; and marketing.

Children's responsibilities can be divided into two, depending on whether or not they are going to school. Within each group, further divisions can be observed. Children who are in elementary school will have a greater farm workload in anticipation that they may drop out. Those who are close to finishing high school are given less of a work burden. Children aged 15 or less are mostly engaged in herding, fuel wood collection and helping their parents. Children's responsibilities also differ by gender. Girls aged 13 or more are usually assigned to household duties or the wife's share in farming. Girls help their mothers and boys, their fathers. Girls who are close to puberty spend most of their time close to the house under supervision by parents. Boys are expected to stay outside the house to experience what men are expected to perform.

The changes in village organization and the structure of authority in village-level institutions have also brought changes in the patterns of responsibility among members of household. The youth in the pre-1974 period had no formal age-specific institutions. Their labour-time was allocated between working for the family, going to school or participating in social activities with youths in their age-group. Women's activity was also confined to the house, farm or markets.

Since 1975, youth and women associations have been established that stretch from village to national level. The youth are organized under the Revolutionary Ethiopian Youth Association (REYA) and the women under the Revolutionary Ethiopian Women Association (REWA). These institutions shared the labour pool of the family through meetings, group activities and festivities. As a result, the head of the household is involved in some house work such as taking grain to milling plant, looking after cattle, baby sitting and collecting fuel wood. Although men are still the head of the household, their activities are challenged by the wife through her political institution which promote equal participation of women at all levels. The traditional hierarchical decision-making and functional responsibilities have been loosened by the socialist institutions.

Since villagization, children in pseudo-cities have been exposed to the behaviour of children from other households. Social activities of children have increased because of greater proximity of residential units. Thus, the supply of children's labour-time to perform household activities have been reduced.

3.5.5.3. The Symbolic Element

The symbolic elements of a household are those which are physically visible and comprehensible, and help to distinguish between households. In this study, the symbolic elements include wealth (identified by clothing, cattle holdings, valuables, etc) and number of houses in one compound.[20] Furthermore, religious practices such as witchcraft and priesthood symbolize the status of a household in a village (see Bauer, 1985; Pankhurst, 1961, 1966, 1967).[21]

The type of household and the degree of inter- and intra-household cooperation in Ada and Selale varies primarily by wealth and prestige (Netting, et al., 1984; Wilk, et al., 1984, 1989; Bauer, 1985) . In addition to the husband, wife, children, and grand parents, a household may include adopted children and close relatives as its members.[22] Wealth and prestige attract and hold the members of large households while the poor can usually sustain only smaller groups of co-residents (Netting, 1982). The tasks of reproduction, feeding, child care and education modify production, distribution and transmission.

Poor households cooperate in production and consumption as a form of social insurance, thus spreading the poverty around. This is more typical of Ada than Selale households. In the pre-1974 period, large households that pooled in production and distribution tended to be stable and have generational continuity. While this behaviour of households is still observed in the study sites, differences in wealth or income are less visible in the post-1974 period. Wilk (1984, 1989) argued that households that cooperate in scheduling labour or in pooling for distribution tend to be less stable and often fragment. However, this is true of households that are more integrated into the market system and where village institutions emphasize economic cooperation (Ada) more than redistribution (Selale) (see chapter VI of this thesis).[23]

Poor households can not support many members. Therefore, these households include only a husband, wife and unmarried children. If households are classified by the degree to which they are at or below subsistence, those which are close to abandoning farming are often young families including only the husband, wife and infants as members. Children between the ages of 1 to 10 may in these cases be given for adoption. Those who are at least 10 years of age may be hired out for herding, farming or may migrate to urban centres. Households which are wealthier and have high social status include about ten members.

The household passes through certain stage of expansion and contraction. In the stage of high reproduction, the number of members of a household increases faster. This is a stage where household heads assert their existence and claim status as responsible persons. It is a stage where they require respect not only from unmarried but also from all age groups. As children grow and reach adulthood, they will get married and leave the household. The size of the household will continue to decline if households are poor (Table, 3.2). If they are rich, however, there is a tendency to adopt children of close relatives, friends or poor households to fill the deficit in the labour pool. Prestige comes from supporting a large family. Thus, when members of the nuclear family leave the household, parents, relatives or friends of parents are requested to live with them, even if they are old. However, as parents grow old, the size of the household declines and it is the children's obligation to request parents to live with them. This cycle is circular and may be obstructed by policies such as villagization. The labour pool reaches its peak, stays there if households are rich and declines among the poor families.

In general, households in the Selale region tend to have larger number of relatives than Ada farmers (Table 3.2). In both regions, wealth, expense, feed area, off-farm income, crop area, worker: consumer ratio and experience increase with increases in family size. Significant differences, however, are observed with respect to feed area and wealth. Selale farmers had larger feed areas while the value of wealth of Ada farmers is greater than that of Selale farmers. Table 3.2 also indicates that while larger family sizes in the Ada region are associated with older farmers (farming experience of 40 or more), larger family sizes in Selale region are associated among farmers with less than forty years of farming experience.

One way to identify a household in a family cycle or distinguish one household from the other is by the number of small houses or huts built adjacent to the main house. Male members greater than 15 years of age show to their parents and villagers that they are capable of building their own houses (like their fathers). Starting at this age unmarried male members of the household are considered to enter the stage of manhood. The more the number of small houses, the more the number of males of working age. This indicates a large labour pool. In some areas, manhood is manifested by the ability of a male member to pair oxen and plow without the father being present. Villagers think of him as a possible candidate (groom) for their relative's or friend's daughter. The ability of a household to raise children who have entered adulthood, demonstrated through the houses they build, their activity in farming, their discipline, and their passing school examinations, brings prestige to the family.

Table 3.2.: Mean Value of Selected Features of Households by Region[1/]

	Family Size	Rela.	Wealth	Expen	Feedare	Offinc	Areacro	Work:Con	Exper.
Selale	2	8	2180	431	12000	154	14950	0.66	26
	3	7	2121	589	10784	174	20492	0.67	25
	4	8	2728	597	18270	98	24395	0.73	33
	5	9	3360	632	11789	191	22209	0.67	37
	6	8	3157	592	16313	138	25445	0.77	32
	7	9	3167	676	20707	189	25451	0.78	32
	8	9	3190	708	19759	149	24495	0.69	32
	9	8	3331	739	22337	146	25909	0.78	35
	10	10	3230	788	22095	149	27775	0.81	27
	11	10	3714	985	29986	109	26871	0.79	32
	12	9	3492	897	20182	249	32125	0.88	38
	13	10	3780	985	28246	217	30175	0.84	41
	14	18	3801	871	20452	274	32242	0.89	35
	15	14	3890	886	21594	284	38450	0.85	36
Mean	9	10	3224	741	19608	180	26499	0.77	32.93
SD	4	3	554	167	5628	58	5695	0.08	4.60
CV	49	30	17.18	22.54	28.70	32.05	21.49	10.16	13.97
Ada	2	4	4568	356	4500	57	22375	0.73	26
	3	5	4131	476	4167	45	22583	0.69	34
	4	6	3911	431	6600	78	25600	0.56	33
	5	7	5132	493	4480	98	23830	0.78	32
	6	8	5635	498	9008	89	28929	0.76	35
	7	7	5350	627	9287	98	21433	0.87	36
	8	6	5989	644	9519	142	22733	0.89	38
	9	7	5409	684	8750	182	26125	0.87	41
	10	4	5885	672	9200	165	27613	0.88	44
	11	9	5719	679	9375	200	30013	0.91	40
	12	7	6224	889	10625	198	38750	0.89	41
Mean	7.00	6.36	5268	586	7774	123	26362	0.80	36.36
SD	3.32	1.57	763	151	2377	57	4996	0.11	5.12
CV	47.38	24.62	14.48	25.75	30.58	46.05	18.95	13.70	14.09

[1/] Where Rela- is number of relatives in nearby area; Wealth- is measured as the sum of the market value of crop and livestock outputs (in Ethiopian birr); Expen- is total expenses for farming, purchase of non-farm produced food products, expenses for school, clothes and taxes (in birr); Feedare- is the area grazing land and stubble extensively used as a source of feed (hectares). Offinc-is income from off-farm employment (birr); Areacro-area under crop production (hectare); Work:Con.- is the ratio of household members between the age of 16 and 60 to those who are below 16 and above 60; Exper.- is the number of years of farming experience.

2/ CV- refers to coefficient of variation [{standard deviation (δ)/mean}*100].

3.5.6. Household Integration: Householding

The household is integrated into the wider community through involvement of its head in several fields of social relations including parish, wards, neighborhoods and feasting associations (Bauer, 1985). Householding is a process through which ties of households with other people, villages and institutions are expanded or contracted. However, this process is forced to be less functional in the Ethiopian highlands because some of the essential resources are fixed or have no market. Land is a public property which cannot be sold, bought or leased by individuals. Farmers were not allowed to hire labor until 1991. Thus, the degree with which households integrate with other systems may be limited. When land is fixed, even in the absence of property rights, an alternative means to perpetuate the cycle of householding is intensification. This may take the form of adoption of new production technologies, increasing the "quality" of labor through schooling, use of available infrastructure such as roads and markets, and alternative investment decisions intended to counteract forced intervention of government into the system.

Households within a village can establish relationships with other households because of their common membership in a parish, social networks or village (Bauer, 1985; Hoben, 1973; Levine, 1974). Neighborhoods (defined in residential terms) are the locus for most social networks. Neighbors have reciprocal rights and duties in supporting one another's lives in times of crisis and ceremonies.

Householding as a process of integrating households into larger systems can be examined with respect two factors, among others: markets and land tenure. These factors influence the social and economic organization of households and villages.

3.5.6.1. Householding and Markets

By its very establishment, the villagization program created pseudo-cities that are close to main roads, cities, schools and markets. Consequently, households could easily be integrated to higher political and marketing systems.

The way markets affect farmers' behavior, and how households see and utilize opportunities offered by markets can change production and consumption decisions (Polyani, et al., 1957). According to Messing (1962), unlike Arabs, the Amhara are too proud and not so intent on economic "maximizing" as to resort to bargaining with a customer. An Amhara refuses to admit that their wares on display are for sale, thus avoid having to carry unsold goods home, they sell cheaply when the markets begin to close. The same marketing behavior is observed among households of the study sites. Thus the proverb advices the buyer,

"To church (go) early,
to Market (go) late"

To maximize the use of family labor for farm production, most market days are set on holidays or weekends (e.g., Saturday).

In the rural Ethiopia, there are several village-level markets. Most of these markets are located close to large villages or cities. On average farmers have to walk about five kilometers to these markets. Traders travel to these village markets to purchase crop and livestock outputs. These traders sell their purchases to large wholesalers or retailers in major cities, or directly to consumers.

The activities of traders were curtailed between 1978 and 1991 as the result of the policy of the socialist government that was intended to eliminate middlemen from the marketing channel. However, they were not completely removed from the marketing system. Some traders obtained authorization from the government to purchase grain from selected regions. In return, they had to sell part of their output to the state marketing agency. Since, 1990, however, any individual could trade without any restriction.

Ada is located closer to the capital city (50 km) and other major cities such as Debrezeit, Nazareth and Mojo. Ada has been the focal region for several development projects since the 1960's. These projects have constructed schools, roads, water wells and clinics. Thus, farmers in this region have more infrastructures compared to those in Selale. Ada farmers have easy access to credits and crop production inputs from the Ministry of Agriculture, city merchants and development projects. Selale, on the other hand, is located about 130 km from the capital city, and closer to a small city called Fitche. Most farmers do not have access to means of transport and roads compared to farmers of Ada region.

Villagization has stimulated the integration of households into the marketing system by facilitating access to traders, merchants and government marketing bodies. Furthermore, it facilitated travelling of household members who planned to sell their crop or livestock outputs in major markets. Consequently, few farmers shifted to trading (i.e., became part time farmers). This may have reduced the number of traders or middle men travelling to rural or village markets.

The formations of socialist institutions (e.g., PAs) have given credibility to people involved in special skill-requiring activities (e.g., small scale craft producers, iron-smiths and other talents). Skilled manual labor is traditionally regarded as a misfortune (Messing, 1962). It often led, during feudal times, to long term military conscription. Hence the proverb,

"From cleverness of hand, (you get) serfdom;
from cleverness of mouth, (you become boss)!".

Indirectly, farmers are integrated into the larger marketing system through these skilled farmers who also depend on outputs of industries (e.g., metal for making plow and hoe).

3.5.6.2. Householding and Land Tenure

Ethiopian farmers have been subjected to semi-feudal or feudal regimes since the 4th century A.D. Ever since, taxes have been extracted for running the administration. The reciprocal relationship between the state and farmer has, however, been unbalanced thus creating hostility, fear and anguish (Pankhrust, 1961, 1966; Hyden, 1980). In fact, since taxes are still generally viewed as a sort of tribute to the feudal chief, a deposit of money is regarded as surrender.

Non-secured tenancy has been cited as one of the most important structural obstacle to agricultural development. Farmers associate non-secured tenancy with the absence of ownership right. Appropriate agrarian reform should include land markets, access to credits, roads, inputs and output markets as its components. Ada and Selale farmers did not receive the full benefit of the land reform, except becoming the sole user of their production, because of the absence of many components of an agrarian reform.[24] Compared to farmers of Selale region, however, Ada farmers have benefited from the land reform because of the relatively good infrastructure in this region. Group discussion with farmers of both study sites indicated that Ada farmers rank the effect of non-secured tenure on patterns of resource allocation relatively lower than Selale farmers.

In pre-1974 period, most peasants were tenants who had to share part of their produce with absentee landlords, and land owned by small farmers was purchased by mechanized farmers. In the post-1974 period, peasants have had only user rights. There has been no land market. Their land was reshuffled more frequently within or between PAs (Belete, 1989). Therefore, most farmers did not invest in short-term fertility-increasing inputs (e.g., inorganic fertilizer) or long-run soil conservation methods (e.g., terracing). The way households feel about non-secured tenancy may affect resource allocation and variability in production (see Chapter VIII).

A relatively greater number of components of an appropriate land tenure reform are available in Ada compared to Selale region (see section 3.5.6.1). Ada is also easily accessible to private traders who are authorized to operate in the region. Consumers from the capital and nearby cities travel to Debrezeit (capital city of Ada district) to purchase grain directly from farmers. Thus, they can receive better prices for their products compared to Selale farmers. Nevertheless, Ada and Selale farmers are equally affected by the land reform policy and changes associated with the use of land and village-level institutions of peasants.

Householding may endanger the ability of households to secure subsistence requirements because of involvement of family members with city-type life style and not much time on farming. Moreover, farmers have to travel up to three hours to their crop fields from these pseudo-cities.

3.5.7. Economic Characteristics of Ada and Selale Regions

3.5.7.1. Introduction

Previous sections described the social and institutional characteristics and forms of integration of households in the study sites. Equally important aspects of households are their economic characteristics. Before examining the structure of crop and livestock production, some basic features regarding the economy of Ada and Selale households are described below.

Prior to 1974, most fertile arable lands in Ada and Selale regions were owned by landlords. Since the overthrow of the imperial regime, however, two major changes have taken place. Firstly, every individual capable of farming was entitled to obtain land. Secondly, the average farm size was frozen at 10 hectares and varied by family size (see Chapter II).

Selale farmers own larger numbers of livestock than do Ada farmers. Historically, the Oromos are herders or pastoralists (Levine, 1974). They regard livestock husbandry more highly than crop farming. However, the necessity to produce grain on the farm, their versatility in adopting the Amhara culture and increases in human population resulted in sedentary farming, involving crop and livestock husbandry. In the Ada region, while the majorities are Oromos, acculturation took place earlier than Selale. Moreover, proximity to the capital city, the fertile nature of the soil, and high density of population resulted in sedentary plow farming earlier than in the Selale region.

Selale has more grazing land than Ada. Selale farmers regard livestock husbandry more highly than crop farming, and believe that it is their sphere of expertise. Ada farmers face shortage of feed and regard crop farming more highly than livestock husbandry. The historical, geographical and cultural developments that took place as early as the 14th century contributed to differential resources, experience, goals, and preferences of farmers in the study sites. That is, Selale farmers prefer livestock over crop production. The reverse holds for the Ada region.

Selected social and economic characteristic of the study sites are presented in Table 3.3. Significant differences are observed with respect to the number of people living in the rural and urban areas between the study sites. Ninety percent of households in the Selale region are rural while only seventy percent of Ada population lives in rural areas (Table 3.3.).

Ada farmers allocate most of their land to crop farming, while arable and grazing lands occupy almost the same amount of land in the Selale region (Table 3.3.). With respect to climate, both Selale and Ada regions belong to the high altitude geographic zone. Nevertheless, most of the study sites in the Selale area are located at an altitude of 2200 or more meters above sea level (a.s.l.). The study sites in the Ada region are located at an altitude ranging between 1800 to

Table 3.3.: Major Socio-Economic Features of the Study Sites

Categories	Sub-Categories	Selale	Ada
Population	Male ('000)	237	145
	Female ('000)	239	144
	Rural (%)	90	71
	Urban (%)	10	29
Land use ('000 hectares)	Arable	287	156
	Grazing	215	5
	Forest	36	15
	Other	67	37
Climate (percent):	Lowland	33	34
	Woinadega	44	57
	Dega	23	9
PAs (numbers):		270	193
PA Members ('000)		70	40
SCs (numbers)		65	40
SC Members ('000)		62	30
PSc (numbers):1/		31(7)	40(1)
Soil Type (percent):	Black	35	70
	Red	41	25
	Brown	9	5
	Grey	15	-
Livestock Population ('000):	Cattle	342	150
	Sheep	247	13
	Goat	78	49
	Draft	66	2
	Chicken	187	104
Yield/ha (quintals):	cereal	10	14
	pulse	6	9
	oil	6	4
Milk yield/yr (litres):	local cows	390	250
	cross-bred cows	2550	1950
Input use:	Fertilizer (qts/ha)	0.06	0.6
	Improved Seed	0.004	0.04
	Pesticides (kg/ha)	0.04	0.05

1/ It was difficult to obtain accurate figure regarding the number of households in these PCs. However, the average number of households per PC is about 46. The numbers in bracket indicate those that were functional until April 1991.

Source: MOA, 1990; Base-line Information from the Current Research.

2200 meters a.s.l. Climatically, relatively more places in the Selale region experience the cold temperature of highlands. That is, the temperature ranges between -5^{0}C and 25^{0}C.

The size of PAs in the Selale region seems to be relatively larger than Ada. That is, the average land size per PA may exceed 800 hectares. The average grazing land per household in the Selale region is greater than that in Ada region. The size of grazing land per PA in Selale is larger than that in Ada

(Kebede, 1993). Furthermore, some areas in the Selale region are dominated by gorges and valleys, thus making measurement of land difficult. These factors may have contributed to inflated land size per PA. Nevertheless, various sources from MOA indicated that farmers in Selale region own larger crop and grazing lands than those in Ada (MOA, 1984, 1990).

The soils of Ada are mainly black (vertisol) while that of Selale black and red. Reddish brown soils are in general less fertile than black soils. Crop yield per hectare is larger in the Ada than in the Selale region. In addition to natural soil fertility, yields in Ada may be higher due to greater use of crop production technologies (e.g., fertilizer) per hectare of crop land.

Selale farmers own a larger number of livestock than those in Ada. Milk yields per cow for both local and cross-bred cows are higher in the Selale region compared to Ada.

3.5.7.2. Statistical Comparison of Selected Features of Study Farmers

In addition to differences and similarities discussed in previous sections, statistical analysis (analysis of variance) was carried out with respect to selected parameters (Table 3.4.). Graphical presentation of information contained in Table 3.1 is not possible because the units of measurement are not the same for all variables. Therefore, tabular presentation was chosen. The results suggest that the two regions exhibit statistically significant differences with respect to: i) the number of household members within the age groups of 11-15 and 16-60; ii) the education of the head and number of household members who have attended sacred, elementary and junior schools; iii) the degree of experience of the household head as an independent farmer; iv) the frequency of market and city visits; v) the number of livestock owned; vi) the number of relatives; vii) the average income received from the sale of grain and fuel wood; viii) the level of expenditure for seed and oxen input; ix) the farm size; x) the amount of milk produced per household; xi) the size of grazing area; and xii) the amount of grain produced (Table 3.4.).

In general, Ada farmers had more years of schooling, more years of farming experience and fewer relatives. They gain most of their income from the sale of grain while that of Selale from livestock and livestock products. The productivity of livestock (milk/cow) is higher among Selale farmers while Ada farmers produce greater crop yields per hectare. Most of these differences can be attributed to the socio-economic organization of villages and households described earlier.

Table 3.4: Selected Characteristics of Selale and Ada Farmers

Variables	Selale		Ada			
	N	Average	N	Average	F-Value	Prob>F[1/]
Household members who are who						
<11 years	173	2.01	41	1.99	0.412	0.469
11 15 years	137	1.45	25	1.3	4.46	0.03*
16 60 years	207	1.75	48	1.5	4.52	0.03*
>60 years	55	1.02	13	1	0.292	0.591
Education of HH head (yrs)	55	2.5	23	3.6	5.671	0.001*
Number of children who attended:						
Sacred	117	1.72	17	1.12	15.25	0.001*
Elementary	187	1.01	38	1.37	14.28	0.002*
junior	81	1.37	37	1.16	4.002	0.05**
high	76	1.04	19	1.11	1.396	0.24
quit	110	1.39	23	2.18	5.6	0.010*
Experience (years)						
dependent	176	11.24	50	13.44	0.044	0.834
independent	176	24.58	50	27.88	4.173	0.04**
Visits per month (day)						
market	184	3.31	47	2.04	21.38	0.003*
city	187	3.71	44	2.2	21.13	0.001*
No. of relatives	203	9.8	48	7.4	21.52	0.005*
Income from:						
sale of grain	203	230.27	49	828.6	65.46	0.006*
Sale of Livestock prod.	194	451.4	22	203.11	1.09	0.098
sale of fuel wood	169	343.58	31	63.97	13.84	0.004*
weaving	6	306.41	1	173	0.442	0.535
daily labor	20	126.73	4	123.75	0.003	0.956
remittances	18	139.88	3	73.33	0.354	0.559
Expenses for:						
purchase of food	214	268.2	50	228.14	2.366	0.125
clothing	205	114.49	39	106.09	0.309	0.579
tax	217	43.14	51	37.03	15.584	0.008*
school fee	153	22.26	38	28.62	3.049	0.04**
farm tools	216	35.89	52	36.81	0.109	0.742
seed	122	66.41	11	34.39	4.17	0.04**
ox rent	12	8.25	4	30.25	13.92	0.002*
labor	182	91.41	41	105.53	0.819	0.366
Milk production per Month (liters)						
local cows	193	56.9	35	42.6	6.79	0.05**
cross bred cows	66	320.35	14	186.29	5.76	0.011*
Area under (hectares)						
crop	217	25470	52	23115	19.562	0.001*
grazing	208	8205.63	37	1641.3	26.29	0.006*
Livestock number	165	10.89	16	5.18	0.69	0.016*
Crop Production (qts)	217	14.88	52	21.41	2.98	0.05**

1/ * and ** refer significance at 1 and 5 percent respectively

3.5.7.3. Livestock and Crop Production Systems

3.5.7.3.1. Livestock Production

The functions and structures of a household are influenced by the process of householding, especially by government policies. The technologies investigated by this study have not brought radical changes in the social organization of households in the study sites. The outcomes of activities performed by households are reflected by the structure of crop and livestock production. Since a household in the Ethiopian highlands is defined by virtue of both harvesting crops and raising livestock, the next section discusses the production system, and elements of household income and expense. The survival of households in the Ethiopian Highlands is determined by access to the services of land and cattle. A household without working oxen is not regarded as a farmer but as a wage worker. By virtue of being native to an area, he/she is entitled to receive land from PAs. If a household does not own oxen, friends and relatives may hesitate to lend their working oxen to him/her to prepare a plot for planting.

Livestock are a means to crop production. They are complementary and competitive enterprises (Gryseels, et al., 1986). Peasants have to use oxen to prepare land for crop production. Income obtained from crop production helps to purchase cattle, feed or medicine for livestock disease.

The herd structure of Selale varies by altitude (Finnida, 1989). At lower altitudes, where farming activities constitute the largest part of the smallholder's income, the herd structure is characterized by a larger number of oxen kept for traction purposes. At an altitude of 2800-3000 mts a.s.l., the percentage of cows in the herd and their contribution to household income increases. The presence of large numbers of cattle across different altitude regions is the result of variation in feed resources and degree of concentration in crop or livestock enterprises.

The proportion of cows in the herd for the Selale and Ada regions are respectively forty-five and twenty-five percent. This suggests that Selale farmers have the resources necessary for raising large numbers of cattle and that livestock husbandry is a more important enterprise in this region compared to Ada.

Milk production per local cow of test and control farmers is 57 and 51 litres per month for Selale and 43 and 38 for Ada respectively.[25] The average lactation period is ten months. The introduction of cross-bred cows has produced significant changes in milk production. Milk production levels of cross-bred cows indicate 316 and 190 litres per month with a lactation length of 9 and 8 months for Selale and Ada areas respectively (Table, 3.4). From Table 3.5, it can be seen that the proportion of fresh milk sold is higher among test farmers while milk processed is higher among control farmers. Selale farmers who own local breeds of cows tend to process a larger proportion of the milk compared to farmers in the Ada region. Farmers of both regions sell most of the butter compared to what they consume. Furthermore, the percentage of butter sold is lower among

Table 3.5. : Percentage Distribution of Livestock Products and Crop Disposal

	Selale		Ada	
Livestock Products	Test	Control	Test	Control
Milk -Sold	80	30	75	58
- Consumed	10	10	10	13
- Processed	10	60	15	29
Total	100	100	100	100
Butter - Sold	75	90	88	75
-Consumed	25	10	12	25
Total	100	100	100	100
Crop - Sold	10	13	20	20
- Consumed	75	74	75	71
- Others[a/]	15	13	5	9
Total	100	100	100	100

a/ refers to crops used for exchange and feasts.

test farmers when compared with control ones. The difference in crop production and disposition between Ada and Selale region is smaller. However, Ada farmers sell a relatively larger proportion of their crop output compared to Selale farmers.

The degree of concentration on either livestock or crop enterprises is influenced by availability of grazing land or sources of feed. An average Selale farmer owns 0.8 hectares of grazing area whereas an Ada farmer owns only 20% of an average Selale farmer's holdings. The size of grazing area and the capacity to produce sufficient feed are equally important. Farmers supplement livestock feed with stubble grazing. Farmers in Selale and Ada feed their livestock on 1.3 and 1.7 hectares of stubble respectively. Moreover, farmers use end-products of beer brewing as a source of feed. On average, test and control farmers of Selale brew about 35 and 26 times per year while Ada farmers brew 26 and 35 times per year. The higher frequency of brewing among test farmers may be partly because of the high feed consumption of cross-bred cows.

The success rate of new crop and livestock intervention strategies is influenced by the willingness of farmers to sacrifice one enterprise for another (e.g., crop for livestock). Farmers in Selale are willing to allocate 0.4 hectare of crop land for growing livestock feed, Ada farmers commit only 0.01 hectare. Moreover, study farmers of both sites are willing to spend about 30 birr/cow per month for the purchase of feed.[26]

Livestock, in addition to contributing to crop production, are a repository of wealth and/or savings. A farmer's livestock holding is influenced by the number he possessed at the time of marriage. Selale and Ada farmers, respectively, reported having 5 and 4 livestock at the time of their marriages. In 1991, the average Selale and Ada farmers had 10 and 5 livestock, respectively.

3.5.7.3.2. Crop production

Both regions are located in the high potential cereal livestock zone (HPCL). Nevertheless, the preference of peasants for new production methods differs. In Selale, farmers prefer methods of increasing livestock production and productivity, whereas crop production is the preferred enterprise among Ada farmers. Nevertheless, livestock for traction purposes are a requirement in both sites.

The average distance travelled by farmers to their plots ranges from 30 to 40 minutes. The average area cultivated by test and control farmers is 2.6 and 2.3 hectare for Selale and 2.3 and 2.1 hectare for Ada region. Aggregate production (cereals and pulses) for test and control farmers is 19 and 13 quintals and 25 and 22 quintals for Selale and Ada areas respectively. The cropping pattern for the study sites is given in Table 3.6. An important difference between Ada and Selale regions is with respect to the cultivation of teff. Teff is both a staple and a cash crop. Ada is known for growing high quality teff and their teff commands higher prices compared to teff grown in the Selale region.

Table 3.6.: Percentage Allocation of Area among Different Crops

	Selale		Ada	
Crop	Test	Control	Test	Control
Teff	30	32	49	50
Barley	23	21	19	20
Wheat	13	15	11	10
Bean	18	19	16	13
others*	16	13	5	7
Total	100	100	100	100

* - include oats, chickpeas, lentils, peas and safflower.

Farmers employ various kinds of physical production inputs. The most important are oxen, seed and labour. In addition, commercial fertilizer, pesticides, improved seed and manure are applied to selected crops. Adoption of production augmenting inputs embodies risk. In subsistence agriculture where marketable surplus is limited, allocating a portion of the revenue from farming to investment in new inputs is a risky decision. Thus, farmers prefer to use purchased inputs for crops or classes of livestock which have relatively larger direct or indirect market value.

Average fertilizer consumption varies between 40 and 96 kilograms for the Selale and Ada areas respectively. An average Ada farmer applies 30% of the commercial fertilizer on wheat and 60% on teff. Selale farmers apply 10, 20 and 50% of total fertilizer purchased on wheat, barley and teff, respectively. The average pesticide use per household is 5 and 2 kilogram for Ada and Selale areas, respectively.

The average annual oxen and labour use per hectare are 25 oxen days and 43 man-days in the Selale region. For Ada, the average oxen and labour use are 34 oxen days and 48 man-days. The difference in the amount of oxen and labour use between Selale and Ada farmers is mainly due to the fact that Ada farmers grow teff on a relatively larger area. Teff requires greater expenditure of oxen and labour inputs per unit area compared to other crops. Moreover, Ada farmers are relatively more knowledgeable about crop farming compared to Selale farmers. Thus, the amount of oxen and labour inputs used by Ada farmers may be closer to what may be "required" for maximum crop output per unit area.

3.5.7.3.3. Household Income and Expense

Households in Ethiopia are integrated into the market system through its members purchasing non-farm produced goods, selling outputs and participating in trading. This linkage is not restricted to the local level but occurs at the regional and national levels. The household may be influenced by what has happened in the world market through the prices of inputs it purchases. For instance, Selale producers understand the value of artificial insemination or of introducing exotic genes into local breeds. Changes in these techniques would be felt at the household level via increases in prices paid to acquire these services.

Both study sites can be characterized as self sufficient in terms of food production. Sales of crop and livestock products provide the major sources of income. Male and female members of a household engage in daily (wage) labour, while only male members perform weaving. Income and expense items are given in Table 3.7. The major categories of expenditure items include production inputs (tools, seed, labour, and ox rentals), consumer items (salt, sugar, coffee, soap, gasoline, and oil), clothing and land tax.

Six percent of the sample farmers in both sites respond that they rent oxen. In many cases, peasants make use of traditional systems such as "debo" or reciprocal arrangements whereby a household will work on another person's plot with his oxen or labour. Thus, it is not common to rent an ox for cash but rather in kind. In this study, the amount of grain paid was converted into monetary equivalents.

Income from the sale of livestock is higher among Selale than among Ada farmers, and the reverse is true for sale of crop products. In both cases, test farmers have higher incomes than control farmers.

Table 3.7.: Pattern and Amount of Expense and Income

Category	Selale		Ada	
	test	con-	test	con-
Expense				
Farm tools	35	38	37	40
Consumer goods& Cloths	293	245	250	206
Tools	40	34	41	31
Seed	74	59	53	15
Ox rental	8	10	40	21
Labour	90	50	99	108
Land tax	40	40	40	40
Other expenses	122	72	88	28
Income				
Value of grain sold	447	373	880	720
Sale of Livestock & Livestock	493	208	277	109
Sale of fuel wood	332	355	54	71
Other Income	150	75	80	90
Gross income	1422	1011	1291	990
Total expenses	702	548	648	490
Net income	722	463	643	500

3.6. Summary

This chapter pointed out that difficulties in identifying the unit of analysis are caused by conflicting arguments regarding the structure and function of households, and differences in the socio-economic environments within which households operate. Thus, a working definition of the unit of analysis is a necessary step in the study of households in a specific society.

Households in the Ethiopian highlands should be the primary target of development strategies designed to increase food production. Food production in these highlands can be increased through better management of existing resources. More importantly, however, this can be accomplished with the help of new technologies. To examine the feasibility of this proposition, three crop and one livestock related technologies are selected. The research sites are typical of the Central Ethiopian highlands. Mixed farming is the dominant system of production in these regions with specialization either in crop and/or livestock production.

Appropriate research designs are identified. Households who adopted cross-bred cows are grouped as test farmers. These farmers are compared with farmers who did not adopt cross bred cows (control group).

The social and economic organization of peasants has been altered by the villagization program. Elements of the pre-1974 values, behaviours, norms and forms of organizations of households have been modified. Interpersonal relationships became more impersonal after farmers started living in pseudo-city type residential units. Moreover, psychological values related to land and sacredness of residential units have been loosened. But, interaction among households themselves and with external institutions has increased. This is primarily due to changes in residential arrangements from scattered to city-type villages that are accessible by vehicles and other means of communications. However, the introduction of new technologies has not brought about significant changes in the social organization of households. The structure and functions of households in Ada and Selale regions do not show substantial differences. Thus, the findings from these study sites can be presumably applied in other regions with similar socio-economic organization with little modification.

Visible and measurable characteristics of households such as houses, incomes, crops and livestock production systems reveal that Selale farmers prefer livestock related technologies while those of the Ada region prefer crop related technologies. This preference pattern follows regional advantages such as availability of feed and pasture in the Selale region compared to that of Ada. Furthermore, it also reveals ethnic-enterprise specializations. The Oromos of the Selale region believe that their experience and "soul" is intertwined with livestock husbandry. Ada producers prefer to specialize in crop production. Nevertheless, farmers of both regions prefer to secure subsistence requirements through intensifying crop production.

Farmers in the Ada and Selale regions, despite similarities with respect to certain parameters, may follow different rules and behaviours of decision-making regarding adoption of technologies and responses to intervention strategies. To examine the manner in which households select their goals and their norms and patterns of decision-making, and the consequences of the decisions for the adoption of technologies and the efficiency of production, "appropriate" methods of analysis should be utilized. Chapter four outlines methods of decision analysis selected for this study.

3.7. Notes

1. Table 2.1., in Chapter two, includes some portion of the lowlands to group the total population into the three agro-climatic zones.
2. One study site in Northern Shoa includes the Selale and Debrelibanos districts. These two regions were under one district until 1989, the old Selale. Preliminary comparison of data pertaining to the characteristics of study farmers showed statistically insignificant differences between the two sites. Therefore, the two sites were treated as one under the Selale district.
3. Teff is a local name of a crop that belongs to a grass species known as *Eragrostis Abyssinca*. It is cultivated as staple crop only in Ethiopia.

4. Compared to bullocks of Asia and oxen from cross-bred cows, the local zebu breeds provide less traction power. It requires relatively long periods of time and heavy investment to obtain cross-bred oxen. Moreover, traction power of local breeds can be increased either through adequate and balanced feeding or modifying plowing implements (Astatke, et al., 1986).
5. This figure refers to the state of the technology in 1985.
6. CIMMYT stands for "Centro Internacional de Mejoramiento de Maiz y Trigo". It is an International Agricultural Research Centre, under the Consultative Group on International Agricultural Research (CGIAR), located in Mexico.
7. Prior to selection of the control group, the sample size was determined according to the following procedure. The sample size (N) is given as: N= (KV)2/D2 , where D is the largest acceptable difference (in percent) between the estimated sample and the true population parameters which are calculated using sample information. K is a measure of confidence (in terms of the number of deviations from mean) with which it can stated that the result lies within the range represented by plus or minus D and V is the coefficient of variation of yields (Casely, et al. 1982).
8. This applies only to cross-bred cows. Some farmers were over-burdened by debt and others were identified as less diligent in managing animals by officials of Pas. The project has to insure that farmers will pay the loan on time. Therefore, some kind of discrimination was applied in the selection process.
9. Evaluating the economic well-being of other farmers would make farmers as if they are intruders into private life of others. This is not a socially acceptable norm. However, options were explored with groups of farmers and they suggest this method as a "better" alternative.
10. A general meeting of all members of twenty PAs in the Selale region was conducted. Members were briefed about the objectives of the research and how long it might take to complete the questionnaire.
11. These policies were effective until 1990/91 crop year.
12. See next section for historical, socio-cultural and economic characteristics of farmers in the Selale and Ada regions.
13. The title of this commentary is given by the researcher.
14. A minimum of 24 pieces are required to build a small house of about 30 meter square (i.e., 480 birr). This size of house is not large enough to accommodate more than three persons. Considering an average family size of five, it is costly to build a new house.
15. Personal communication with students who recently came to McGill (in 1993) indicates that most farmers are still living these pseudo-cities.
16. If there were a land market and no restrictions on land holdings, we might see most Selale farmers becoming livestock raisers rather than their current crop-livestock combination. This limitation on the expansion of land area for large scale livestock production means that they receive less income from livestock for purchases of grain and consumer goods. In the face of increases in grain prices, therefore, they may be forced to expand crop farming. Crop farming in Selale is also linked to acculturation.
17. This doesn't imply that the economy of affection described in Hyden (1980, 1983) doesn't exist. Rather, the political structure of the pre-1974 regime tends to reflect authority structure of village institutions (Korten, 1972; Pankhrust, 1961, 1966; Hoben, 1973; Levine, 1974).
18. Exceptions to these are married couples who may sometimes temporarily live with parents of the groom until they build their own house.
19. Less proximity to cities and absence of off-farm work limited the movement of

household members outside the homestead. Moreover, openness of household head to ideas from spouses could reinforce delegation of more responsibility to members in the Selale compared to Ada region.

20. In a situation where there is market imperfection, prices may not reflect the "actual" value of outputs or inputs. However, the best approximation to calculate what a producer will receive for his output is the nominal market price. Thus, in calculating the value of wealth, the amount of grain and each class of livestock are multiplied by their respective market prices.

21. Identification of the self among the Amhara is with a place name of a parish or region, while lineage is used by Oromos. The Oromos in the study sites, however, use region as a symbol of identification.

22. The Oromo household is normally a small and nuclear family unit while the Amhara household contains members who are not spouses or immediate kin. The later characterizes household structure in the study sites.

23. Village institutions that emphasize economic cooperation tend to embody impersonal relationship among members and could dissolve easily. It can be argued that institutions in Selale exhibit organic solidarity while that of Ada mechanical solidarity (see Durkheim, 1945).

24. Between 1978 and 1990, however, farmers had to sell a portion of their production to the government market at a low market price, thus reducing benefits from the land reform.

25. These differences are statistically significant (at 5%) between regions but not significant within region.

26. Until 1992, 1 U.S.$ = 2.07; since 1993, 1 U.S.$ = 5 birr.

CHAPTER IV

REVIEW OF METHODS OF HOUSEHOLD DECISION ANALYSIS AND MODEL SELECTION

4.1. Analytical Problems to Household Decision Analysis

Studies of household decision-making in developing countries can be categorized into two types. The first derives its root from neoclassical economic theory. This approach assesses the rationality or efficiency of peasants in developing countries using classical theories of profit maximization (Webb, 1988). This is referred to as a "rationalist model". The second type of study (here called a "logic model") investigates the manner in which households choose among alternative strategies and the reasons underlying the hierarchy of goals or preferences of households. This line of study emphasizes the rules that households apply in discriminating among a set of possible decisions.

Rationalist models examine the marginal value of a production input as a method of analysis. Specifically, the value of each additional unit of output produced per unit of input is used as a basis to determine remuneration for resources used in the process of production (Menger, 1950). This method assumes that market prices provide the only catalyst that ensures efficiency in allocation of resources.

In line with marginal value analysis, mathematical psychologists and other social scientists developed the concept of utility. It is a composite function which combines an individual's preference for income in a risk-less situation and an individual's attitude towards risk or chance-taking.[1] In the process of translating Menger's concept into mathematical models, values are replaced by utility, value of inputs by costs of inputs and values of outputs by prices. In addition, marginal utility is changed from an ordinal to a cardinal measure. In a peasant economy, the value or utility vis-a-vis the drudgery of raising children, planting a particular combination of crops and so on have been investigated extensively (Barlett, 1980; Schultz, 1964; Ellis, 1988).

Other researchers have singled out labour as the most important input of human actions and have concentrated on examination of various aspects of labour in decision analysis. For instance, the value of output is translated into the quantity of labour required to produce each good (e.g., Ortiz, 1980).[2] The problem in this approach arises from difficulties in distinguishing levels of remuneration for labour according to its quality. Studies in decision analysis approximated the quality of labour by the techniques involved in the production of goods. This approach attributed distinctions between developed and developing nations to techniques of production and the skills that accompanied them. Labour in developed countries received a higher share of the value of output, while peasant's labour in LDC's is considered surplus and its marginal product is equated to zero (see for example, Lewis, 1954, 1984; Fei and Ranis 1964; Mellor, 1966). This implies that the amount of output produced in the agricultural sector of LDCs would not decline if labour is pulled out of agriculture. This approach has led to strategies that focus on the establishment of large scale farms in the agricultural sector, resulting in the eviction and impoverishment of peasants.

Douglas and Isherwood (1979), on the other hand, examined consumption in order to understand the value of goods produced. The value of a good embodies information such as what went into its production, where it was produced and who produced it, thus reflecting the relationship between the consumer and his social world. However, this approach cannot be used in quantitative decision analysis because not all the information embodied in a good are quantifiable, nor even identifiable.

Another method related to the approach adopted by Douglas and Isherwood is the use of labour accounting or the labour theory of value. There is a controversy in this method because of two assumptions. There is an assumption which states that what gives value to a commodity is how it relates to the individual and how it symbolizes social obligations. The second assumption states that what gives value to the commodity is how labour transforms nature and raw materials into other commodities (see Ortiz, 1983).

In contrast to economic rationality models, a distinction can be made between what is called socially necessary and surplus labour (Marx, 1970), the institutional framework which governs the allocation of labour (see Meillassoux, 1972 for example) and the rational allocation of labour among peasants (Norman, 1980; Geertz, 1973, 1983; Berry, 1980, 1984, 1986). These approaches to decision analysis believe labour is the most important single factor without which neither households nor societal production and reproduction are possible.

The logic model emphasizes rules that households apply in discriminating among a set of decisions. It differs from the rationality model because it does not set the objectives of households a priori. It concentrates not only on the rationality of decisions but more importantly on how these decisions are selected and executed (Webb, 1988; Gladwin, 1980).

Both types of studies of household decision-making face a common problem in that the characteristics of the household are generalizable. These studies assume that the response of households to opportunities and constraints are similar (Webb, 1988; Isaac, 1989; Netting, et al., 1984). Given the sharp division of labour and responsibility by age and sex within a household, members of the domestic unit do not necessarily use the same criteria for evaluating the feasibility of a specific decision (e.g., adoption of a new technology). However, several studies consider the male head as the only decision-maker (Barlett, 1980; Schultz, 1964, 1981; Wharton, 1969; Gladwin and Zabawa, 1987). Consequently, policies that are based on these kinds of studies may not be successful.

The theoretical definition of decision-making relies on the principle of "self-satisfaction". The components of decision analysis defined in psychology, economics or related disciplines cannot be applied to peasant's household decision-making without modification. In reality, there are no: i) generally accepted ways of measuring utility, ii) logical patterns of aggregating preferences of individuals and iii) smooth movements from one set of preferences to another.[3] Moreover, not all decision criteria and their consequences are measurable or quantifiable. This is especially true when producers operate in what has been called "the economy of affection" or "the moral economy" rather than in the market economy (Scott, 1975; Hyden, 1983; Rubenson, 1984; Netting, et al., 1984).

Models of decision-making at a household level assume the following for all practical purposes: i) the objectives, strategies and consequences of decisions are accepted by all members of the household, ii) the outcomes of activities pursued by households are shared by all members of the household, iii) there is essentially no contribution from relatives and friends outside the household, iv) all household decisions are made by a single person and v) the head is assumed to survive without its sub-units or members, thus member's contribution to the household income or survival is not properly valued or assessed (Webb, 1988; Ellis, 1988).

The framework of decision analysis utilized by several development studies assumed only one decision-maker (e.g., Ellis, 1988). This framework may be defended on the grounds that in every household, decision-making emerges from negotiation, disagreement, conflict and bargaining. Often times, the head of the household has a veto power over the ideas of his/her members if compromise cannot be reached.[4] If members of a household have divergent interests, they are brought to the attention of the head of the household. In many circumstances, these interests are accommodated. It is, therefore, possible to obtain information reflecting the interests of household members from interviewing the head of the household. Accordingly, this framework makes feasibility studies about the effect of public policies or projects easier because a large number of households affected by interventions can be studied.[5]

Public policies can be designed to satisfy Pareto optimality criteria.[6] Nevertheless, at a micro level, there are losers and gainers. If most households consent to the goals of a village or district, then it is possible to design policies that may bring more gains than losses. This assessment can be made if information obtained from each member of a household is evaluated prior to implementation of a project or introduction of a technology. Analytically, however, it is impossible to formulate production, consumption, reproduction and other decision models for every member of a household.[7] If these arguments are correct, statistical analysis of information from a household head, though not complete, is the best analytical tool for the study of household decision-making. All decision-makers, for instance farmers, consumers and firms, have commonality in the sense that they have to identify their goals, gather information, assess the choices and foresee the consequences of executing decisions. Executing decisions involves several factors specific to each decision-maker.

In between the rationalist and logic approaches to the study of households described above, several methods of analyzing decisions regarding intra-household resource allocation strategies have been proposed. Most of them, however, are linked to contributions from economists (Becker, 1981; IFPRI, 1992). These methods include the separate spheres bargaining model, an event-history approach, the altruistic model, the collective model, the new home-economics and the reciprocal claim model (IFPRI, 1992). The contribution of most of these methods to policy design, however, is limited (for details see IF-PRI, 1992).

In general, there are numerous objectives of households that govern their decision-making processes. Often times, these objectives are not captured by mathematical models. Instead, observation of what producers do, or how they allocate their resources and draw implications may be an appropriate approach to decision analysis in peasant economies.

In chapter I, it is stated that examination of household decision-making involves two aspects: the processes that take place before executing a decision, and the consequences of a decision. The first part involves identification of goals and strategies, information gathering and processing, production knowledge, intra-household decision-making and the decision-making environment. The second part involves assessing the consequences of decisions to adopt innovations and of risk-taking behaviour, the contribution of material inputs and know-how to the production process and the efficiency with which they are used, and production risk.

Modelling and analysis of the first part involves multiple regression and descriptive analysis using frequencies, percentages, means and standard deviation of selected parameters. Analysis of the second part also involves relatively rigorous statistical analysis. Selected equations are presented in the text. Derivations and additional explanations for the mathematical and statistical models are presented in the appendix to this thesis.

4.2. Household Goals and Resource Allocation Strategies

Analysis of household decision-making requires, among other things, identifying the objectives of household resource allocation strategies, and the kinds of production knowledge and information processed in making those decisions. Knowledge of farmers' goals or objectives provides an important basis for understanding their preferences for, and choices among, various farm adjustment strategies. Such information is valuable in estimating the acceptability of various intervention strategies to farmers.

The guiding principle of a household's decision-making is its objectives set in hierarchical order (Barry, 1984; Berry, 1986; Barlett, 1980; Cary and Holmes, 1982; Webb, 1988; McCann, 1986; Norman, 1980; Hyden, 1980). Household goals may include securing family survival, maximizing farm income, safeguarding income of the future, gaining recognition and prestige as a farmer, building a big house, expansion of livestock husbandry, belonging to a farming community, feeling pride of ownership, gaining respect for doing a worthwhile job, and independence and freedom from supervision (Eisemon and Nyamete, 1988; Warren, et al., 1988; Cary and Holmes, 1982). Studies in LDCs (e.g., Eisemon, et al., 1988; Barlett, 1980) identified family survival as the primary objective of households.

The objectives or goals of households vary depending on resource endowments, access to non-farm produced inputs, authority and support services from relatives. To attain their goals, households employ several strategies. These include, among others: i) planting more than one crop and raising different classes of livestock, ii) purchase of household equipment and livestock, iii) engaging in off-farm work, iv) storing, saving or selling various kinds of liquid or fixed assets, v) sending children to school and vi) investing in livestock or crop production increasing technologies.

The goals of decision-makers may determine the choice of strategies or vice versa (see Cary and Holmes, p.144-145). Other studies argue that clear separation of means (strategies) from ends (goals) is difficult (Lutz and Lux, 1979). The problem of distinctiveness of goals and means is further complicated by the social and economic decision-making environment. The study in Selale and Ada regions of Ethiopia implemented interview and group discussion methods to determine whether goals shape strategies or vice versa. Once the direction of causality is established, the structural relationships between goals and strategies of households can be examined. Intervention strategies can make the best use of financial and physical resources by focusing on those strategies which are strongly related to the goals of households.

To examine the degree of association or structural relationships between household resource allocation strategies and goals, a multivariate regression model with dummy variables is employed (Cary and Holmes, 1982). The model can be specified as:

$$Y_{ijk} = \alpha_{j1} + \sum_{c=1}^{q} \beta_{jc} O_{ijc} + \sum_{d=2}^{v} \theta_{jd} S_{ijd} + \sum_{e=1}^{r} \delta_{je} R_{ije} + \upsilon_{ijk} \text{ ...... (4.1)}$$

for $i = 1 \ldots n, j=1 \ldots p, k=1 \ldots w, c=1 \ldots q, d=2 \ldots v, e=1 \ldots r, (c, d, e) \in K$

where Y_{ijk} is individual "i's" ranking of jth goal on k^{th} strategy, α_{j1} (intercept) is the rank on strategy 1 for j^{th} goal, β_{jc}, θ_{jd} and δ_{je} are weights, O_{ijc} is a dummy variable (0,1) categorizing opportunistic strategies (c), S_{ijd} is a dummy variable categorizing strategic or long-range strategies (d), R_{ije} is a dummy variable categorizing risk management strategies (e) and υ_{ijk} is residual error term (see Chapter V).

Statistical analysis using equation (4.1) may not reveal the importance of all strategies that may be associated with the goals of households. Therefore, descriptive analysis of goals and strategies of households is carried out. Farmers are requested to answer questions in the following format:

> *After you have planted barley, the government announces that it will decrease or increase the price of barley. What will you do in response to this policy measure?*

This type of question reveals the resource allocation behaviours and objectives of smallholders that are reflected in changes of cropping plans, marketing strategies and livestock possessions.

4.3. Production Knowledge

The goals and resource allocation strategies of households depend on the production knowledge and information possessed by households. There is no hard and fast rule to measure or quantify production knowledge. Studies in cognitive psychology have demonstrated the usefulness of measuring knowledge using problem solving tests or comprehension ability (see Eisemon, 1988; Bransford, et al., 1983). Causal attribution or explanation of reasons for a specific problem has also been found useful (Eisemon, 1988).

To assess and measure agricultural knowledge, two approaches are employed in this study. First, problem solving tests are constructed to measure agricultural knowledge and skills related to current production technologies and practices. The tests are intended to examine the kinds of solutions households provide to crop and livestock production problems based on their agricultural knowledge. For instance, farmers who plant barley are presented with the following problem solving task:

> *Your barley plants are stunted exhibiting yellowish colour and do not grow tall enough to produce good seed. What are the possible causes of this problem? How may it be prevented?*

The information used in solving such problems comes from schooling, contacts with extension agents, friends, relatives and experiential knowledge.

Secondly, participatory evaluation of indigenous production knowledge and discussions are carried out. The major researcher and senior field assistants participate in plowing, planting, weeding, cultivation, harvesting and threshing of major crops for about five to ten days per study farmer. The researchers provoke conversation related to production knowledge of farmers. For instance:

Opening a dialogue on why households destroy specific weed species first; what do they believe are the causes of stunted plant growth; why do they plow a given plot many times before planting; and how deep do they want the seed bed to be before planting, say, barley.

Answers for these kinds of questions are also analyzed using flow charts involving for instance the type of crop disease, its causative agents and solutions or remedies.

Answers obtained from problem solving tests are scored to compare variations in knowledge of farmers within and between regions. The basis for scoring is answers obtained from discussions with farmers of different age groups. The premise behind this basis for scoring is that experience and indigenous knowledge vary by age. For the answers or solutions to reflect actual problems of farming in the regions, therefore, it is important to have a consensus from group discussions with farmers of different age groups.

4.4. The Decision-Making Environment

Kinds and patterns of decision-making vary depending on whether or not they are shared among households, kin or government agencies. Under this situation one can look for shifts in the locus of decision-making patterns of producers. Accordingly, one should ask, what decisions are left to them and how do they respond to the loss of decision-making power? For example, suppose the government is taking a portion of household's decisions regarding resource allocation or marketing. The decision strategies of households may change from marketing in the legal to the illegal markets and from growing crops that are under marketing quotas (e.g., cereals) to those which are not (e.g., vegetables or small ruminants). In economies where the law and the market are not the only institutional frameworks within which people have access to resources, studies of individual choices must specify the nature of other social institutions (Berry, 1984).

Adoption decisions are influenced by access to new technologies and support services, the status of households within a peasant association (e.g., politician or priest) and level of education. Households may depend on the nuclear family, relatives, city people, markets and government to obtain information, to gain access to resources and to maintain minimal subsistence requirements. The degree of dependency can take various forms such as reliance on credit and insurance institutions, membership in social networks, demand for goods produced outside the farm and access to information about issues affecting the livelihood of farmers. These dependencies influence resource allocation strategies, technology adoption and consequently the level of production.

Ownership of resources, access to various support services, agricultural knowledge and information influence social relationships and vary by family cycle. Production processes and the economic relationships are central to the formation of many, if not most, important social relationships (Scott, 1975).

Social relationships, in turn, influence economic relationships by introducing non-economic factors into the processes of exchange and production (Wilk, et al., 1989).

If changes in production incentives lead to reinforcement or alteration of social relationships, the changing social relations may also affect the organization of production and/or the allocation of productive resources among alternative uses. For example, the rate and patterns of agricultural intensification may depend on the division of labour between men and women (Guyer, 1984, 1986; Henn, 1983) and changes in the size or structure of farming units may affect cropping patterns or methods of cultivation (Norman, et al., 1981; Norman, 1978).

In this study, dependencies or influences with respect to labour, wealth, market, government and social networks are investigated. In examining the decision-making environment, the following methods of analysis are employed: i) changes in economic organization as a result of changes in government policy, ii) variations in wealth and worker:consumer ratio by family cycle and iii) various forms of self-help groups or social networks that affect production activities.[8]

4.5. Adoption Decisions

4.5.1. Introduction

Adoption of innovations is a process composed of learning, deciding and acting over a period of time. The adoption of a specific practice is not the result of a single decision to act but of a series of actions and thought processes (Rogers, 1983). It is dependent on the prior existence of certain factors and conditions. Decisions to adopt innovations are assumed to be associated with social, economic and psychological characteristics of the potential adopters (Brown, 1981). Therefore, variations in the individual characteristics of the study group are focused upon to explain decisions regarding adoptions of different innovations.

In adoption studies, it is necessary to understand not only the decision-making processes but also the characteristics of the innovations and of farmers. The type of innovations, their characteristics and complexities are studied to understand why people do or do not adopt innovations. Moreover, the kinds of information households gather influence adoption decisions. Innovations are evaluated with respect to their relative advantage, compatibility, complexity, divisibility, "trialability", observability and communicability (Rogers and Schomaker, 1971; Rogers, 1983).

Producers do not adopt new innovations before gathering information and evaluating their social and economic merits. It is argued that decision-makers pass through five stages in the adoption process (Rogers and Shoemaker, 1971).

These include the awareness, interest, evaluation, trial, and adoption stages (see also Gartell, et al., 1979; Mason, 1963). At each stage, information is gathered, processed and given meaning. Several factors intervene at each stage which may result in acceptance or rejection of innovations. Selected results of adoption studies are reviewed below.

4.5.2. Adoption of Agricultural Technologies

In order to alleviate some of the constraints facing the agricultural sector, a number of innovations are introduced into the mixed farming system of LDCs. Due to the complexity and diversity of the economic, social, cultural and technical characteristics of the household and the farming unit in LDCs, the success rate of adoption of innovations is low (see Feder, et al., 1982; Molnar, et al., 1983; Hayami and Ruttan, 1985; Jamison and Lau, 1982; Warren, et al., 1988). The reasons for low success rates range from irrational to rational behaviour of peasants.[9] In fact, most studies argue that economic feasibility is the major determinant of adoption decisions. Household decisions, however, are delivered provided that they are meaningful within the environment of the decision-making process. Decisions can be reasonable, socially, culturally or logically, but not economically. Studies that evaluate the feasibility of innovations in a holistic perspective are scant. This remark is demonstrated by a synopsis of the findings of adoption studies presented in the following pages.

Exposure to mass media and inter-personal communication were found to have a significant and positive effect on adoption of agricultural innovations in Brazil (Longo, 1990). Other studies argued that personal contacts in a face-to-face situation and access to information were very important in decisions regarding adoption of agricultural technologies in Ethiopia (Ayana, 1986; Aklilu, 1980). Adoption of technologies was found successful when a two-step flow of communication is implemented (i.e., extension agents to opinion leaders and then to farmers) (Longo, 1990). In adoption decisions, identifying the sources of information and how they are transferred is part of the process. The degree to which this information is processed or comprehended is an important aspect of adoption decisions.

Gladwin and Murtaugh (1980) incorporate a cognitive process in decision-making using attentive and pre-attentive models. This distinction is made to see how past decisions are integrated into the behaviour of choice patterns. It helps to show why some farmers cannot always discuss the reasons for their decisions, as the choices and aspects of choices may be pre-attentive. For cognition to occur, there must be a framework (of knowledge and experience) to process the information gathered. Information received from various sources is comprehended differently by different households. Knowledge gained through experience greatly influences the cognition or comprehension ability of decision-makers (Warren, et al., 1988; Eisemon, 1988).

Several studies have demonstrated that knowledge of innovations, other things being equal, significantly influences adoption decisions (Jamison and Lau, 1982; Roumasset, et al., 1979; Eisemon, et al., 1988; Barlett, 1980). Variations in knowledge are attributed to social, cultural and economic factors (e.g., Toborn, 1978; Aklilu, 1974; Mason, et al., 1980; Eisemon, et al., 1988; Eisemon, 1988). Moreover, experiential knowledge influences willingness to take risks (Warren, et al., 1988).

Homans (1961) argued that people of low status and very high status are likely to be innovators. The logic behind this argument is that low status people cannot lose much by deviating from the norms of the system, while the upper class group have already established themselves in the system. By acting differently from others or pursuing activities which are not in conformity to the system, the upper class group do not lose their status. Cancian (1979), contended instead that the upper-middle rank group is more unwilling to take risks than the lower-middle rank. There is a greater desire among the lower rank group to move upward whereas the upper class does not want to risk their relatively secured socio-economic status (Dewalt and Dewalt, 1980; Barlett, 1980).[10]

Besides information, knowledge and status, resources such as land, labour and capital influence adoption decisions. For instance, Hicks and Johnson (1974) have argued that a larger rural labour force in Africa leads to greater adoption of labour intensive technology (see Ayalew, 1986 for a contrary view).[11] Harris (1972) and Helleiner (1975) suggested that not only shortage of labour but also of resources such as capital and land have contributed to non-adoption of high yielding varieties in India and Africa. Besides the supply of labour, the age of household members was found to influence adoption of crop production innovations in the Ethiopian highlands (Ayana, 1985; Aklilu, 1980). Furthermore, enhancing skill or human capital through education has positively influenced adoption decisions of divisible technologies (Lockheed et al., 1980; Eisemon, et al., 1988; Schultz, 1964).

Often, new innovations require complementary inputs in order to realize their maximum potential. For instance, the introduction of new technologies such as high yielding varieties was found to be successful because farmers had access to pesticides and fertilizers (Gafsi,1976; Ruttan, 1977). Availability of credit and subsidies may help to reduce the financial risk of farmers. With increases in availability of technologies, the quality of information or experience regarding better management of agricultural production increases. These conditions may contribute to increase in risk-taking behaviour of households in LDCs (Feder, et al., 1985). That is, the degree of risk aversion is hypothesized to decline between exposures or learnings and with increases in the number of agricultural technologies adopted (Perrin, et al. 1976; Moscardi, 1979; Roumasset, 1979; Feder, 1981). Nevertheless, the degree of risk aversion of smallholders negatively influences adoption of new innovations (Colmenares, 1970; O'Mara, 1983; Cutie, 1976).

Rogers (1971, 1983) provided empirical evidence that the wealthier a farmer is the more likely that he will adopt innovations first.[12] The value of livestock owned, shortage of cash, the price of outputs, farm size, farm and non-farm income, profitability and degrees of farm fragmentation were identified as factors explaining decision behaviours of the Ethiopian smallholders regarding adoption of improved seed, pesticides and fertilizer (Ayana, 1985; Aklilu, 1980; Waktola, 1980).

Farmers in Arssi and Shoa regions of Ethiopia were found to be very receptive to improved practices and ideas which had economic and technological advantages over traditional practices (e.g., Waktola, 1980; Tecle, 1975). Adoption studies regarding livestock technologies in the Central Ethiopian highlands are non-existent except for the results from on-farm verification trials conducted by ILCA. These studies concentrated only on the economic impacts, thus leaving out the social and cultural aspects of innovations (see for example Mekonnen, 1983; Gryseels, et al., 1986).

The findings discussed above and those documented in Feder, et al (1985) indicate that there is no single or dominant factor influencing the adoption of innovations nor a single method of examining adoption decisions. Several technical, socio-cultural and economic factors influence adoption. What is required, therefore, is to employ methods (qualitative and quantitative) that will help illuminate influences from these complex set of factors on adoption decisions. Assuming that households choose to adopt innovations if they are satisfied with their outcomes, and to enable comparisons of findings of this research with studies conducted in other LDCs, probabilistic models of adoption decisions are selected for this study.[13]

4.5.3. The Empirical Model

Adoption decisions are influenced by various social, cultural, economic and technical factors (Barlett, 1980; Warren, et al., 1988). To examine the relative impact of production knowledge, information, wealth, age and other socio-economic factors on adoption decisions, a multivariate logit model is recommended by several studies (Maddala, 1983; Jamison and Lau, 1982). An alternative to the logit model is the probit model. However, Amemiya (1981) has demonstrated that statistical results obtained from the probit and logit models do not show significant differences. The present study uses a multivariate logit model to examine adoption decisions of households in the Selale & Ada regions.

Each household is confronted with a decision to adopt the new technology, partly or fully, in combination with other technologies or to continue using traditional practices. Denoting the probability of adoption (P), new technology (N), traditional technology (T), the relative probabilities of adopting versus not adopting a new technology are given by:[14]

$$P_i[(N \mid X_i)/(T \mid X_i)] = \exp(\beta X_i)*[\{1+\exp(\beta X_i)\}/\{1+\exp(\beta X_i)\}]$$
$$= \exp(\beta X_i) \quad \text{............ (4.2)}$$

Taking the logarithms of both sides,

$\ln [P_i(N \mid X_i)/P_i(T \mid X_i)] = \beta X_i$ .... (4.3)

Where ß is a vector of unknown parameters which can be interpreted as the net influence of the vector of independent variables on the probability of adoption of new technologies (Rahm and Huffman, 1984; Kebede, et al., 1990). Hypothesis testing for individual and joint influences of socio-economic, technical and physical factors is carried out using the likelihood ratio (Jamison and Lau, 1982).

Separate adoption equations are estimated for the Ada and Selale areas. The independent variables include soil type, plot characteristics, family size, education, expense, farm and off-farm income, wealth, awareness, number of relatives, years of farming experience, farm size and so on.

4.6. Risk-taking Behaviour

In his seminal work on risk, uncertainty and profit, Knight (1921) distinguished between risk and uncertainty on the basis of the information available about the likelihood of outcomes of action choices. Knight associated objective probabilities with risk and subjective probabilities with uncertainty. According to Arrow (1971), when uncertainty as to the state of nature is introduced, the decision problem becomes more complicated because of the multiplicity of outcomes which may occur with probability greater than zero (Arrow, 1971; Heath, 1976). Roumasset (1979), notes that uncertainty is a state of mind in which the individual perceives more than one possible outcome to a particular act. Risk, on the other hand, is the probability or chance that the variable in question will fall below some critical level. Risk aversion may be defined as the willingness to accept a lower expected return in order to obtain a lower risk (Roumasset, 1979). The decision-making environment, the nature of production technologies as well as production knowledge of households greatly influence household's risk-taking behaviour.

Risks are not just taken or avoided: they have costs which are passed on in predictable ways. Gender and intra-household relations affect how risk is taken, avoided or shared (see Guyer, 1984, 1986; Conti, 1979). Distributional aspects of risk should also be given due consideration. Cultural practices provide a means of reducing risk. For instance settlement types, social networks and agricultural practices are designed to minimize not only economic but also social risks (Wiessner 1982; Braun and Plog, 1982; Scott, 1976; Cancian, 1980).

Peasants employ various methods of minimizing risk (Roumasset, 1976; Wharton, 1971). The primary risk minimizing unit is the household, taking action through wage labour and through establishing ties with the larger market economy when the social mechanisms of kinship networks are too few to reduce risk (Barlett, 1982; Firth, 1969). Social relationships and interdependence, involving sharing and exchange, provide both spatial and temporal buffers in a form of social insurance. Techniques such as mobility and scattering fields serve as spatial buffers and storage as a temporal buffer (Jodha, et al., 1983; Wiessner, 1982). Despite alternative methods of reducing risk, households exhibit varying degrees of risk-taking behaviour.

Reactions of peasants to development strategies are, in part, explained in terms of risk. Conservatism of peasants and reluctance to adopt innovations are not due to irrational behaviour, as often assumed, but to peasants' desire to maximize their security by minimizing their risk (Ortiz, 1980; Scot, 1975; Roumasset, 1976; Wharton, 1971). New techniques are often adopted on a small scale and therefore entail little risk (Barth, 1967).

Cancian (1980) argued that peasants' willingness to adopt new practices depends on their wealth and social standing. The relatively wealthy peasants are more willing to take a highly variable technique that might produce greater yields in some years, because their wealth buffers them from falling below the threshold in poor years. Poor peasants are sometimes willing to take unknown techniques involving high uncertainty and possibly high risk because they have little to lose (Cancian, 1980; Dewalt and Dewalt, 1980; Roumasset, 1976).

Shahabuddin and Mestelman (1985) argued that farmers in traditional agriculture are exposed to numerous sources of risk. The consequences of these risks for resource allocation decisions may result either in a decline or an increase in income and production (Dillon and Anderson, 1971; Herath, Hardaker and Anderson, 1982; Weins, 1970; and Wolgin, 1975). Furthermore, in peasant agriculture, where farms are extremely small and cultivation is dependent upon highly variable rainfall, the farmer may not be as preoccupied with maximizing income as with maximizing the chances of survival. Safety-first patterns of decision-making are based on this disaster avoidance behaviour (Roy, 1952). The behaviour of farmers in maintaining subsistence requirements or some threshold level of production has been found to be the primary objective among farmers in Bangladesh and the Philippines (see Shahabuddin et al., 1985; Roumasset, 1976).

Once a problem is identified as requiring an analytical decision, the risk-taking attitude of decision-makers plays an important role on the efficiency with which decisions are made. Risk-taking behaviour is influenced by several factors. Methods for analyzing these risk-taking behaviours of households range from those utilizing hypothetical gambling to those conditional on actual production decisions (Anderson, et al., 1977). Recent contributions to the theory of decision-making under uncertainty include theoretical multi-attribute utility functions (Barry, 1984). These models are based on the assumption of rational economic or utility maximizing behaviour of households or firms.

Methods of studying risk-taking behaviour are divided into direct elicitation of utilities (DEU), observed economic behaviour method (OEB) and experimental methods (Binswanger, 1980; Fleisher and Robinson, 1985; Anderson, Dillon, and Hardaker, 1977; Barry, 1984). Alternatively, these models can be grouped into econometric, optimization and experimental methods. Variations in methods of risk analysis have contributed to variations in conclusions about the risk-taking behaviour of households. Unfortunately, these inconsistent results are adopted by policy makers in the design of development projects.

In this study, risk-taking behaviour is analyzed using information obtained from actual decisions made by households. It is hypothesized that the primary objective of farmers is to secure subsistence for reproduction of producers and the farming unit. Therefore, households follow the strategy of minimizing the risk associated with securing subsistence requirements. Data on crop yield variability, cropping pattern, prices of inputs and outputs, debts, loans, farm and non-farm incomes and related variables are collected. Risk sensitivity and crop diversification indices are calculated (Hardaker, et al., 1984; Valdes, et al., 1979).

The risk sensitivity index or disaster level (i.e., maximization of the negative of gross profit) for crop farming is calculated as :[15]

$$RSI = \frac{(X_1 + X_2 + X_3) - (X_4 + X_5 + X_6 + X_7)}{Ha} \quad \text{..... (4.4)}$$

Where X_1 is expense for household necessities for the past year, X_2 is expense for sending dependents to school, X_3 is urgent debt, X_4 is off-farm income, X_5 is liquid assets, X_6 is saving, X_7 is the amount of loan and Ha the number of hectares planted to cereals (Valdes, et al., 1979). A ratio of the difference between expenses and gross household income to hectares allocated for cereals can serve as a risk sensitivity index. This ratio increases with increases in expenses and declines with increases in income. It also shows to what kind of crops a farmer allocates a larger area when he/she is indebted or in cash surplus. Similar calculation is made for livestock production.

The crop diversification index is calculated as :

$$I = 1 - \sum_j L_j^2 \quad \text{........... (4.5)}$$

Where L_j is the proportion of land allocated to crop j. The index "I" has an upper bound of 1.0 and takes a value of zero for monoculture. Similar calculations are made for the composition of livestock. This measure generally increases with the degree of risk aversion (Hardaker, et al., 1984).

4.7. Production Function Analysis

4.7.1. Introduction

Analysis of production decisions requires identifying variables that influence the amount of production and the efficiency with which crop and livestock outputs are produced. Information from these kinds of analyses facilitates the design of agricultural policies aimed at increasing food production. Several methods are used to examine production decisions. The most important ones are production functions and optimization. Production functions are used to evaluate the "rationality" of production decisions and the consequences of the efficient or inefficient use of resources (Ortiz, 1983).

The production function is a mathematical specification of a rule that defines outputs given a set of inputs and a known way of organizing production. It has been used, apart from economists, by anthropologists, sociologists and biological scientists. It represents the technical relationship between physical outputs and limitational and limitative factors of production.[16] The inputs include traditional (e.g., land and labour) or new technologies. Adoption of a single (e.g., fertilizer) or mixes of production technologies (e.g., fertilizer and pesticides) are anticipated to have a differential impact on the production structure (inputs used, crops grown and milk produced) of smallholders (Eisemon and Nyamete, 1988). Of particular interest is whether or not selective mixes of production technologies produce different results in farm efficiency compared to a single technology and traditional inputs. To estimate the impact of traditional and new techniques on the amount and efficiency of production, data are collected on production inputs (e.g., expenditures on seed, fertilizer, labour, oxen, and so on), the amount of grain and milk produced for the 1990-91 season.

4.7.2. The Average Production Function Model

Inputs used in the production process cause variations in outputs. The response of outputs to the amount of inputs used can be represented by different kinds of mathematical or statistical models. There are controversies on what kinds of functional forms to use in production function analysis (see Peterson, et al., 1977; Chambers, 1988). Because of simplicity, computational convenience and ease of interpretation, the modified Cobb-Douglas production function is selected as a functional form to evaluate the impact of inputs on the level of production realized by households (Jamison and Lau, 1982; Ward and Zahalka, 1983; Ortiz, 1980, 1983).

The linear in logarithms Cobb-Douglas form of the function is given as:[17]

$$\text{Log } Y = \delta_o + \Sigma\beta_i \text{ Log } (X_i) + \Sigma\, \alpha_j Z_j + U \qquad .....(4.6)$$

Where Y is output, X's are continuous inputs, Z's are dummy (0-1) variables, δ_o is the intercept, β and α are unknown parameters. Equation (4.6) is estimated by the ordinary least squares (OLS) technique for the control and test groups separately using cross-sectional data. The significance of individual inputs and structural changes (shifts) in the production function parameters are tested following Chow (1960).

The dummy variable in equation 4.6 is assumed to capture the effect of factors for which measurement is not available or is qualitative. It indicates the percentage impact of `Z' variables on `Y'.

The Cobb-Douglas production functions are estimated for six crops commonly cultivated in Selale and Ada areas.[18] These include four cereal crops (teff, wheat, barley and oats), two pulse crops (beans and field peas) and one aggregate function for cereals. The dependent variable is the amount of production for each crop.[19] The inputs include area (in square meters), seed (in kilogram), labour (in mandays), oxen (in oxen days), soil characteristics recorded as (0-1) variable, and so on.

Production function analysis using equation (4.6) indicates the impacts of inputs on the amount of outputs obtained. To investigate the efficiency with which inputs are used by the test and control farmers of the Selale and Ada regions, other methods of production function analysis are selected. The data used for average production function analysis are applied to the estimation of parametric (stochastic frontier) and non-parametric (data envelopment analysis) production functions (Taylor, et al., 1986a and b). The reasons for the selection of two methods are: i) methodological comparison and ii) to obtain reliable results through analysis of aggregate (data envelopment analysis) and dis-aggregate (frontier production function) data. The theoretical background and specification of the empirical models are presented below.

4.7.3. Measurement of Efficiency

Early efforts in the investigation of efficiency and its measurement were made by Koopmans (1951, 1957) and Debreu (1951). These studies dealt only with technical inefficiency. Koopmans (1951) defined a feasible input-output vector to be technically efficient if it is technologically impossible to increase any output or to reduce any input without simultaneously reducing at least one other output or increasing at least one other input.[20]

Farrel (1957) obtained a partial decomposition of efficiency into technical and allocative components. He also provided indexes of technical, allocative and overall private efficiency. However, Farrel's specification was confined to a single-output production technology with strong scale, monotonocity and curvature properties. Deficiencies in the Farrel method led to two developments. First, Farrel's two-way decomposition of efficiency was retained, and the measure of technical efficiency was generalized to allow for disproportionate input reductions and/or disproportionate output increases. This generalization was modified in a single-output context by Fare and Lovell (1978). This approach was extended to multiple-output technologies by Fare, et al. (1983). The second development augmented Farrel's measure of technical efficiency with the adoption of a separate measure of structural efficiency (see Fare and Grosskopf, 1983, 1985).

The idea of gauging the performance of a producer by comparing actual to optimal farm or firm size, and the association of optimal size with input-output vectors satisfying constant returns to scale is attributed to Frisch (1965). An index of scale efficiency based on this notion of technically optimal size was proposed by Forsund and Hjalmarsson (1974), Banker (1984), and Banker, Charnes, and Cooper (1983).

Fare et al. (1983, 1985) pointed out that the results obtained from data envelopment analysis and stochastic frontier production functions are not consistently the same, or one may not be approximated by the other. In the first approach the best practice frontier is constructed from observed inputs and outputs as a piecewise linear technology. In the second approach, the production function should be well-behaved satisfying strong disposability of inputs, differentiability and strict monotonocity. This study employs the two approaches for the purposes of methodological comparability and reliability of the results.

4.7.3.1. Data Envelopment Analysis (DEA)

Charnes and his colleagues (1978) formulated DEA which was an extension or modification of the Farrel measure of efficiency (Farrel, 1957; Farrel and Fieldhouse, 1962). Data envelopment analysis (DEA) was developed to measure the relative efficiency across individual production units (see Farrel and Fieldhouse 1962; Charnes, Cooper and Rhodes, 1978; Charnes and Cooper, 1980; Sengupta, 1987; Silkman, 1986). These units could be schools, hospitals, parks, crops or livestock farms. This method assumed that the production units produce one or several homogenous outputs through transforming a set of homogenous inputs.

The objective of this research is to measure the efficiency of individual producers who are engaged in the production of crops and livestock simultaneously. The production units that attain a given level of production from a certain amount of inputs are given scores ranging from 0 to 1. All production units are compared relative to the units with a maximum efficiency (i.e., score of 1). This method is similar to the engineering production function in a sense that it does not restrict individual decision-making units from using various levels of inputs and attaining different levels of outputs (Wibe, 1984; Heathfield, et al., 1987). Thus, it is possible to compare units producing different levels of outputs from different levels of inputs.

In Ethiopian agriculture, crop and livestock production are inseparable. The assumption of homogenous products does not hold if physical units of measurements are used. Therefore, physical outputs are multiplied by their respective market prices (see Appendix 4c).[21] The physical inputs required to produce milk are cows, labour, feed, veterinary services and related expenses. Farm size, seed, labour, fertilizer, pesticides and oxen inputs are required to produce cereals and pulses. Let X_{ij} be the amount of input i used by farmer j in the production of milk or crops. Similarly, let Y_{rj} be the level of output r produced by farmer j. Each producer is assumed to use some amount of input to produce certain level of output, i.e., $X_{ij}, Y_{rj} > 0$. The decision variables of the problem are the unit weights to be attached to each of the inputs and outputs by farmer j. Each producer possesses a different level of managerial skill. Therefore, weights (e.g., prices, skill) are attached to both inputs and outputs. For instance, U_{rj} is the unit weight of output r produced by farmer j. Similarly, V_{ij} is the unit weight of input i used by farmer j. The unit weights are defined via the constraints which are specified in the problem.

The linear fractional program is formulated for 270 producers. The linear program to perform DEA by means of the simplex algorithm is given as:[22]

$$\text{Maximize } Z_j = \sum_{j}^{n} U_{rj} Y_{rj} \qquad \text{............ (4.7)}$$

Subject to

$$\sum_{r=1}^{n} U_{rj} Y_{rj} - \sum_{i=1}^{m} V_{ij} X_{ij} \leq 0 \; ; j = 1,2,3, ..., n$$

$$\sum_{i=1}^{m} V_{ij} X_{ij} = 1;$$

$$U_{rj}, V_{ij} \geq 0; r=1, 2,\ldots m, i=1,2\ldots .q$$

4.7.3.2. Frontier Production Function

Traditional econometric methods for estimating stochastic economic relationships have implicitly assumed that all economic agents are successful in reaching the efficient frontier. If, however, the economic agents are not equally efficient, then the average relationships estimated by ordinary least squares methods might not reflect the frontier relationships. The maximum output sets the frontier. The amount by which a firm lies below its production frontier can be regarded as a measure of inefficiency. Detailed discussions on frontier production functions can be found in Frosund, et al., (1980), Greene (1980, 1982), Stevenson (1980), Schmidt (1985-86), Lee (1983), Waldman (1984) and Kumbhakar (1987).

The frontier functions can be categorized into full frontier and stochastic frontier models. The two models differ with respect to factors contributing to technical inefficiency. The full frontier production function assumes that all random variations in output because of factors exogenous to the firm, e.g., weather, are attributable to technical inefficiency.

The need to consider factors outside the control of the firm and statistical noises that contribute to inefficiency led to the development of a stochastic frontier function (Aigner, et al., 1977; Meeusen, et al., 1977). The stochastic production frontier assumes that the disturbance term is composed of two parts, a symmetric and a one-sided component. The symmetric component captures the random effects outside the control of the decision-maker including the statistical noise contained in every empirical relationship. The one sided-component captures deviations from the frontier due to inefficiency.

The stochastic frontier production function, assuming the Cobb-Douglas form, can be written as:[23]

$$\ln Y_i = \ln\alpha_0 + \Sigma\beta_i \ln X_{ij} + \varepsilon_j \quad \text{..............} \quad (4.8)$$

Where $\varepsilon_j = V_j - U_j$ is a composed error term with $V_i \sim iid\ N(0,\sigma v^2)$ and $U_j \sim iid|N(0,\sigma u^2)|$. V is the two sided random error, U is the one-sided efficiency component distributed half-normally, Y is output, α is the intercept, β is an unknown parameter, X's are inputs (Aigner, et al., 1977; Meeusen, et al., 1977). V and U are assumed to be uncorrelated and ε is assumed to be independent of X. The symmetric component, V, permits random variations in output due to factors outside the control of the firm such as weather.[24]

Estimation of the parameters of the stochastic frontier production function was accomplished by the maximum likelihood technique using LIMDEP (Greene, 1990). The algorithm is based on an iterative solution to the likelihood equation and yields the appropriate maximum likelihood estimates. The advantage of stochastic production frontier model is the introduction of a disturbance term representing noise, measurement error and exogenous shock beyond the control of the production unit, in addition to the efficiency component. Thus, technical efficiency measures obtained from stochastic frontier functions are expected to be higher than those obtained from deterministic models such as programming (Bravo-Ureta, et al., 1990).

4.8. Production Risk Analysis

In production risk analysis, the research interest is either to examine the sources of variability in production because of i) the amount of decision inputs or factors ii) the importance of, regardless of the amount, selected inputs or factors. When production is unpredictable, the parameters of the production function across decision-making units vary. Production risk in Ethiopian agriculture because of variability in essential inputs such as land, labour, seed and oxen are examined by Belete (1989). One of the objectives of the present study is to investigate whether or not the use of new technologies contributes more to the variability of production than essential traditional inputs (e.g., labour, land). The method employed to attain this objective is similar to that used for examining the relationship between goals and strategies, i.e., multivariate regression model with dummy variables. [25] To examine the contribution of new technologies and other non-essential inputs or factors to variability in production, standard deviations of crop and milk yields (as an indicator of production risk) are regressed on binary variables (coded 0-1). The model is allowed to have an intercept which captures the effect of essential inputs that are not included in the model. The coefficient of the independent variables shows their impact or contribution to yield variability relative to the impact of the intercept term (Walker, et al., 1982; Maddala, 1977).

The least squares binary regression, assuming fixed effects, is specified as:

$$Y_{ij} = \Omega_{j1} + \sum_{d=1}^{n} \beta_d X_{ijd} + \sum_{p=1}^{z} \delta_p X_{ijp} + \sum_{f=1}^{v} \theta_f X_{ijf} + \sum_{s=1}^{q} \pi_s X_{ijs} + \sum_{t=1}^{k} \Phi_t X_{ijt} + U_{ij} \quad \text{............ (4.9)}$$

where Y_{ij} is standard deviation of yield of plot i cultivated by farmer j, X_{ijd}, X_{ijp}, X_{ijf}, X_{ijs} and X_{ijt} are information, policy, financial, socio-economic, technical/environmental factors, and U_{ij} is a disturbance term. [26] In equation (4.9), the intercept term captures the effect of essential inputs such as labour, oxen, seed and land. The logical estimate of say β_d is the difference between the expected yield from the use of essential inputs not included in the model and that of impact from socio-economic variables. It must be noticed, however, that this model assumes no interaction between the dummy variables (see Kennedy, 1985).

The variables of interest in this study and hypothesized to be major sources of production risk are identified from group discussion with farmers in five peasant associations in the Selale and Ada regions. These variables include information (extension education, schooling), policy (price variability, land tenure), institutions (grow feed), financial (fertilizer, pesticide, improved seed, cross-bred cows, concentrate, stall feeding and veterinary service), socio-economic (experience, production Knowledge), and technical/environmental (use of atela, number of crops planted, slope of a plots, soil type, location of a plot, rainfall and region) (see chapter VIII)

4.9. Linear Structural Relations (LISREL)

Structural equation models have been found useful in solving many substantive research problems in the social and behavioural sciences. Such models have been used in the study of macroeconomic policy formation, racial discrimination in employment, housing and earnings, studies of antecedents and consequences of drug use, scholastic achievement, evaluation of social action programs, voting behaviour, and many other phenomena. These models are useful in statistical modelling techniques such as simultaneous equation systems, linear causal analysis, path analysis, structural equation models and dependence analysis (Joreskog, et al., 1989).

A structural equation model is used to specify the phenomenon under study in terms of cause-effect variables and their indicators. Each equation in the model represents a causal link rather than a mere empirical association, thus structural parameters do not, in general, coincide with the coefficients of regressions among observed variables. Instead, structural parameters represent relatively unmixed, invariant and autonomous features of the mechanism that generates the observed variables. To serve these purposes, structural equation models require statistical tools that are based upon, but go well beyond, the conventional regression analysis and analysis of variance (Joreskog, et al., 1989).

LISREL is the most general program that is widely available for estimating structural equation models. It can be used to analyze data from surveys, experiments and longitudinal studies (for details see Joreskog, 1973, 1978 and 1981). The LISREL model chosen in this study is used to examine linear causal relationship (path analysis) between independent (exogenous or cause) and dependent (endogenous or effect) variables.

Path analysis is a technique used to assess the direct causal contributions of one variable to another in a non-experimental condition. The system involves two kinds of variables: independent or cause variables (x's), and dependent or effect variables (y's). The classic technique consists of solving the structural equations for the dependent variables in terms of the independent variables and the random disturbance terms to obtain the values of structural parameters. Models of this type and estimation techniques in econometrics can be found in Theil (1971) and Goldberger (1971).

If the data set contains only observed variables and assumed zero measurement error, then the LISREL model can be formulated as:[27]

$Y = \beta y + \Gamma x + u$ ...... (4.10)

This is a structural equation model or a path analysis for directly observed variables. The y's are to be explained by the model, i.e., variations and covariations among the y-variables are to be accounted for by the x-variables. The x-variables may be random variables or a set of fixed values. This model assumes that i) $(I-\beta)$ is non-singular, $E(u)=0$, where E is the expected value operator, and u is uncorrelated with the x's. The u's represent an aggregate of all known and unknown influences on y's which are uncorrelated with the x's. The parameter matrices involved in this model are β, Γ and $\Phi = \mathrm{cov}(u)$.[28]

Estimation of path analysis for directly observed variables using LISREL can be carried out using a system of equations and estimate all structural parameters directly. The structural equations include: specification of the data type (raw, covariance, correlation), the model (the number of x and y variables and the form of the matrices of data) and an output statement (for the details see, Joreskog, et al. 1989).

Specification of all kinds of relationships between x's, x's and y's, and between y's for all conceivable variables may result in a lack of convergence even with increases in the number of iterations (Joreskog, et al. 1989; Saris, et al. 1984; Hayduk, 1987). In the study of Ada and Selale farmers, based on regression analysis and prior results from group discussions, x-variables whose effect on the y's are relatively low are excluded from the analysis. Two models are estimated: one with five and three, and another with six and three, exogenous and endogenous variables respectively. The exogenous variables include wealth, worker:consumer ratio, experience, extension education, secular education and farm or feed area. The endogenous variables include livestock or crop production knowledge, number of technologies adopted and production efficiency. Analysis is performed by region, and within a region by control and test groups.

4.10. Summary

Analysis of decision-making involves two parts: the processes involved before executing specific decisions and the consequences of these decisions. This chapter reviewed methods of household decision analysis and their findings. Combinations of various methods that range from observation or participation to those involving rigorous statistical models are selected for this study.

This chapter examined the use of tools such as multivariate regression model, frequency distribution and means to analyze the goals and strategies of households. Information processing and agricultural knowledge are assessed using tools such as frequencies, percentages, problem solving tests, causal explanation and participant-observer methods. The decision-making environment is investigated using changes in socio-economic organization as a result of changes in government policy, variations in wealth and worker:consumer ratio by family cycle and various forms of social networks or institutions that affect production activities.

Decisions regarding adoption of technologies are analyzed using responses to qualitative variables, observation and logistic regressions. Information from actual decisions made by households is examined to calculate crop diversification and sensitivity indices that are used to assess the risk-taking behaviour of households. The consequences of decision-making are analyzed using average and frontier production functions, and linear programming technique (Data Envelopment Analysis). Binary least square regressions are used to assess production risk. Finally, LISREL analysis is selected to test the conceptual model presented in chapter one and establish the magnitude of causal linkages among variables that influence the processes and consequences of production decisions (Sarris, 1984). The results of the application of these methods to cross-sectional data collected from the Ada and Selale regions of Ethiopia are presented in subsequent chapters.

4.11. Notes

1 Models of risk attitude or risk-taking behaviour and resource allocation assuming profit maximization as the only goal of peasants have been applied to test the efficiency of peasants in LDCs (Schultz, 1964, 1978; Barlett, 1980; Ortiz, 1980; Cancian, 1980; Ellis, 1988). The results of these studies are mixed and incapable of assisting in the design of appropriate policies that help to attain sustainable growth.

2. Ortiz (1983) classified methods of decision analysis into the categories of utility, game theory, programming, information processing and production function models.

3. For the theoretical exposition of preferences and choices, see Anderson, et al. (1977).

4. In fact, Janeflid (1980) argued that sources of income available to family members are subject to the judgement of the head of the household, thus constituting a rude but effective social security system.

5. This approach, however, is by no means a perfect substitute for information that can be obtained from each member of the household (see also IFPRI, 1992).

6. Pareto Optimality is obtained when policies result in at least some being better-off while no one is made worse-off (Just, et al., 1982).

7. Take as an example an average household size of five persons for the Ethiopian highlands. If members of the household have divergent interests and strategies to achieve their goals, i) financial and time requirements may not allow detailed analysis of information from each member and may limit the coverage of household studies to small geographic zones, and ii) overlapping of objectives or strategies and use of resources commonly managed by members may pose difficulty in selection of suitable analytical models.

8. With respect to point (ii) examination of wealth and worker:consumer ratio is carried out for younger or smaller family size (three or less persons), average family size (4 to 6 persons), larger family size (at least seven persons), widows and old couples (see also Barlett, 1980).

9. These characterizations of peasants for low rate of adoption are those of researchers not peasants themselves. In fact, for peasants farming under uncertain socio-economic system and environment, their primary goal is to secure adequate food for their family. Therefore, given the resources and constraints peasants face, the decision they made are rational. However, because these decisions do not fit into the neoclassical or western models of rational behaviour, peasants' behaviour is labelled as irrational.

10. According to Cancian (1980), therefore, the lower rank group is the most innovators.

11. Socio-cultural factors such as observance of religious holidays and meetings in the Ethiopian highlands influence adoption of technologies by reducing the available labour supply (Ayana, 1985; Kiros, 1976).

12. This finding, also supported by several others, is the opposite of the argument provided by Cancian (1980). However, Cancian's argument is based on a study in Latin America. Therefore, it cannot be argued to hold in every society.

13. Several studies of adoption decisions in LDCs have employed probabilistic models (see Feder, et al., 1982; Jamison and Lau, 1982).

14. The derivation of equation 4.2 and 4.3 are presented in Appendix 4a.

15. The risk of disaster is defined as $Pr(\Omega ij < d)$, where Pr is the probability level, Ωij is gross profit and d is a disaster level of gross profit below which household may not survive. The equation in 4.4 assumes that households maximize the negative of the disas-

ter level or gross-profit.

16. An input is said to be limitational if an increase in its usage is a necessary, but not sufficient, condition for an increase in output. An input is said to be limitative if an increase in its usage is both a necessary and a sufficient condition for an increase in output (Ferguson, 1971:9-10).

17 For derivation of equation 4.6., see Appendix 4b.

18 Oats are not grown in Ada region. Therefore, the number of crops for which production functions are estimated for this region are five.

19. In LDCs, measurement problems and memory biases of households (see Coleman, 1983), result in unreliable yield per unit area. Consequently, this would result in biases of regression results (Kebede, 1988). Therefore, total production is used as a dependent variable in this study.

20. Alternatively, a producer is said to be technically efficient if production occurs on the boundary of the producer's production possibilities set, and technically inefficient if production occurs on the interior of the production possibilities set (Fare, Grosskopf and Lovel, 1985).

21. In a situation where markets are fragmented and pricing policy is inappropriate, use of market prices may not be an appropriate measure to value inputs and outputs. This study was conducted during the year in which market was liberalized and that farmers could sell their output either to the government or free market. When the policy was announced, the inputs for the 1990-91 crop year were committed. Moreover, certain input markets are not yet liberalized (e.g. land). Thus, only outputs are converted into monetary equivalent to standardize the units of measurement of different kinds of crops and livestock outputs.

22. The derivation of equation 4.7 may be found in Appendix 4c.

23. Kopp and Smith (1980) argued that the choice of functional specification has a very small impact on measured efficiency. The Cobb-Douglas production function is a compromise between a complex production process and a complex estimation technique.

24. The derivation of equation 4.8 may be found in Appendix 4d.

25. Walker and Rao (1982) call this method least squares binary regression model (see also Maddala, 1977; Johnston, 1984).

26. A value of 0 is assigned to a variable if it, based on group discussion with farmers and findings from previous research, doesn't increase variability in production.

27. Detailed specifications of the path models are provided in appendix 4e.

28. A special case of this model is that when β is sub-diagonal and Φ is diagonal, the structural equation for observed variables model is called a recursive system or path analysis.

CHAPTER V

COGNITION AND DECISION-MAKING: GOALS, STRATEGIES, INFORMATION AND KNOWLEDGE

5.1. Introduction

This chapter examines the structural relationship between the goals and strategies defined by households. It assesses types and sources of, and access to information on agricultural production. The objective of this chapter is to show differences in goals, strategies, information processing skills and knowledge among producers and the manner in which these differences may affect household decision-making and its consequences.

Applied decision analysis should identify the goals that subjects aspire to achieve through implementation of calculated actions or strategies. A strategy is a rule or schedule that specifies in advance for future information signals the action that will be taken in response to it (Anderson, et al., 1977). Goals may be defined as ends, objectives or states that an individual wishes or plans to achieve (Cary and Holmes, 1982). While attempts have been made to argue the distinctiveness of goals (ends) and strategies (means), these distinctions are by no means certain. Strategies affect goals, the latter only becoming clear in retrospect, and goals may also affect strategies.

Identification of goals and strategies in a peasant economy is difficult because of the complex nature of the criteria used by households to rank goals and strategies. Certain goals and strategies not only re-appear every year but also are of a long-term nature. Moreover, several factors outside the control of decision makers influence the selection of goals and strategies. Nevertheless, goals and strategies are the critical components of the decision-making process. Assessing household decision-making is possible only if goals and strategies are identified.

There are many field methods for identifying goals and strategies of producers (see Barlett, 1980; Ortiz, 1983; Ellis, 1988; Netting, et al., 1984; Barry, 1984): through observation, participation, visual representations (e.g., cards that bear pictures of goals) and interviews. Each of these methods have their own advantages and disadvantages. Gladwin (1980) contends that it is possible to identify the goals and strategies of households with the help of decision trees and argues that elimination by aspects is too simplistic an approach for this purpose (see also Tversky, 1972).[1]

Decision trees require subjects to demonstrate the ability to retain and recall information. Coleman (1983), however, has shown that non-literate farmers cannot remember all conceivable goals, strategies and selection criteria applied because of memory biases. Group discussion with farmers in the Ada and Selale regions revealed that with the passage of time it is difficult for households to recall all the criteria they used in the selection of goals and strategies. Thus, application of sophisticated decision trees and probabilistic calculations in identification and stepwise selection of goals and strategies of households in the Ethiopian highlands may not be a feasible method. Instead, this study concentrates on those goals and strategies that can be recalled by decision makers. This was accomplished through a combined approach that made use of cards, participatory, observatory and interview methods to identify the goals and strategies of households.

To design agricultural policies, it is essential to know whether goals or strategies should be targeted. The choice depends on whether goals shape strategies or vice-versa. Empirical tests of causality using cross-sectional data are difficult (Granger, 1969). In this study, to obtain evidence regarding causality, 25 producers were asked questions of the following format:

> "*What kind of goals do you plan to achieve this year? How do you think you will attain these goals?*"

Answers to this question were taped. Goals identified by the producers and strategies they will implement were tabulated. Moreover, whether they examined the resources available to the households and factors external to them (e.g., policies) or set their goals and selected strategies is identified from their responses.

To minimize biases coming from this method of investigation, responses to the above question were compared with responses to the base line questionnaire dealing with household decision-making.[2] Furthermore, the results from the above type of question were compared with notes from the participatory study that was undertaken on selected households in the Selale region. The result of this method of assessment indicated that eighty-four percent of households select their goals before choosing strategies. The average age of these household's heads was 47 years. The average age of household's heads who selected strategies prior to goals was 32 years. Thus, it can be concluded that most farmers chose goals before strategies, and this pattern of decision-making was observed among farmers who were 40 or more years of age. Once this simple method of assessing causality was accomplished, the next step was to identify strategies which played significant roles in attaining the goals of the decision makers.

Before presenting the results of statistical analyses of the relationships between goals and strategies of the study farmers, outcomes of the participatory-observatory method from two case study farmers in the Selale region are discussed. Analysis of the relationship of goals and strategies is followed by examination of differences in access to information and knowledge between farmers and regions. Furthermore, the effect of information and knowledge on choices of enterprise combination and decisions regarding the selection of goals and strategies is investigated.

5.2. Household Goals and Objectives

5.2.1. Participatory-Observatory Method

In field research it is very important to distinguish the researcher from individuals whom peasants see as representative of institutions designed to extract their produce. The researcher should identify a farmer with good social status or reputation as his/her person of contact in a village. He/she should explain the reasons for the visit to farmers. He/she must try to use a means of transport similar to that used by farmers or different from that used by government officials. He/she must be willing to spend sufficient time with farmers. Most importantly, he/she must have ability to communicate with farmers in order to identify goals, resources, constraints, opportunities and strategies of producers (Warren, et al., 1988).

It was difficult carry out the participatory-observatory method for a large number of producers because of time and financial constraints. Therefore, two producers were randomly selected for this purpose: "Ato" Tesfaye Kebede of Wayu and Ato Abebe Dilnesaw of Gulele Biro peasant associations, both from the Selale region.[3] A short description of each of these farmers and their family is presented in Appendix 5a.

Producers may have different goals to attain in any particular year, but securing food production is the principal goal of all producers. Therefore, identification of goals among peasants of the Ethiopian highlands can be accomplished best if examined during a season of the year that is critical to their survival. In this study, the plowing season is chosen to serve this purpose.

Farmer # 1: Ato Tesfaye Kebede

The time of plowing is an important agricultural season to observe and identify the goals and resource allocation strategies of producers in the Ethiopian highlands. A typical routine may be as follows: Every day the husband, wife, Abebe (the eldest son) and Alemitu (the eldest daughter) wake up early in the morning (6 a.m. at the latest) and go to bed late (10 p.m.). The rest of the children wake up by 7 a.m. or earlier depending on the amount of time required to walk to schools or to respond to requests for help from other members of the household. The mother milks the cows. Daughters help prepare coffee to be served after breakfast. The husband, wife and eldest children have breakfast between 7 and 8 a.m. The eldest brother or sister looks after the livestock while the youngest children have their breakfast. The mother and daughters stay within the compound and carry out activities such as cleaning of the house and seeds for planting, packing goods for market, preparing lunch or supper, cleaning the barn, washing clothes and making dung cake.

The father and the eldest son alternate duties; if the father is going to work on the farm, he has to make sure that his oxen are well fed. He separates the working oxen from the rest of the stock and provides them with the best feed available for that day. If he is not working on the farm, he may stay around the homestead and help supply fuel wood to prepare breakfast and lunch. Milking and pregnant cows are kept in a special section of the barn for "special feeding" by the wife.

To complete the plowing, the father applies a number of passes to the plot before it is considered ready for planting. While working with Ato Tesfaye, I initiated discussions of the following type:

Question: "Why do you have to plow so many times and overwork your oxen? Would not it be better to fatten the oxen for sale?"

Answer: "You see prices of food crops are high for most parts of the year except at times of Gena (Christmas) and Timket (Epiphany). If I put my ox on a fattening scheme, I would not be considered a farmer. I was born to be a farmer and my ancestors were farmers. Nothing will make me more proud than to be a farmer. However, I do not want my children to toil like me."

Answer: "Why not?"

Answer: "What are they going to farm? I used to own sufficient farm land during Haile Sellasie's regime (the monarchy that lasted until 1974). Now, only 3 hectares for 7 people and probably less in the future. Besides, the times are good. Unlike in my youth, there are schools everywhere. The young generation knows more than when we were their age. We are getting old. Probably, they can support us (parents) if we send them to school today. Besides, even if I sell beef cattle, it may worth only 800 birr (400 U.S.), if I am lucky."

Question: "Why?"

Answer: "Cattle die because of disease or accidents and can be stolen. God knows what may happen. I may lose everything. With 800 birr, I can purchase 800 kg of grain without allowing for purchases of salt, oil, gasoline and cloth, and school expenses for my children. How could I survive? If I stay on the farm and grow crops, even if there is a crop failure people will be sympathetic and help me survive difficult years."

Question: "If you really want to stay on the farm, why do you not use fertilizer, pesticides, improved seed and cross-bred cows?"

Answer: "Son, it is not easy. The Ministry people know only how to talk. I do not know what they are writing, they come and ask if I would like to use [these technologies] and they leave. I would like to use anything that will help increase production without getting into Eda (debt). I have used pesticides and will continue using them. In fact, if you are going to `Sheger' (Addis Ababa, capital city of Ethiopia), bring me medicine for ticks and coughs (for cows and oxen). I will pay you or you will take one of the sheep. I will try to get you the medicine but I will not take money."

Ato Tesfaye: "In that case, leave my place now!"

Question: "Why?"

Answer: "I am not a beggar. I do not want people's money as long as I am alive and God's with me."

We worked on a quarter of a hectare. The wife and the daughter brought us food and drinks at the place of work. We looked for shade under the few remaining trees and sat. While we were eating, I asked Weizero Tenaye (the wife):

Question: "What would you like to have as a change in the economy of your household?"

Answer: "Cows for my children! And to purchase coffee any time I want."[4]

Ato Tesfaye exclaimed, "Well, I need a powerful ox from the dam. They are very strong, but less resistant to disease."

Later in the evening, I talked to Abebe who was in his final two years of high school.

Question: "Why did you decide to go to school?"

Answer: He laughed and said, "I did not. My parents decided whether I should attend school or not."

Question: "What would have been your choice?"

Answer: "To raise a lot of cross-bred cows and be rich."

Question: "If you do not attain this goal, what would interest you?"

Answer: "I think I want to be a doctor."

Question: "Why?"

Answer: "Because when my parents get sick I can cure them."

Farmer #2, Ato Abebe Dilnesaw

Question: "What did you plan to accomplish in your lifetime?"

Answer: "Many things. First of all to get married and have children. I work hard to get more land, but the "derg" took it.[5] I even worked as a guard of the satellite dish under the Ministry of Telecommunication. I have always wanted to learn more and be a minister. I could not achieve this goal because when my parents migrated to this place, they were old and I had to take care of them. My two brothers died during the Italian invasion. I was the only remaining child. That was why I quit schooling at grade 4.

"Now, I wanted my children to be well educated. All of them have completed high school but could not get a job. Times have changed. Spending 12 years in secular plus 2 to 3 years in sacred school should have been enough to guarantee them a job. Instead, they are toiling like me. In fact, I am better off than they are." (He was the vice-chairman of the peasant association and runs the service cooperative's shop).

Question: "You told me that you took training in management of dairy cows nearly 10 years ago. Why and how?"

Answer: "I can read well. I am a good and respected farmer and I can afford to purchase cows. The staff of the Ministry of Agriculture used to visit my farm often and recruited me for the training."

Question: "Why did they visit you often?"

Answer: "They gave me advice on what kind of pesticides to use, and provided me with veterinary control and related things. In return, I did a very good job.

Thus, they kept on coming."

Question: "You have planted a lot of land with teff compared to other farmers. Why was that? You also sold most of the milk from your cross-bred cows. Would not it be feasible to process it into butter?"

Answer: "These thieves, you know the "AMC", they collaborate with people who prosper with their tongue (he meant politicians) and allocate a larger amount of grain marketing quota than what is expected from a PA. They sell the extra grain in the free market, and collect the difference between the value of grain at the AMC and market price for themselves. If I do a good job, I get good production. I sell the teff and purchase barley or oats to satisfy the quota. If I get 8 quintals of teff per hectare, I can sell one quintal for 100-120 birr. Then, I can purchase two quintals of barley for 100 birr and drop it to the `hyenas' (marketing board authorities). If I plant barley, I may get 12 quintals but remain with 10 quintals after the quota. The market value of 7 quintals of teff (700 to 840) is higher than 500 birr worth of 10 quintals of barley. I save at least 200 birr. Besides, I also raise sheep. If I sell one sheep, I can purchase barley and give it to AMC. I sell fresh milk because the amount of butter produced per liter of milk is small. I feed the calf well and sell the rest. The calf grows fast and sells at a high price. Which one do you think is better?"

He has a big house with a roof made of galvanized aluminum. He prefers to die rather than being forced not to visit church, at least every Sunday. According to Ato Abebe all the crops and livestock, all the children and members of his family, all the village and the country are in the hands of the Almighty God.

"Ignoring God is to forget who we are. What I tried to work is because of God's blessing. No matter how hard I worked I will not succeed if I don't respect the church. Thus, I contribute more to the church than to AMC."

Discussions with Ato Tesfaye and Abebe reveal a number of goals and strategies. The goals of the two producers include: i) being a good farmer, ii) being wealthy or rich, iii) securing subsistence, iv) helping parents and v) attending church. The strategies followed by the two producers are sending children to school, adopting technologies, raising milking cows and ruminants, avoiding risk of losses and being friends with extension agents.[6] Nevertheless, they encounter constraints such as declining farm size, accessibility of resources, limited employment opportunities, biases of extension activities towards well-to-do farmers, unfavorable terms of trade for agricultural products, and risk embodied in lump-sum investments (e.g., cross-bred cows).

The discussions also reveal that education as a means of self-employment is not considered a worthwhile investment. The only option is to use their education to run for political office or look for employment in institutions within a peasant association. In either way, education may help them to gain access to resources or to obtain more income.

5.3. The Relationship Between Household Goals and Strategies[7]

5.3.1. Introduction

Fifteen cards bearing most of the goals of households were prepared and presented to farmers for ranking (Table 5.1.). Producers were instructed to suggest goals which they felt were left out. Goals that were chosen by most farmers in the Selale region relate either to livestock production or affiliation with social networks (e.g.,, cattle husbandry, producing more milk and butter, becoming leader of self-help group, gaining respect and prestige and raising more horses). Goals that were chosen by most households in the Ada region relate to crop production, trade or marketing (e.g. increasing grain production, becoming a business man, purchasing valuables, fattening of cattle or small ruminants, maintaining current income and creating intimacy with traders or city people).

The percentage of test farmers corresponding to the ranking of most of the goals is higher than control farmers for most goals. Fewer control farmers in the Selale region chose gaining respect and prestige, or raising horses, as their preferred goals compared to test farmers. A relatively high percentage of control farmers prefer the goal of becoming a businessman compared to test farmers that chose this goal. This may indicate the need to compensate for foregone income because of not receiving cross-bred cows through participation in non-agricultural activities. In the Ada region, most control farmers did not prefer the goals of producing butter and milk, becoming a business man and maintaining current income compared to test farmers that chose these goals. However, more control farmers would like to attain the goal of purchasing valuables compared to test farmers.

Goals ranked first to fifth were chosen by about 80% of the producers in both areas (Table 5.1). These goals were selected for detailed statistical analysis. Similarly, strategies employed by producers to achieve these goals were gathered through an open-ended questionnaire. To study factors explaining variations in the choice and ordering of these goals, multivariate regression analysis with dummy variables was employed. The goals chosen for the analysis were to: i) secure family survival, ii) expand cattle production, iii) build a big house, iv) rear small ruminants (sheep and goat) and v) be a good farmer. Why are these goals important to households? Before presenting statistical results, the implications and significance of these goals are discussed below.

Goals ranked first to fifth were chosen by about 80% of the producers in both areas (Table 5.1). These goals were selected for detailed statistical analysis. Similarly, strategies employed by producers to achieve these goals were gathered

Table 5.1. Rankings and Percentage Distribution of Goals by Region

	Selale		Ada	
	Test	Control	Test	Control
Rank of Goals	N=89	N=128	N=26	N=271
	Percentage (%)			
1. Securing family survival	98	97	97.5	99
2. Cattle husbandry	91	85	55	54
3. Building big house	71	71	78	81
4. Raising sheep and goat	75	79	74	71
5. Be a good farmer	81	83.5	91	93
6. Producing more milk and butter	69	61	51	38
7. Going to Church	66	61	57	54
8. Increase grain production	43	35	55	62
9. Seeing children married or getting good job	45	49	51	49
10. Becoming leader of self-help group	31	33	24	21
11. Gaining respect and prestige	55	22	23	17
12. Be business man or open Shops	24	39	59	45
13. Purchasing valuables such as radio/watch	15	19	22	31
14. Start fattening of oxen and sheep to increase income	37	34	45	49
15. Raising more horses	51	38	14	12
16. Maintaining current income	22	15	31	17
17. To create intimacy with traders or city people	19	18	28	31

1/ where N= number of farmers

through an open-ended questionnaire. To study factors explaining variations in the choice and ordering of these goals, multivariate regression analysis with dummy variables was employed. The goals chosen for the analysis were to: i) secure family survival, ii) expand cattle production, iii) build a big house, iv) rear small ruminants (sheep and goat) and v) be a good farmer. Why are these goals important to households? Before presenting statistical results, the implications and significance of these goals are discussed below.

5.3.2. Implications and Significance of the Selected Goals

Goals are selected by households only if they are given meaning by the culture. Thus, goals can be seen as cultural values. On the other hand, attainment of goals serves as a means or instrument to attain status. Hence, goals may serve both as cultural values and instruments. Nevertheless, applied household decision analysis should distinguish goals from strategies, even though their distinction is by no means clear.

Securing subsistence is the goal of all producers (Table 5.1). The household, however it is defined, cannot reproduce itself or the farm without securing food. This goal does not, therefore, require further explanation.

Why is raising cattle the second major goal of many producers? What do cattle do for the household economy? What do they represent? The household is sometimes used as a metaphor to describe cattle. For the farmer, cattle as a class of livestock include the ox (father), the cow (the mother), the steer (the son), the heifer (daughter) and calf (the child). These divisions within the class of cattle resembles those of humans. They are like a family. A cow with good milk production and reproduction has the privilege of forming her own family. That is, all her calves (young and mature) will be kept as husbandry animals for a relatively longer time. In addition to providing metaphorical correspondences between humans and animals, it is important to know the values and belief systems pertaining to cattle.

Farming as a culture in the Central Ethiopian highlands is based on livestock and land. Each class of livestock fulfils different purposes. From the informants' point of view there is more than one reason for keeping a specific class, breed, type and color of livestock. But one can broadly categorize these reasons into those based on "utilitarian" or "value" perspectives, the latter based on cultural attributes or categories. The following paragraphs illustrate the value perspective while revealing utilitarian aspects of the farming system and its sub-systems.

The view of livestock among informants can be described as follows;

> "The commercialization of livestock raises particular problems, since livestock represents not just stores of particular `use values' for which certain demand might exist, or even `exchange values' of certain worth assessed in terms of the local currency, but in Schneider's words function them-selves as `money' for which there is intrinsic demand, in pastoral and non-pastoral societies alike (Galaty, 1988:9)."

To an Orthodox Christian farmer, a cow is considered as physically weak because of its reproductive trait, a symbol of good will and generosity. It is intended for reproduction of a required trait. As the rich feed the poor, so does the cow feed her calf and the child of the house. Cows, like calves, are not resistant to water deprivation or under-feeding. Thus, exposing the cow to hardship is considered insane and counterproductive.

The sacred cow, according to the cultural values of the Christian farmer, is the mother of the "earthly God" (i.e., the ox) or like Virgin Mary, the mother of the Savior. The cow is the mother of the ox who saved mankind from using hand tools in order to plow and helps him to harvest enough for the survival of the family. The ox is capable of resisting nutritional stress, and is considered a saint sent by God to help the farmer produce and live. Yet, in the peasant economy, the ox is considered a "fool" or humble creature. For example:

"berie hoy berie hoy ante mognu hoy
Sarun ayehina gedelune satayie

elim Kalew gedele gebteh Kereh Woyie"

This proverb is interpreted as:

You, foolish ox, in search of grass
and without noticing your surrounding
fell off the cliff and never returned.

Producers who cannot feed their family adequately are equated to an ox with weak traction and a lazy person. Songs and metaphors related to the ox are a common feature of the Ethiopian highlanders and cover each phase in the crop cycle (plowing, planting, harvesting and threshing).

The steer and heifer are replacement for the ox and the cow, respectively. Therefore, they are cared like a "child" until they become an ox or a cow. The calf should never be beaten or hurt. Farmers will never eat calves before they reach a certain age. Sheep are considered "holy" and goats are associated with snakes or "Satan".[8] They help to stabilize household income. Draft animals are not eaten but sent by God to serve humans as a means of transportation or exchange.

For ecologists and economists, the logic behind the classification of animals is correlated to adaptability or the notion of utility. That is, only their extrinsic value as farm animals counts while their intrinsic value and importance as cultural categories does not. However, the producer's classification is not only rational but carries with it an important value connotation.

The socio-cultural organization of peasants in the Ethiopian highlands associates a farmer with ownership of cattle. A producer without cattle is not a farmer. He is not very different from a migrant worker who travels to an area where crop production is the dominant enterprise where a shortage of peak season labor is common. The household head and members of his family are given lower social status. Members of a village seek the labor of these poor households for the price of feeding them. This kind of labor acquisition carries with it impersonal social ties. Nothing is more disgraceful than working for food in a peasant society. It can, therefore, be concluded that cattle form the basis for the economic, social and cultural status of peasants. A number of producers have difficulty in ranking family survival and owning cattle separately. It is virtually impossible to have one without the other.

The third goal of households in the study sites is building a big house. Building one's own house is a very important accomplishment for the household head. The house is a public statement of the household head's hopes and ambitions with regard to status (Bauer, 1985; Levine, 1974; Hoben, 1973; Pankhurst, 1961, 1966). It is not only the ability of a producer to support his own family that accords him a status from within and outside his village. Equally important is the image that he inculcates in the minds of many who pass by his house or look at his house from a distance.

A man's prospect for a good life lies not only in how good his crops look but also in how well he and his family members dress and the size of their house. This is strengthened by the ability of a producer to hold occasional feasts. No matter how good the feast is, the "smile" and the size of the house are accorded much of the credit for organizing the feast. Producers use metaphors such as "kefitfitu fitu" for the smile and "Enkuwan Sew Feres Yasgalibal" for the house. Respectively, these mean that "the joy and smile of people who provide the food and drinks are the essence of a feast rather than the meal itself", and that "the house can accommodate `horse riding', let alone movements of people". Therefore, an image of a good person and a good life is established if a household head satisfies the goal of building a big house.

The house inhabited by an individual is a sacred place. An individual would be more reluctant to enter the compound of a big house compared to a small or poor man's house. Everybody murmurs "yeklie lij bet", which means "Ato x's son's or daughter's house". A respect for privacy is given not only to those who own big houses but also to their parents as well. It is a means to establish closer ties with people from outside the village. For instance, extension agents and staff of development projects are invited or given appointments at "Ato x's house".[9] Other members of the village supply the drinks and food to the lady of the house to entertain these guests. People with big houses obtain respect from extension agents either because they were invited in their house or assessment of what is required to be a good farmer. Consequently, they gain access to credit, innovations and related benefits from agencies.

The fourth major goal of farmers is to successfully pursue small ruminant husbandry. The category of small ruminants includes sheep and goats. The study sites are in highlands favorable for sheep husbandry. Goats are lowland animals while sheep perform well in the highlands. Availability of feed and leaves of trees for goats plays a significant role in the distribution of goats. The highlands are cleared of many trees and bushes. The lowlands are less suitable for crop production and are characterized by an abundance of shrubs and bushes. Moreover, goats are resistant to water stress or are considered to be hardier than sheep. Many producers believe that the meat of sheep rots more quickly in a hot climate than that of the goat. The highlands are characterized by cold or mild temperatures thus enabling mutton to stay fresh longer without applying the traditional methods of preservation. Besides their religious significance, sheep and goats provide quick income that enables households to pass through hard times without resorting to assistance from friends, relatives or the government. Owning sheep or goats, which serves as collateral, makes borrowing easier and reduces the chances of indebtedness.[10]

No matter how poor a producer is, it is socially desirable to reciprocate invitations. For Christians, in addition to major fasting days, no fatty foods are permitted on Wednesdays and Fridays. Feasts should normally be on non-fasting days. Sheep constitute the major source of meat for feasts. Not only will people be happy but they will spread news that "Ato X" killed a sheep from his herd. This commands respect. People who are invited to feasts are expected to respond positively to requests regarding help at times of plowing, weeding, harvesting, weddings and deaths. In addition, children in their twenties and thirties often leave their parents for job-related reasons. Parents visited by their children are expected to kill a sheep. Small ruminant husbandry, therefore, serves important social and economic functions in the organization and behavior of households.

The fifth goal of Selale and Ada producers is to be good farmers. Who is a good farmer? What does this imply for members of households, villages and of outsiders? A good farmer should: i) be hard working or diligent person in either crop or livestock husbandry; ii) be able to help relatives; and iii) have stable and happy marriage (See Korten, 1972; Levine, 1974; Pankhurst, 1966).

5.3.3. Results of Statistical Analysis

5.3.3.1. Introduction

The choice of goals determines the selection of strategies (see section 5.1 and 5.2). Interviews and group discussions revealed that the strategies employed by producers of Selale and Ada regions fall into three categories. They are: i) maintaining or increasing current food production levels or standard of living (opportunistic), ii) securing subsistence or risk minimization and iii) long-range or strategic planning.[11] Opportunistic strategies include purchases of improved seed, fertilizer, pesticides, medicine to control animal disease, cross-bred cows and expansion of farm size. Risk minimization strategies include off-farm work, planting more than one crop and raising more than one class of livestock. Strategic or long-range planning include saving, sending children to school, storing grain, running for political office and increasing asset holdings. The statistical analysis is discussed according to these categories.

Five regression equations with dummy variables were performed corresponding to the five goals (section 5.1). In using dummy variables as explanatory variables, an equation with an intercept term (which is the value of the reference category) is more convenient because it facilitates the interpretation of the coefficients (Kennedy, 1988).[12] The choice of strategies is given meaning by the socio-cultural categories and institutions. Households cannot produce subsistence food requirements unless they receive support from their family members and social networks. Therefore, goals and strategies are chosen in terms of their ability to satisfy the strategy of continuing the family tradition of active involvement in village life or social networks. Most study farmers indicated that maintaining family tradition or belonging to social networks is an important part of their existence and they evaluate their actions relative to this goal. Therefore, farmers were asked to rank other goals relative to it. Thus, continuing family tradition or involvement in social networks is used as the reference or intercept with which the importance of other strategies are compared.

The dummy variable's coefficients are interpreted as the extent to which behavior in one category (e.g. planting more than one crop in the category of risk minimization) deviates from the reference base. If the variable's importance to the ranking of a particular goal is greater than participating in social networks, then the coefficient will be negative. Conversely, if the contribution of a strategy to the ranking of a goal is lesser than participating in social networks, then the coefficient will be positive.[13] In addition to t-test of the statistical significance of individual coefficients, the importance of all strategies to the overall ranking of a goal is tested using the F-ratio.

In all regressions, the dependent variables are the five goals discussed above. Furthermore, attainment of one goal can serve as a means to satisfy other goals. For instance, securing subsistence may provide a buffer or an incentive for households to take actions to satisfy the goals of building a big house and pursuing livestock husbandry. The goals which enter as regressors with other strategies indicate sequentiality of decisions. The result of this sequentiality is presented in a separate table. The analysis is carried out for each region by test and control groups.

5.3.3.2. Opportunistic Strategies

In chapters two and three, increasing the productivity of resources such as land, livestock and labor, and expansion of land were suggested as a means to increase food production. In identifying strategies implemented to satisfy the goal of securing subsistence requirements these two methods of increasing food production problems were considered. Specifically, opportunistic strategies include methods to increase productivity of resources (purchases of improved seed, fertilizer, pesticides, medicine to control animal disease and cross-bred cows) and expansion of area under crops. The result of the relationship between these strategies and goals is presented in Table 5.2. Analysis is performed to compare Ada and Selale regions, and farmers in each region.

Most of the strategies are used to achieve all the goals selected for statistical analysis. Compared to participation in social networks (the intercept term), most strategies satisfy the goal of securing subsistence as indicated by the low or negative signs of the regression coefficients (weights). Nevertheless, only fertilizer and pesticides show a consistent and statistically significant importance in both regions and for both groups of farmers.

Expansion of farming area seems to have its greatest importance on securing subsistence food requirement in an area where crop farming is the dominant enterprise (e.g. Ada). Purchases of improved seed and cross-bred cows are ranked highly on the goal securing subsistence requirement in the Ada and Selale regions respectively.

Table 5.2. Relative Importance of Opportunistic Strategies[1/]

	Securing family survival				Cattle husbandry			
	Selale		Ada		Selale		Ada	
Categories/Farmer Group	Test	Cont.	Test	Cont.	Test	Cont.	Test	Cont.
Opportunistic Strategies								
Intercept	1.85*	1.85*	1.67*	1.89#	1.61*	0.51	1.38*	2.26*
Improved Seed	0.03	-0.04	-0.36*	-0.44#	-0.03	-0.12	-0.05	-0.04
Fertilizer	-0.47*	-0.46*	-0.32*	-0.49*	-0.04	-0.24*	-0.75*	-0.48*
Pesticides	-0.67#	-0.67*	-0.22#	-0.69#	-0.44#	-0.37	-1.47*	-0.13
Crossbred Cows	-0.45#	-0.26#	-0.02	-0.13	-0.76*	-0.34*	0.2	-0.13
Veterinary Service					-0.01		-0.01	
Farm Size	-0.48#	-0.42#	-0.54*	-0.62*	-0.67*	-0.33#	-0.61*	-0.08
Risk-Minimization Strategies								
Trading	-0.46*	-0.43*	-0.68*	-0.89*	-0.04	-0.02	-0.13	0.11#
Off-farm work	-0.04	-0.04	-0.35*	-0.24*	-0.04	-0.01	0.11*	-0.08
>1crop	-0.03	-0.03	-0.26*	-0.54*	0.55*	-0.88*	-0.21	-0.32
>1livestock	-0.46*	-0.45*	-0.04	-0.06	-0.77*	-0.67*	-0.07	-0.08
Long-Range Planning Strategies								
Saving	-0.46*	-0.25#	-0.94#	-0.77#	-0.47#	-0.55#	-0.06	-0.10#
Sending Children to School	-0.04#	-0.14#	-0.36#	-0.17*	-0.04	-0.09	-0.06	-0.05
Store Grain	-0.06*	-0.16*	-0.08*	-0.07#	0.23#	-0.27*	-0.58*	-0.31#
Running for political Office		-0.06#	-0.07*	-0.01	-0.01	-0.31#	-0.35#	-0.02
Asset Accumulation		-0.22	-0.33	-0.21	-0.52	-0.01	-0.67*	-0.95*
N	88	127	25	26	88	127	25	2688
R^2	0.94	0.93	0.97	0.97	0.7	0.6	0.9	0.9
F-Ratio	5.8*	4.5*	7.1*	4.1*	7.9*	6.2*	2.1#	1.5

1/ The values are partial regression coefficients and lower weights indicate higher importance.

- * and # refer to significance at 1 and 5 percent levels respectively. And, Test and Cont. refers of test and control farmers respectively.

Table 5.2. Relative Importance of Opportunistic Strategies[1] (Continued)

	Selale		Ada	
Categories/Farmer Group	Test	Cont.	Test	Cont.
Opportunistic Strategies				
Intercept	0.19*	0.35	1.07*	1.30*
Improved Seed	-0.12	-0.01	-0.08	0.91
Fertilizer	-0.67*	-0.73*	-0.37#	-1.05*
Pesticides	0.21	-0.11	-0.86*	-0.32*
Crossbred Cows	-0.46*	-0.19#	-0.08	-0.17#
Veterinary Service				
Farm Size	-0.09	0.23	-0.22#	-0.59#
Risk-Minimization Strategies				
Trading	-0.01	-0.07*	-0.01	-0.11
Off-farm work	-0.19*	-0.14*	-0.28*	-0.30*
>1crop	-0.14	-0.12	-0.19	-0.16
>1livestock	-0.32#	-0.16	-0.51*	-0.35
Long-Range Planning Strategies				
Saving	-0.38*	-0.28#	-0.29	-0.31*
Sending Children to School	-0.46*	-0.88*	-0.19	-1.09*
Store Grain	-0.02	-0.02	-0.03	-0.04
Running for political Office	-0.03	-0.18#	-0.40#	-0.34#
Asset Accumulation	-0.84#	-0.76*	-0.42#	-0.49#
N	127	25	26	88
R^2	0.58	0.49	0.88	0.8
F-Ratio	5.8*	7.4*	6.6*	4.8*

1/ The values are partial regression coefficients and lower weights indicate higher importance.

- * and # refer to significance at 1 and 5 percent levels respectively. And, Test and Cont. refers of test and control farmers respectively.

Planting improved crop seed is not widely practiced in Selale area. The performance of improved seed in demonstration sites is poor. The average yield per hectare of local cereal variety is 8.4 qts.[14] The average yield of improved variety is 8.2 qts/ha (MOA, Selale office). There is little wage employment. Thus, these strategies are not significantly associated with the goal of securing subsistence. Producers recognize opportunities open to them and attempt to exploit them usefully. A Selale farmer, Ato Abebe Dilnesaw, explains why he does not intend to adopt new cultivars as follows:

> " New crop varieties have not performed well. We have seen them on the plots of extension agents (demonstration plots). Purchasing these varieties and getting only a one-fourth increase in production is not worth it. You see I could afford to take risks before I got married. Once I had children, I couldn't live for myself but for my children. You know our grandparents say `Even a mother does not love her poor son'. Everybody likes you, including your own family, as long as you can support them and be better off."

Expansion of area is valued highly by test farmers of both regions in satisfying the goal of animal husbandry. This may imply that test farmers would like to increase their land holdings for the purposes of increasing grazing and hay lands. Fertilizer and cross-bred cows are ranked highly in attaining the goal of cattle husbandry in the Ada and Selale regions respectively. In the Ethiopian highlands, crop and livestock production are intertwined. They compete for the limited resources owned by households, and the income generated from one enterprise helps in the expansion of the other. Their importance, however, varies depending on the comparative advantage of the regions. Crop production increasing strategies may be ranked highly among the goals of Ada farmers while that of livestock production increasing strategies among the goals of Selale farmers. The strategies of purchasing fertilizer and pesticides, and expansion of area are relatively more important and statistically significant in attaining the goal of building a big house relative to other strategies among farmers of Ada region. However, fertilizer and cross-bred cows are ranked highly in attaining this goal among farmers of the Selale region.

Purchases of cross-bred cows and fertilizer in the Selale region, and expansion of area in both study sites are valued highly and are statistically significant in satisfying the goal of expanding the small ruminant husbandry.

Cross-bred cows, fertilizer, pesticides and expansion of area are important in attaining the goal of being a good farmer in the Selale region. In the Ada region, however, purchases of fertilizer, pesticides and improved seed, and expansion of area are the strategies that are valued highly in attaining this goal.

5.3.3.3. Risk Minimization Strategies

All strategies of risk minimization are ranked highly on the five goals relative to participation in social networks (Table 5.3.). Risk minimization strategies are influenced by availability of employment opportunities and regional advantages regarding crop or livestock production. Sharply contrasting differences exist across regions. In the Selale region, all strategies are statistically significant in satisfying the goal of securing subsistence except planting more than one crop. In Ada region, however, all strategies are important and statistically significant except raising more than one class of livestock. Ada producers do not have adequate feed resources to support livestock husbandry. Therefore, raising more than one class of livestock is not the best-bet solution to minimize risk.

With respect to the goal of livestock husbandry, planting more than one crop and raising more than one class of livestock are statistically significant for the Selale region but only off-farm income is significant for the Ada region. Selale farmers have a large number of livestock compared to Ada farmers. With the decline in the size of grazing area, crop by-products offer an alternative source of livestock feed. This may be the importance of planting more than one crop in the Selale compared to Ada region.

Building a big house requires a substantial amount of monetary expenditure. Given the small volume of grain that households harvest, alternative sources of cash are sought that include off-farm employment and selling livestock. The result indicates that off-farm employment and raising more than one class of livestock are ranked highly in attaining the goal of building a big house in both regions, and trading is ranked highly in the Ada region.

Raising more than one class of livestock and off-farm income are consistently ranked highly (and statistically significant) on the goal of small ruminant husbandry compared to participation in social networks. Most of the regression results indicate that both cattle and small ruminants are more important in Selale when compared with Ada region.

Trading, planting more than one crop and raising more than one class of livestock are statistically significant in attaining the goal of being a good farmer in the Selale region. The magnitude of the weight of trading and planting more than one crop in attaining the goal of being a good farmer are higher among Ada compared to Selale farmers.[15] Nevertheless, most strategies are valued more important in attaining this goal relative to the reference strategy or base (i.e., maintaining family tradition and participation in social networks).

Table 5.3. Relative Importance of Risk-Minimization Strategies

	Risk-Minimization Strategies			
Goal/Region/Farmer	Trading	Off-farm Work	> 1 crop	> 1 Livestock
Goal: Securing Subsistence				
Selale				
Test	-0.456*	-0.036	-0.03	-0.46*
Control	-0.427*	-0.038	-0.031	-0.457*
Ada				
Test	-0.688*	-0.346*	-0.262*	-0.043
Control	-0.89*	-0.238*	-0.543*	-0.059
Goal: Cattle husbandry				
Selale				
Test		-0.037	0.551*	-0.77*
Control		-0.01	-0.888*	-0.666*
Ada				
Test		0.105*	-0.212	-0.07
Control		-0.08	-0.321	-0.08
Goal: Building a big house				
Selale				
Test	-0.044	-0.192*	-0.139	-0.315#
Control	-0.022	-0.135*	-0.116	-0.162
Ada				
Test	-0.129	-0.282*	-0.19	-0.506*
Control	0.105#	-0.304*	-0.158	-0.353
Goal: Small Ruminant hus-				
Selale				
Test	-0.007	-0.199*	0.032	-0.448*
Control	-0.07*	-0.135*	-0.081	-0.333*
Ada				
Test	-0.014	-0.004	-0.063	-0.247
Control	-0.107	-0.002	-0.041	-0.322*
Goal: To be a good farmer				
Selale				
Test	-0.227#	-0.154	-0.596*	-0.804*
Control	-0.207#	-0.14	-0.43*	-0.508#
Ada				
Test	-0.357*	-0.151	-0.75*	-0.474
Control	-0.335#	-0.431	-0.828	-0.321

* and # refer to significance at 1 and 5 percent levels respectively.

5.3.3.4. Strategic or Long-term Resource Allocation Strategies

Strategic resource allocations include saving, sending children to school, storing grain, running for political position and accumulating assets (Table 5.4.). Similar to risk minimization and opportunistic strategies, these strategies are important to attain the five goals selected for analysis. Large differences are observed with respect to the impact of saving in the Ada region when compared with the Selale region. In societies where social institutions influence the forms of cooperation to be more distributive, saving as a strategy to attain food security may be less important (See chapter VI on distributive [Selale] and economic [Ada] forms of institutions).

In regions where the major source of income is livestock (Selale), saving provides a means to purchase small ruminants. However, in regions where farmers own few livestock (Ada), purchases of small ruminants can be achieved with income obtained from selling stored grain.

Sending children to school is ranked highly on the goal of building a big house among Selale farmers compared to Ada farmers. This may be because an average Selale farmer receives 139 Ethiopian birr per year in remittance from children with urban jobs while an average Ada farmer receives 73 Ethiopian birr. Accumulation of asset is ranked highly on the goal of building a big house in the Ada region when compared with the Selale region.

Saving and sending children to school play important roles in attaining the goal of small ruminant husbandry in both study sites. However, the importance of sending children to school in attaining the goal of small ruminant husbandry is ranked highly among control farmers compared to test farmers in both regions.[16] Accumulation of assets is more important in satisfying this goal among test farmers compared to control farmers in both regions. Saving and sending children to school are ranked highly on the goal of being a good farmer in Ada compared to Selale region. However, accumulation of assets is ranked highly in the Selale region compared to Ada.

Saving and sending children to school show relatively more importance and are statistically significant in attaining most of the goals, compared to other strategies. Running for political office is valued highly in attaining most of the goals among Selale compared to Ada farmers. The latter result may indicate the fact that the largest proportion of asset owned by Selale farmers is embodied in cattle and that ownership of large number of cattle accords the status of being a good farmer in this region.

The F-ratio indicates that most strategies are significant predictors of the rankings on all the goals of households. This significance is high for test farmers, and for the Selale region when the goals are livestock related, and for the Ada region when the goals are crop related.

Table 5.4.: Relative Importance of Strategic or Long-Range Strategies

	Strategic or Long-Range Strategies							
Goals/Region/ Farmers	Saving	Sending Children to School	Store Grain	Run for Political office	Asset	N	R2	F-Ratio
Goal: Securing subsistence								
Selale								
Test	-0.466*	-0.039#	-0.069*	-0.067#		88	0.94	5.8*
Control	-0.245#	-0.141#	-0.169*	-0.074*		127	0.93	4.5*
Ada								
Test	-0.939#	-0.355#	-0.08*	-0.003		25	0.97	7.1*
Control	-0.776#	-0.17*	-0.067#	-0.007		26	0.97	4.1*
Goal: Cattle husbandry								
Selale								
Test	-0.47#	-0.036	0.228#	-0.11#	-0.22	88	0.7	7.9*
Control	-0.551#	-0.093	-0.266*	-0.052	-0.33	127	0.6	6.2*
Ada								
Test	-0.056	-0.061	-0.588*	-0.021	-0.21	25	0.9	2.1#
Control	-0.104#	-0.054	-0.314#	-0.03	-0.52	26	0.9	1.5
Goal: Building a big house								
Selale								
Test	-0.375*	-0.465*	-0.017	-0.18#	-0.01	88	0.58	5.8*
Control	-0.281#	-0.883*	-0.017	-0.002	-0.67*	127	0.49	7.4*
Ada								
Test	-0.285	-0.195	-0.031	-0.04	-0.95*	25	0.88	6.6*
Control	-0.309*	-1.09*	-0.043	-0.53	-0.84#	26	0.8	4.8*
Goal: Small Ruminant husbandry								
Selale								
Test	-0.147	-0.572#			-0.76*	88	0.73	10.0*
Control	-0.296#	-1.099#			-0.02	127	0.66	7.1*
Ada								
Test	-0.812*	-0.442#			-0.09	25	0.88	2.5#
Control	-0.442#	-0.921*			-0.03	26	0.92	1.1
Goal: To be a good farmer								
Selale								
Test	-0.395*	-0.143	-0.031	-0.76	-0.98*	88	0.62	7.8*
Control	-0.41*	-0.01	-0.007	-0.58#	-1.51*	127	0.54	2.5#
Ada								
Test	0.574#	0.373*	-0.118	-0.75*	-0.25	25	0.85	6.1*
Control	-0.511#	-0.316#	-0.122	-0.23	-0.08	26	0.73	3.1#

* and # refer to significance at 1 and 5 percent levels respectively.

Table 5.5 shows that goals ranked high in the hierarchy are important in attaining goals that are ranked low in the hierarchy. Securing subsistence is the first goals. This pattern of relationship between goals is similar to sequentiality decision-making. The first, second, third and fourth goals (i.e., securing subsistence food requirement, cattle husbandry, building house and ruminant husbandry) influence the chances of satisfying the second, third, fourth and fifth goals compared to the base (i.e., maintaining family values and participating in social networks).[17] All the weights of goals that enter in the regression (listed on the columns) are negatively signed (Table 5.5.). It means that, relative to the base, all the goals ranked high in the hierarchy are ranked highly, most of them statistically significant, in attaining goals ranked low in the hierarchy. These sequential patterns of goals do not show significant variation across regions or across farmers in each region. However, most of their effects are statistically significant among test compared to control farmers.

5.3.3.5. Summary

From the statistical results of the structural association between strategies and goals, the following implications can be drawn. Most strategies are relatively important in satisfying household goals in both regions. Strategies which are proven to be useful by other producers prior to this study tend to have a stronger relationship with goals (e.g. pesticides and fertilizer). The importance of strategies that are directly related to livestock husbandry is higher in the Selale than in the Ada area. Strategies that help boost crop production are ranked highly on goals of Ada farmers than on goals of Selale farmers. Because of the desire to reduce income differentials between test and control farmers, control farmers engage in alternative sources of income. Thus, strategies that involve non-agricultural activities tend to have greater importance in attaining the goals of control compared to test farmers. Producers rank strategies hierarchically, and goals ranked high in the hierarchy are valued highly on the attainment of subsequent goals.

5.4. Information and Knowledge

5.4.1. Information and Cognition

In examining information, three important points ought to be considered, concerning the types, sources and accessibility of information. The types of information gathered depend on the goal of the subjects. Information can concern production, consumption, marketing, reproduction or other goals. Within these broad categories, subjects search for specific sets of information. If the goal of the subjects is production, the information sets may include input and output markets,

Table 5.5: Sequentiality of Goals

Goals/Region/Farmers	Securing Food	Livestock Husbandry	Building a Big House	Ruminant Husbandry
Goal: Cattle husbandry				
Selale				
Test	-0.132#			
Control	-0.17*			
Ada				
Test	-0.109#			
Control	-0.118			
Goal: Building a big house				
Selale				
Test	-0.216#	0.194		
Control	-0.008	-0.107		
Ada				
Test	-0.008	-0.157#		
Control	-0.641*	-0.077#		
Goal: Ruminant Husbandry				
Selale				
Test	-0.12#	0.247*	-0.057	
Control	-0.028	-0.35#	-0.014	
Ada				
Test	-0.286#	-0.041	-0.103	
Control	-0.228*	-0.165#	-0.157	
Goal: To be a good farmer				
Selale				
Test	-0.013*	-0.402#	-0.023	-0.241#
Control	-0.008	-0.542#	-0.032	-0.401*
Ada				
Test	-0.019	-0.928	-0.374*	-0.091
Control	-0.146*	-0.021	-0.067	0.112

* and # refer to significance at 1 and 5 percent levels respectively.

management, soil type, family food requirements, land, oxen and seeds. Depending on their goals and strategies, therefore, households look either for general or specific information.

Verner and Gubbels (1967) distinguished several sources of information: government (Ministry of Agriculture), commercial (information from business agents or establishments dealing with farmers), farm organizations, and personal (family, friends, neighbors, own observation and experience). The sources of information vary depending on the degree to which subjects articulated themselves with the larger economy. For instance, a peasants' major source of information could be friends, relatives, extension agents and members of the household; a traders' source of information could be producers, local, national and international markets. Once the types and sources of information are identified, households have to examine the method of gathering and processing information before taking actions.

Access to information depends on a farmer's intimacy with and knowledge about the sources. The information gathered has to be processed and given meaning. Most households have access to information from friends and relatives. The skill with which information is processed determines the outcomes of decisions. Given experience and skill, the amount and quality of information processed depend on the ability to think, recall, memorize and comprehend.[18] Information that cannot be comprehended or given meaning enters the body of experience as a "black-box".[19] The value of information, regardless of its source, is analyzed according to a pre-constructed context or framework of thinking. Information which is found to be meaningful will be added to the body of knowledge and can thus enlarge the framework of thinking or comprehension ability of subjects.

A subject first comprehends sentences and then makes cognitive contributions such as elaborating spatial implications and inferring probable consequences.[20] For information to be comprehendible, subjects must have a working knowledge of the language in which information is transferred. Cognitive processing can therefore take over after linguistic comprehension takes place (Bransford and McCarrel, 1983; Cole, 1983). Linguistic knowledge alone, however, is not enough for comprehension. Decision makers need to relate information contained in an object or entity to experience and to external information. Comprehension results only when the comprehender has sufficient non-linguistic information to use the cues specified in the linguistic input to create semantic content that allows a subject to understand utterances. The ability of a subject to comprehend depends on the knowledge of the language, the world derived from experience and how the information is processed.

Comprehension involves assimilating the information presented in the individual sentences to existing lexical and non-lexical knowledge schemas. The variety as well as the amount of practice with a particular subject matter an individual has is crucial to a wider application of cognitive skills. Humans learn generalized problem solving skills through repeated experience with different problems of the same type. The ability of subjects to make use of information depends on the types of processing mechanisms.

There are two types of information processing mechanisms, the conscious and the unconscious. Pre-attentive or unconscious processing refers to any information processing that is outside of a decision maker's ordinary attention and awareness. This processing underlies routine decisions. Attentive or conscious processing involves long-range planning and critical thinking. The ability to suddenly shift attention suggests that humans are continually monitoring their environment for matters of immediate importance, unconsciously processing to some degree a variety of information. For example, a peasant may be plowing his plot when he suddenly sees a dark cloud and feels cold, which are interpreted as indicators of rain. He immediately pulls the working oxen to the side and runs to his homestead to cover the piles of hay and straw before the rain causes seepage of water and rotting. He is consciously acting or plowing when he unconsciously processes the information related to the coming of rain. The larger the amount of information subjected to critical or conscious thinking, the more likely that it will form part of experiential knowledge. It is, therefore, necessary to design ways of assessing the types of information that are subjected to conscious and unconscious processing to differentiate farmers by the quality of knowledge they process.

Prior knowledge and experience are identified as criteria for acceptance or rejection of information conveyed in the form of premises. Sources of new information from which to derive conclusions and proof or verification of a conclusion reached are also linked to knowledge and experience (Bransford and McCarrel, 1983; Cole, 1983). Cognitive contribution of information and its comprehendability are linked to experience and knowledge. The information which is processed and given meaning will add to this body of knowledge and experience of subjects. Thus, it is possible to infer that knowledge possessed by subjects is the driving force behind most human actions. Before discussing issues related to knowledge, data relating to study sites is used to show the three aspects of information outlined above.

5.4.2. Types of Information

Information obtained from extension agents is similar to the content of agricultural courses offered in high schools, junior agricultural training centers, colleges and universities. The kinds and sources of information households receive are presented in Table 5.6. The same kinds of information are also obtained from social experience. If an important set of information cannot be obtained from social experience, it creates a mechanism of dependency on extension agents and "modern" or science education. The problem arises when actions pursued by subjects produce unexpected consequences. That is, outcomes not implied by the information set. Regardless of the outcomes of action, processed information will be added to the experience and knowledge of subjects.

5.4.3. Source of and Access to Information

Local knowledge includes a repertoire of various kinds of information and action-oriented experience. Various studies have demonstrated that information necessary to guide action-oriented behavior can be obtained either from formal sources or informal sources (Belay, 1977; Bell, 1972). Formal sources include

Table 5.6. Types and Sources of Information (N=90)

	Sources of Information			
	Extension	Families	Friends	Other[1]
Crop Production				
Pesticides Use	X	X	X	X
Rotation & fertilizer	X	X	X	X
Crop Protection	X	X	X	X
What to plant		X	X	
Soil type		X	X	
Texture and depth		X	X	
Improved seed	X		X	X
Livestock Production				
Better management	X	X	X	X
Forage production	X			
Disease control	X	X	X	X
Use of A.I.	X		X	X
Marketing				
When to sell		X	X	X
How to sell		X		X
What to sell		X		
Where to sell	X	X	X	X
Home Economics				
Few children	X			
Clean house	X	X		
Use of latrine	X			X
Child care	X	X		X
Dividing the house	X	X		
Soil & Water Conservation				
Fallowing		X	X	
Terrace	X	X	X	X
Furrows		X	X	X
Overturning green		X	X	

X- indicates source from where households gather information.

1/ This group includes city people, newspapers and magazines, markets, other government agencies and development projects.

schooling, extension agents, radio, newspapers and magazines. Informal sources include experience, interaction with friends, relatives, children, city people and markets (Table 5.6). It is not identification of the source which is difficult but obtaining access to it (Warren, et al., 1988). Most farmers in Ethiopia have not attended secular school and don't own a radio or have access to magazines attended secular school and don't own a radio or have access to magazines and newspapers. Therefore, the only formal source of information is from contacts with extension agents.

The experience of extension programs in LDCs has shown the complex and difficult nature of transferring information to peasants. Over the past three decades, it was assumed that transferring information from experimental stations to farmers would help to increase adoption of innovations. Actual figures, especially in Africa, suggest otherwise (Moock, 1986; Hayami and Ruttan, 1985; Molnar, et al., 1983; Jamison and Lau, 1982; Feder, et al., 1985). Recently, a training and visit scheme was started in India, Ethiopia and other developing countries (Feder, et al., 1982). However, this approach is questionable particularly because of the huge cost involved in reaching a large number of farmers in the Ethiopian highlands.

Besides issues regarding transferability of information, the activities of extension agents are biased towards types of farms favored or run by the state (see Chapter II; Table 5.7). Furthermore, producers who are favored by projects and political institutions receive more attention from extension agents.

Table: 5.7. Frequency of Visits of Extension Agents by Type of Farm

	Frequency of Contact per Year[1/]	
Types of Farm	Selale	Ada
Individual Peasants (in PAs)	24(81)	42(41)
Service Cooperatives	84(10)	80(6)
Producers Cooperatives	96(6)	112(4)

[1/] The figures in bracket refer to sample size.

The average frequency of contacts with extension agents per farmer per year ranges from 8 to 28 (Table 5.8). This range is influenced by various factors. For instance, political elites and test farmers have more access to extension agents than farmers who neither receive cross-bred cows nor hold a political office (see Table 5.8.). However, access to information from informal sources is more widely dispersed than from government or project sources.

Berry (1988) demonstrated that producers with status or authority are more likely to capture more resource than others. The results from this study indicate that farmers who hold political office own more grazing and arable land than those who do not hold a political office (Table 5.8).[21] Moreover, politicians are able to escape from the unfavorable effects of policies such as marketing quota, taxes and contribution, and recruitment of household members to war zones.

Group discussion, as a source of information, was carried out to assess how information is diffused and whether or not there are variations among households and villages. At the district level, differences in access to information is the result of government policy. In Chapter II, it was noted that the Ministry of Agriculture has divided the agricultural regions into surplus and deficit producing districts.

Table 5.8. Sources of and Access to Information Among Study Farmers

	Categories of Farmers						
	Ada			Selale			
	Politicians	Test	Control	Politicians	Test	Control	Sample Size
Categories	28	19	10	25	17	8	50
Market visit/month[2/]	2	2	2	4	3	3	33
City visit/month[3/]	3	2	2	4	4	4	42
Owning/listening radio[4/]	1	1	0	1	1	1	50
Experience (yrs)[5/]							
dependent	10	12	15	12	11	14	50
Independent	21	26	23	19	23	22	50
Grazing Area (ha)	0.4	0.3	0.3	1.3	0.9	0.8	50
Crop Area (ha)	2.3	2.1	2	2.8	2.5	2.6	50

Source: Computed from field survey, 1990/91.

[1/] Some farmers did not give answers to questions related to market and city visits. Thus, the effective sample size was less than 50.

[2/] refers to the average number of days that members of a household travel to a nearby city.

[3/] refers to the average number of days that members of a household travel to a nearby market place either to purchase necessities, sell farm outputs, or both.

[4/] If a household owns and listens to a radio a value of 1 is given and 0 otherwise.

[5/] Dependent- refers to the number of years a household head spent working for his/her parents, and independent refers to the number of years of farming experience after a household head became an independent farmer or after marriage.

Man-power, services and supplies provided through the Ministry are biased towards surplus producing regions (MOA, 1989).

Information from extension agents, in both surplus and deficit regions, tended to favor peasant associations which are close to cities and service cooperatives whose officials have developed strong friendship with extension agents and which have elected officials at sub-district, district or regional level (Table 5.9.). Most farmers in the Ethiopian highlands live more than five kilometers away from the main roads. A substantial amount of money is needed to cover travelling costs for extension agents in order to reach these farmers. Thus, extension agents visit

Table 5.9: The Effect of Distance and Political Representation on Frequency of Visits of Extension Agents

Pas in Salale	Distance from Cities or SCs	Intimacy with Officials with District DAs	Officials Chosen			Frequency of Visits per Month
			Sub-District	District	Province	
Gulele	1.5	high	no	no	no	21
Anokere	3	high	yes	yes	no	24
Wayu	2	high	yes	yes	no	28
Illukura	17	low	no	no	no	12

farmers who live closer to major roads and cities. Furthermore, producers who have directly or indirectly developed stronger friendship with peasant association officials and extension agents obtain more information than farmers who have not developed such contacts.

Regardless of the variations in the types, sources and access, information is processed and added to the pre-existing body of knowledge through listening, thinking, memorization, teaching (demonstration), recalling, comprehension and trial. In the Ethiopian highlands, the production of crops and livestock form the basis for the survival of households. The knowledge of these highlanders about local farming practices is expected to be substantial. Their knowledge could serve as a basis upon which the generation and dissemination of innovations should be developed (Belay, 1977; Wolde-Mariam, 1984). The results of participatory-observatory and interview methods of investigation related to crop and livestock production knowledge are discussed below.

5.4.4. Indigenous Knowledge

Indigenous/local knowledge refers to skills and experience gained through oral tradition and practiced over many generations. It is this knowledge which serves to solve daily problems and open employment opportunities for peasants. Variations in knowledge can be observed by sex, age, ethnic group, and degree of contact with the outside world (Norem, et al., 1988).

To be indigenous, the knowledge must be functional within a given society. It must be an active part of the culture of the population concerned, preserved, communicated and used by its members to some purpose in relation to a productive activity within the society (Bell, 1972). Thus knowledge, be it orally transmitted or acquired through various means, is used to direct meaningful actions. The manner in which actions or reactions are produced is related to the logic of thinking and acting. According to Piaget (1972), every response, whether it be an act directed towards the outside world or an act internalized as thought, takes the form of an adaptation or, better, of a re-adaptation (see also Geertz, 1983).

The potential of indigenous knowledge to form the basis on which development strategies should be built is discussed in Warren, et al., (1988), Geertz (1983) and Richards (1985). This knowledge can, however, be general or specific to individuals. Despite these differences, access to this knowledge by all subjects and development agents is crucial to the design of development policies. Furthermore, similar to aspects of information, it is essential to examine the types, sources and accessibility of knowledge and its significance for decision-making.

Identifying knowledge should be followed by discovering methods of obtaining and diffusing it to peasants who may benefit from it. How should indigenous knowledge (IK) be diffused and utilized for the benefit of local peasants and peoples elsewhere? Short (1973) notes that one of the requirements of utilizing knowledge is that one must have access to it. With respect to diffusion of knowledge, Guba (1968) noted that diffusion is the acceptance of an idea or practice by individuals, groups, or other adopting units, linked to specific channels of communication, social structure and system of values or culture, over time. Thus, indigenous knowledge is dynamic and changing over time.

Farming system specialists have strongly argued that producers or households are not always in "equilibrium" with the ecosystem (Gilbert, et al., 1980; Hendry, 1987; Shaner, et al.,1982). The moment an element of the system is injected through intervention, the balance changes, leading to another locus of equilibrium or disequilibrium.[22] Thus, what is observed is a moving target. Development projects have not recognized experiential knowledge as the driving force behind the dynamics of household behavior. If knowledge is in dynamic equilibrium, the problem of identifying, collecting, diffusing and utilizing it is as difficult as dissemination of new technologies or ideas. Moreover, indigenous knowledge is locale-specific, such that the same knowledge cannot be used everywhere.

According to Thrupp, perhaps the most promising way to build and legitimize the capacities of rural people is to understand that local knowledge and culture-based capacities can be a source of empowerment (Thrupp, 1989). The process of empowerment requires non-interference in peoples' own decision-making processes and participating at every stage of technology generation. This may involve recognizing the value of informal research carried out by farmers (e.g., Richards, 1985). It is after recognition and legitimation of indigenous knowledge, as the only power peasants have in interacting with outsiders, that constraints on the diffusion and utilization of knowledge are reduced or removed.

One prospect for diffusion of information and knowledge is the use of lateral transfers among farmers themselves. This can be accomplished through self-help groups and local forums whose premise is that not everybody knows the same thing. Applied household- level studies argue that knowledge and differential possession of information are a means of empowerment. However, it is impossible to empower households uniformly. As there are inequalities in ownership of physical resources (e.g. land), so are there differences in experiential knowledge and management styles. Knowledge within and between households is one means through which status or prestige can be gained. Similar to neoclassical theories of the value of information, gaining access to experiential knowledge and information in a peasant economy involves costs and risks. The problem of lateral transfer of information and knowledge, therefore, is the same as that of "disciplinary compartmentalization" amongst the learned circle (Warren, et al., 1989).

The study sites represent the mixed farming systems that dominate the Ethiopian highlands. The livelihood of households is dependent on the knowledge they acquire in both crop and livestock production (Table 5.10). While the types of knowledge presented in Table 5.10 represent those required for day-to-day activities, the quality of knowledge possessed by each individual differs. The focus of this study is production related. Variations in knowledge between regions and farmers are illustrated with respect to crop and livestock production.

5.4.4.1. Crop Production Knowledge

Producers use their knowledge to comprehend ideas and to construct causal linkages and relationships prior to choosing actions. The results of inferential reasoning or actions chosen by households cannot be anticipated a priori for all kinds of problems. For commonly occurring problems, some required actions are part of the norms of the society. For instance, to reduce the problems of soil erosion on steep plots, plowing sideways or making furrows is a solution given by most households.[23] There is, however, a body of knowledge specific to households. This specificity may create inequality in production knowledge and management styles, consequently in the efficiency of production. Several open- ended questions regarding crop production problems were asked. That is, respondents were free to state any response to the questions. Responses that were considered by the largest percentage of farmers were tabulated. The answers given by producers are discussed below. In addition, findings from working with farmers and observations are presented along with the results from interviews.

Most producers indicated that variations in soil fertility are caused by planting the same crop on the same plot every year (mono-cropping), erosion and the absence of fallowing. Farmers of both regions believe that black soil, also called vertisol, is fertile because it is not susceptible to soil erosion. Loss in fertility is associated with a decline in production and prevalence of weeds. Suggested solutions include use of fertilizer, fallowing, furrow making, soil burning and rotation (Table 5.11). There were differences in their responses. While most Selale producers prefer using fallowing, furrows, soil burning and terraces, those of Ada prefer fertilizer and crop rotation. Ada farmers have a lot of experience in the use of fertilizer. They do not want to use methods that reduce existing crop areas (e.g. fallowing). Therefore, most of them chose fertilizer and rotation to prevent losses in soil fertility.

Table 5.10. Every-day Knowledge Necessary for Peasant Farming

Crop production	Livestock Production
Type of crop Planted Determined by:	Class of animal raised is determined by:
consumption debt payment (kind or cash) government tax, quota, etc. soil fertility status available stock, etc.	need cash shortage of oxen need milk have grazing have children
How to allocate area to:	Feed Supply From:
crops fallow/grazing	area fallowed plots for hay
Purchases of inputs:	Purchases of feed:
for which crop (e.g., staple or cash)	for which animal milking, oxen
Crop Diseases:	Livestock Disease:
leaf stalk	internal parasite external parasite
Disease Caused by:	Disease Caused by:
variations in moisture wild animals	lack of feed water stress, age
Disease Control:	Disease control:
Pesticides, Herbicides, weeding, burning, etc.	bleeding, use roots, cutting, use leaves
Weeding Decision:	Milking/Plowing Decision:
which crop to hand weed when and how many times which crop to apply pesticides where to purchase/amount	When to Milk for how long how much for calf
Harvesting Decision:	Breeding Decision:
when to harvest and thresh which crop need more labor	use local bulls use A.I.
Storing/Which one to store	Husbandry for:
at home in Goteras (local name for a storage unit)	to get oxen to increase stock
Marketing Decisions:	Marketing Decisions:
which crop to sell where to sell who should sell for what purpose; etc.	which animal to sell where to sell how much/to whom for what purpose; Sell milk/butter/cheese
Other Decisions/Knowledge	Other Decisions/Knowledge

Most producers indicated that variations in soil fertility are caused by planting the same crop on the same plot every year (mono-cropping), erosion and the absence of fallowing. Farmers of both regions believe that black soil, also called vertisol, is fertile because it is not susceptible to soil erosion. Loss in fertility is associated with a decline in production and prevalence of weeds. Suggested solutions include use of fertilizer, fallowing, furrow making, soil burning and rotation (Table 5.11). There were differences in their responses. While most Selale producers prefer using fallowing, furrows, soil burning and terraces, those of Ada prefer fertilizer and crop rotation. Ada farmers have a lot of experience in the use of fertilizer. They do not want to use methods that reduce existing crop areas (e.g. fallowing). Therefore, most of them chose fertilizer and rotation to prevent losses in soil fertility.

In the absence of access to artificial fertilizer, producers maintain fertility through various means. These include turning-over green plants and residues of previous harvests during plowing, burning the soil and manuring those plots close to the farmstead. Several household studies indicate the value of using manure as fertilizer (Guyer, 1986; Lipton, 1977; Belete, 1989). However, the collection, composting and distributing of manure on plots require a lot of time and labor. Furthermore, manure can be used to make dung cake which can be sold to cover household expenses (Chapter III).

Some soil scientists argue that burning the soil is not an appropriate method of maintaining or increasing soil fertility (Huffnagel, 1961; MOA, 1975). Yet, producers in the Selale region make use of this practice. According to farmers of this region, burning the soil kills soil-borne diseases and weed seeds. It may also kill rodents. Laboratory analysis has also shown that the practice of burning the soil increases the content of soil nitrogen (Alemaya College, 1975). Producers did not need to go to school to learn about this practice but learn from their experience of how crops perform when planted on burned soils.

Divergent responses to the problem of losses in soil fertility were not only obtained from interviewing individuals. Throughout the course of this study, group discussions were also conducted to search for answers to problems of this nature. The following question was asked to a group of householders in Selale region:

Question: "*Why do you think that most producers prefer to use traditional means of maintaining fertility?*"

Answer: "Most plots are located on steep slopes. A couple of years ago Ato Getahun and Ato Belay [friends of extension agents] purchased fertilizer. The production of wheat they obtained was not more than what we obtain. I, among the discussants, used fertilizer but it burned the sprout. We think that the soil and the crop do not like fertilizer."

The same question was administered to Ada producers and I obtained the following answer:

Answer: " We have already used all our arable land. We have used improved seed, fertilizer and pesticides before. If we prepare the land well, fertilizer increases

Table 5.11: Percentage Response to Crop Production Knowledge Oriented Problems

Items	Selale		Ada	
	Test	Control	Test	Control
1. Fertile soils are those which are				
Black	89	82	91	96
Others	11	18	9	4
2. To improve fertility:				
Use fertilizer	22	20	80	84
Fallow/furrow/soil burning	70	70	-	-
Rotation	8	10	20	16
3. Heard alternative to reduce Moisture				
Yes	18	19	52	58
No	82	81	48	42
4. If crop fails then, reduce consumption	18	22	42	54
5. Problems of crop production are:				
Locust, storm, worms, rain	79	81	94	97
Others	21	19	6	3
6. Do not use crop technology because:				
Lack of money	55	51	78	71
Lack of access	22	30	15	24
God's punishment	23	19	7	5
7.To reduce yield variability:				
Use terrace, less seed rate, Furrow and	40	55	71	79
Praying	60	45	29	21
8. Why grow many crops: Survival	61	75	80	92
9. Criteria to choose seed:				
Market value	10	15	20	17
Soil-crop r/ship	15	16	5	5
Crop-plot r/ship	20	14	6	7
Policies	15	10	20	20
Food value	40	45	49	51
10. Apply fertilizer/pesticide on:				
Teff	30	9	80	92
Wheat and barley	70	91	20	8
11. Adopt innovation if:				
It increases production	70	81	98	94
It is less risky	20	11	2	6
Others	10	8	-	-

our production. If it doesn't, we grow chick peas alternately with wheat and teff. The only problem we have is that the price of fertilizer is increasing and the marketing corporation is stealing the little production we obtain. The corporation has not been functional since 1990. We can sell our grain any to anyone and where ever we wish. If We get higher price for our grain, we may afford to purchase fertilizer."

A question was also posed to Selale producers.

Question: "Why do you think that Ada farmers are more interested in using fertilizer than you?"

Answer: "Maybe it worked for them, maybe they need it for teff, or maybe they did not know about it very much."[24]

The answers from group discussions reflect constraints experienced and potentials available in each region. Comparing the answers given by Selale with that of Ada farmers, one striking difference is related to the production of teff (a cash and staple crop). The response indicates that Ada producers would apply fertilizer on crops with a high market value. In the Selale area, there is an abundance of arable land compared to the Ada region. Thus, Selale farmers would not mind leaving a plot uncultivated (fallowing) or construct terraces that reduce the size of the crop farms. The steep plots of Selale region facilitate erosion of fertilizer granules. The response of producers also indicate that fertilizer-crop, fertilizer-soil and fertilizer-crop-soil performance trials should be carried out to obtain reliable information about the effect of fertilizer. Moreover, households closely watch what elites are doing. If those who are intimate with extension agents cannot make the best use of innovations, it is unlikely that other farmers will.

Shortages of rain, infestations of locusts, hailstorms and worms were cited as major problems of crop production. Producers attribute these problems to lack of money to purchase pesticides and to God's punishment. More than half of the study farmers do not know or have not heard of alternatives to control variability in soil moisture (Table 5.11). The water requirements of crops and their growing periods are an important body of production knowledge. Producers plant crops requiring less moisture and a short growing season towards the end of the major rainy season. When producers are asked about their knowledge of soil moisture conservation or water drainage techniques, most of them indicate that they have no knowledge of remedies. In practice, however, most producers make furrows to drain excess moisture. They prepare a plot dominated by soils of high moisture retention capacity early to facilitate drainage of excess rain. Moreover, they plant these soils with crops having a strong root system (stalk crops) and which require a long growing season. Sixty percent of the Selale farmers prefer only to pray as a solution to the problem of production risk or yield variability caused by unpredictability of rain. However, less than 30 percent of Ada farmers suggested praying as a solution to reduce yield variability.

Questions related to the problem of soil and water conservation take the following format:

Question: "What do you think would be a cheaper means of conserving soil and moisture?"

Answer: (Ato Asfaw of Gulele Biro peasant association, Selale):

"I believe that more green plants at the beginning of the cropping season extract most of the soil moisture. Therefore, overturning these plants not only helps to drain moisture but also serve as a fertilizing material. Manure helps keep moisture for a longer time."

Minimizing production risk involves calculated decisions at each point of the production cycle. Households evaluate various points such as soil type, texture, depth and location before making decisions about what crops to plant on a specific plot. Expectations regarding the pattern of rainfall and assurance of survival are key to the choice of crops. Survival accounts for eighty percent of the responses in Ada and sixty percent for the Selale region (Table 5.11). Variations reflect the fact that one region specializes more in crops (Ada) and the other in livestock (Selale).

It was pointed out that technology embodies both material aid and know-how (Chapter I and II). It is argued that if the know-how associated with innovations cannot be given meaning or comprehended by decision makers, it has less chance of being added to the larger body of local knowledge. The occurrence of this situation may result in the rejection of innovations. Furthermore, households discriminate among local crop cultivars. They consider food and market value, the suitability of certain crops to soil, the location of a given plot and government policies, before deciding which crop to fertilize and to apply pesticides to. For example, Selale producers prefer to apply pesticides and fertilizer to barley, wheat and teff in that order. The reasons for this decision are that these crops constitute the daily food menu and production per unit area declines from barley to teff. The percentage of producers who apply fertilizer to crops of high market value (teff) in the Ada region is double that of the Selale region.

Different cropping patterns followed by households are used to minimize crop losses and expenditures on weeding, and to maintain fertility. Access to knowledge regarding cropping patterns is important to researchers, extension agents and development experts. Producers have knowledge of which crops: i) are susceptible to disease, ii) can serve as a wind break, iii) require minimal moisture and iv) grow well on less fertile soils. The kinds of crops to be planted adjacent to plots of other producers, therefore, influence a household's planting decisions.

Adoption of crop production technologies involves a risk of not meeting subsistence requirements. This is especially true when crop production is dependent on rainfall. Studies in the adoption of components of the "green revolution" indicate that for producers to adopt fertilizer, pesticides and improved crop varieties, the increase in production has to be at least twice that of local cultivars (Gafsi, 1976; Perrin, et al., 1976). In the study sites of Ada and Selale, more than seventy percent of the producers identified the ability of a technology to increase production as the single most important attribute for adoption.

5.4.4.2. Livestock Production Knowledge

Similar to crop related questions discussed under section 5.4.4., the study farmers were presented with open-ended questions regarding livestock production. Responses that were considered by most households are tabulated (Table 5.12). Households which have access to specific information (e.g. from extension agents) can have a better knowledge of certain aspects of crop and livestock production compared to households which don't have access to information. In the Ada area, extension efforts towards "improved" methods of production are biased against control farmers. For instance, training in general livestock production methods is given to sixty percent of the test and thirty percent of the control farmers.

Eighty percent of households in both study sites believe that adequate feed and veterinary service are solutions to livestock production problems. Improved forage seeds, veterinary services, artificial insemination, cross-bred cows and management techniques are major livestock innovations being introduced into peasant agriculture. Most farmers in the Ada area evaluate several factors to decide how to use cross-bred animals (Table 5.12., item 3). Furthermore, control farmers of both regions would like to see most of the attributes (good milk production, resistant to disease, good reproduction ability, feed and veterinary services simultaneously before deciding whether or not to adopt cross-bred cows. Test farmers are more willing to take a risk.

Most Ada farmers sell fresh milk because they are close to milk collection centers and major cities or high demand areas for milk and its by-products. The proportion of milk sold by households is greatly influenced by the amount of milk produced and the demand for purchased feed. This is particularly true in a region with an acute shortage of feed (e.g. Ada).

Households not only compare the feasibility of adopting cross-bred animals but also apply various criteria before selection of local breeds of cows, oxen and bulls. In the absence of artificial insemination service or cross-bred bulls, producers select local bulls for breeding purposes by applying the criteria used in making decisions regarding adoption of cross-bred cattle. Of these criteria, traction and reproduction potential are the most important factors in the choice of bulls by the Ada and Selale producers respectively. Study farmers associate the use of local bulls for breeding with increases in butter production and poor calf growth.

Producers were asked questions in the following format:

Question: "What do you think the consequences of using local bulls for breeding cross-bred cows are?"

Answer: (Ato Belete of Wayu peasant association, Selale region):

"Although the level of milk yield may decrease and calves may not grow faster, butter production of local cows is high. Breeding using local bulls will, therefore, increase butter production compared to cross-bred bulls."

Producers were asked questions in the following format:

Table 5.12: Percentage Response to Livestock Production Knowledge Oriented Problems

Items	Selale		Ada	
	Test	Control	Test	Control
1. Gain training in:				
Livestock production	68	45	60	30
Crop production	47	53	53	61
2. Solution to livestock production Prob-				
Feed	9	12	5	16
Feed and veterinary	91	88	95	84
3. Decision criteria on how to using Cross-Bred Animals:				
Height of calves, traction & reproduction	50	82	78	91
Milk Produced	50	61	78	69
4. Which to sell:				
Milk	53	66	81	72
Butter	47	34	19	28
5. How much to sell:				
Milk produced	70	81	60	64
Market price	10	5	10	5
Ability to purchase feed	-	-	-	-
Need cash	20	14	30	31
6. Adopt cross-bred cows if it:				
Increases in milk, good traction, resistant to disease, feed and veterinary	85	91	89	98
7. Choice breeding stock based on:				
Traction	50	55	90	89
Reproduction/milk	45	35	8	10
Colour	5	10	2	1
8. Method of controlling Animal Disease:				
Vaccination	5	5	35	30
Traditional	60	75	45	55
Vaccination and traditional	35	20	20	15
9. Effect of using local bulls:				
Less growth/traction	20	15	50	65
Less butter	80	85	50	35
10. Obtain knowledge of better Livestock				
Extension agents	5	5	4	3
Education	15	10	21	19
Experience	80	85	75	78

Question: "What do you think the consequences of using local bulls for breeding cross-bred cows are?"
Answer: (Ato Belete of Wayu peasant association, Selale region):
"Although the level of milk yield may decrease and calves may not grow faster, butter production of local cows is high. Breeding using local bulls will, therefore, increase butter production compared to cross-bred bulls."

Disease is the major cause of livestock production risk. Unlike technologies such as pesticides, fertilizer and improved seed, investment in cattle is indivisible or lump-sum. To mitigate risk caused by livestock disease, producers not only make use of traditional medicine, but also complement it with modern medicine. For example, thirty-five percent of Selale producers use vaccination in conjunction with traditional medicine involving bleeding of the tongue, back or neck of the animal, the use of leaves and roots of plants, pepper and alcoholic drinks. Forty percent of Selale producers relate use of veterinary services to secular education.

Producers do not measure the milk-fat content of milk from local and cross-bred cows. However, they know how much butter they produce from a certain volume of milk. They compare the amount of time spent in processing "x" liters of milk to produce "y" kilogram of butter. Furthermore, they assess the distance to the nearest milk collection centre to sell fresh milk. They examine the advantages of processing the milk, supplying the skimmed milk to children, and the selling price of butter at the local market before making decisions on how much milk to process.

Selale farmers have an immense knowledge of livestock production. Three important observations in the management of cross-bred cows in the Selale region are provided as examples. The first observation is related to decisions regarding whether or not to sell fresh milk. The second observation is concerned with the effort farmers put into the management of cross-bred cows. The third observation is related to reasons for the susceptibility of cross-bred cows to disease and measures that some farmers implemented.

The farther a farmer lives from the milk collection centers, the less the amount of fresh milk he/she sells. Most of the milk from cross-bred cows is given to calves or to children. However, milk from local cows is entirely processed because of the higher butter-fat content (Table 5.13). Most of this butter is sold at local market or itinerary traders. At this distance, services from breeders and veterinarians are less accessible. Producers, therefore, depend on the revenue from the sale of the offspring of cross-bred cows to repay loans rather than on income from the sale of fresh milk. If a farmer lives close to the milk collection centers, he/she sells most of the milk obtained from cross-bred cows fresh. The largest percentage of milk from local cows is processed with increases in distance from collection centers or highways (Table 5.13).

Table 5.13: Strategic Management of Milk Production and Disposition

	Peasant Associations in Selale			
	Wayu	Segokara	Gendesheno	Illukura
Distance from Milk Collection centre(km)	3	5	7	13
Percentage of milk from Cross Bred Cows:				
Sold	94	71	60	15
Processed	2	20	28	55
Consumed	4	9	12	30
Percentage of milk from Local Cows:				
Sold	15	11	5	1
Processed	77	82	91	98
Consumed	8	7	4	1
Farmers interviewed	8	13	8	6

The further a farmer lives from the milk collection centers, the less care and management effort he puts into cross-bred cows. In fact, some producers feed their local cows equally or better than the cross-bred cows. On the other hand, the closer a householder lives to the milk collection centers, the more effort he puts into proper care of cross-bred cows in comparison to local breeds. This calculated behavior of households is reinforced by the frequent visits of extension agents to households located close to the milk collection centers (see also Tittarelli, 1990).

The further a farmer lives from the milk collection centers, the less care and management effort he puts into cross-bred cows. In fact, some producers feed their local cows equally or better than the cross-bred cows. On the other hand, the closer a householder lives to the milk collection centers, the more effort he puts into proper care of cross-bred cows in comparison to local breeds. This calculated behavior of households is reinforced by the frequent visits of extension agents to households located close to the milk collection centers (see also Tittarelli, 1990).

Finally, participatory study of test farmers revealed that the size and color of cattle influence the degree to which they are attacked by disease-causing organisms. According to the farmers, cross-bred cows are larger and have visible color (black and white). Thus, they attract insects and birds which may carry disease causing organisms. Therefore, they prefer to keep cross-bred cows in stalls.

5.4.5. Problem Solving Tests

Prior knowledge can be obtained from social experience, training or schooling. This prior knowledge is used to interpret information contained in written or oral texts and to assess their cognitive contribution. It is used to construct causal explanations for new information. There are two methods of evaluating the knowledge possessed by subjects used in solving problems they face in their everyday life, or those related to new technologies. The first method uses graphical representations of explanations or reasons to specific production problems. The second method makes use of scoring rules and compares answers from individual decision makers to a bench mark response. The results from the two approaches are presented below.

5.4.5.1. Causal Attribution or explanation

Two villages in the Selale region were continuously monitored with respect to various farming operations. Answers given by one household, despite living in the same neighborhood, are not considered the same as the others. For example: *Question:* "Why do you plant oats every year despite its poor yield, edibility and market value?"

Answer 1 (Ato Degefa): "Well, i) we have no choice and ii) the straw is a very good source of feed to our cattle."

Answer 2 (Ato Mulatu): "Oats i) are a commonly grown crops by our ancestors, ii) have good feed value, iii) are used for making beer and iv) have the capacity to resist frost, which commonly occurs in our area."

Comparing answers of Ato Degefa and Mulatu, it seems evident that the answer given by Ato Mulatu recognizes the real issue, that is, environmental adaptability. Producers resist pressure from extension agents to abandon planting oats.[25]

It is argued that education, as a source of information, helps to understand and provide solution to day-to-day problems (Warren, et al., 1989). To examine the influence of education and age on the ability of households to provide causes, indicators and solutions to production problems, four farmers with different years of farming experience and education were selected. The farmers were asked to provide answers to the following question:

Question: "How do you tell whether the growth of a crop is affected by excess or shortage of moisture, soil borne disease or other factors. Could you discuss indicators of the disease and solutions that you think would solve or reduce the effect of the disease?"

There are two important issues in using causal attribution as a method of assessing variation in knowledge across age and education groups. First, the process of questioning or discussion is time consuming. Therefore, it is difficult to have large sample size. This would make extrapolation difficult. Secondly, responses to a particular question or issues discussed by households differ depending on what they assume about the knowledge of the discussant or interviewer. For example, if a farmer anticipates that the researcher has a good knowledge of crop production, he/she may give short answers. Alternatively, if a farmer perceives that the discussant has little knowledge about farming, the farmers may give short answers because he would expect to be challenged. Because it is difficult to establish an accurate measurement of what farmers know, methods to reduce biases are implemented. The following method of questioning was implemented in this study. Once the question or point of discussion is raised, the researcher tells to the farmers that in the Arssi region farmers do `X and Y' to solve `W and Z' problem. What is the experience of farmers in the Selale region? Since farmers resort to comparative analysis between regions, it is possible to reduce biases because of farmers assumption regarding the discussant or interviewer knowledge. This pattern of discussion or questioning may help to obtain most of the causes, indicators or solutions of a crop or livestock production problems from farmers. Answers given by households are presented in Tables 5.14a and b.

The results of causal attribution indicate that the householder: i) with no schooling and 16 years of experience provided few causes and solutions to the problem, ii) with no schooling and forty-one years of experience noticed several causes and solutions, iii) with two years of secular education and thirty-eight years of experience indicated several different causes and solutions to the production problem relative to answers given by the householder with education but less than twenty-five years of experience and relative to those who have not attended schools. Schooling does not cause significant differences in the answers given by householders with forty years of experience. Two points can be concluded from the above discussion. Farming experience influences knowledge about causes, indicators and solutions to production problems. Education helps a farmer to relate production problems to experience and outside information, and hence provides a more comprehensive set of indicators and solutions to production problems.

5.4.5.2. Quantifying Knowledge

Section 5.2. has indicated that access to information may vary depending on whether a particular group of households is favored by the sources of information. Specifically, test farmers and politicians may gain access to more information than control farmers. This differential access to information may be reflected in their knowledge regarding production and marketing of agricultural products. To illustrate this point, questions related to crop and livestock production, and marketing were administered to fifty producers.

Table 5.14a. Results of Causal Attribution for Crop Production Problems

Caused by	Indicators	Solution
No Schooling and 16 Years of Experience		
If soil borne =====>	The plant wilted & fall because its root rot	Plow many times; Burn soils
If other factors ====>	Deformed leaves, spots and holes on stalk or leaves, plant growth accompanied by thin stalk	Re-planting
If shortage of rain =>	Germination delayed Leaves start to dry	Nothing
If excess rain ====>	Seed don't germinate Plow side-ways Leaves turn into yellowish color	Plow side-ways
No Schooling and 41 Years of Experience:		
If soil borne =====>	The plant wilted & fall because its root rot root rot	Plow many times ; Burn soils; Over-turn/replant
If other factors ====>	Deformed leaves, spots and holes on stalk or leaves, plant growth accompanied by thin stalk	Remove infected plants; Feed to cattle; Replant
If shortage of rain =>	Germination delayed Leaves start to dry	Re-plow; Feed to cattle
If excess rain ====>	Seed don't germinate Leaves turn into yellowish color	Make furrows; Plant stalk crop; Make terraces

Table 5.14b. Results of Causal Explanation for Crop Production Problems

Caused by	Indicators	Solution
Two years of education & 21 years of experience		
If soil borne ===>	The plant wilted fall because its root rot	Plow many times; Burn soils; Extension agents
If other factors ==>	Deformed leaves, spots and holes on stalk or leaves, plant growth accompanied by thin stalk	Re-planting; Remove infected;
If shortage of rain =>	Germination delayed Leaves start to dry	Plant crops with short growing period; Re-planting;
If excess rain ====>	Seed germinate Leaves turn into yellowish color	Make furrows; Make terraces; Plant after water is drained
Three years education and 38 years of experience		
If soil borne =====>	The plant wilted fall because its root rot	Plow many times; Burn soils; Over-turn/replant; Apply rodenticide
If other factors ===>	Deformed leaves, spots and holes on stalk or leaves, plant growth accompanied by thin stalk	Remove infected; Feed to cattle; Re-plant; Apply pesticides or insecticides
If shortage of rain =>	Germination delayed Leaves start to dry	Re-plant; Feed to cattle; Plant crops that require short growing period and less moisture
If excess rain ====>	Seed don't germinate Leaves turn into yellowish color	Make furrows; Make terraces; Plant stalk crop; Plant after water is drained; Plow side-ways

Answers given by extension agents were not found feasible to evaluate production knowledge of producers mainly because the solutions given by extension agents did not recognize some important factors in farmers' decision-making processes. Differences in the content of production knowledge between producers and extension agents can be illustrated with the help of the following question.

Question: "*What is the consequence of inseminating cross-bred cows using semen from local bulls?*"

Answer: (extension agent with 10 years of experience): "The genetic composition of the off-spring will be low relative to the dam and will exhibit characteristics more similar to local stock such as poor milk yield and traction power."

Answer: (Ato Yibeltal, five years of secular education):

"Well, i) cross-bred animals consume a lot of feed which is hard to find and are less resistant to disease, ii) calves of cross-bred cows show high mortality and if we cross them with local breeds their resistance will increase, iii) the meat of the cross-bred animals is not tasty and iv) the butter production and the price for fresh milk are low. Therefore crossing with local increases the butter content and we don't have to sell fresh milk."

Answer: (Ato Tadesse, has attended the literacy campaign):

"The calf will not grow fast, the milk production will decline and feed consumption will be reduced compared to first generation of cross-bred cows."

All answers are acceptable in their own terms, but are not equivalent. Small holders do not want to risk family survival by investing U.S. 425 $ to purchase cross-bred cows. They have to evaluate the social, economic and technical consequences of using the technology. The answer given by Ato Tadesse is close to that given by extension agents. The answer of Ato Yibeltal considers several factors compared to the extension agent or Ato Tadesse. In order to establish a bench mark to compare differences in scores of production knowledge questions, therefore, it seems feasible to make use of answers obtained from farmers rather than from extension agents.

To establish a benchmark for comparing answers given by households, the questions were directed to a group of farmers whose ages range between 18 and 65. A score of 1 to 10 was prepared. Answers given by each farmer were rated relative to those given by the group.

The result of frequency analysis of production and marketing knowledge are presented in Table 5.15. The result indicates that in most cases test farmers and politicians scored higher than the control group. The greatest difference is between Ada and Selale than between politicians and others.

In analyzing production knowledge, it is essential to evaluate how producers understand and incorporate new innovations into the practices of agricultural production and knowledge acquired from social experience. Knowledge possessed by households varies by age. Children are brought up to be good farmers. Household studies in LDCs have found a strong relationship between experience and knowledge (Warren, et al., 1988). To illustrate this case, 42 farmers with different periods of experience were selected and asked to respond to production related questions (see Appendix 3a). A score, as indicated earlier, was prepared (Table 5.16).

Table: 5.15. Differences in Knowledge and Access to Governments Institutions (Political Office and Extension Service)

	Categories of Farmer				
	Selale		Ada		
Categories	Test	Control	Test	Control	Sample Size
Knowledge (score/10)					
Crop Prod. Knowledge	9	9	8.5	7.5	261
Livestock Production Knowledge	7	6	9.2	8.1	265
Marketing Knowledge	8	8	5	6	264
Total	22	21	20	20	

In analyzing production knowledge, it is essential to evaluate how producers understand and incorporate new innovations into the practices of agricultural production and knowledge acquired from social experience. Knowledge possessed by households varies by age. Children are brought up to be good farmers. Household studies in LDCs have found a strong relationship between experience and knowledge (Warren, et al., 1988). To illustrate this case, 42 farmers with different periods of experience were selected and asked to respond to production related questions (see Appendix 3a). A score, as indicated earlier, was prepared (Table 5.16).

From Table 5.16, the score from problem solving tests increases at a faster rate until the number of years of farming experience reaches 25.[26] It means that knowledge gained over a number of years attains its maximum effect when producers are in their 40's (about 25 years of experience). The young generation (18 to 30) visit cities and markets frequently and interact with people who have attended secular schools. Group discussions indicate that this is one of the reasons for the relatively high score of marketing knowledge by the younger compared to the older farmers. One contrasting difference between regions is that crop production knowledge is not only high but also increases at a faster rate among Ada farmers while this pattern holds true only for livestock production knowledge among Selale farmers. Furthermore, marketing knowledge is relatively high among Ada compared to Selale farmers, due to proximity of the former site to the capital city.[27]

Table 5.16: Scores of Problems Solving Tests By Region

	Selale Region			Ada Region			
	Type of Knowledge			Type of Knowledge			
Experience (Years)	Crop Prod.	Livestock Prod.	Marketing	Crop Prod.	Livestock Prod.	Marketing	No. of Farmers
5	5	6	7	6	4	7.5	4
10	5.5	6	7	7.5	5.2	7.9	5
15	6.2	7	6.5	8	5.3	8.1	4
20	6.8	7.1	7.2	9.1	5.5	8.4	12
25	8.1	8.2	8	9.4	5.8	8.9	2
30	8.5	8.5	8.4	9.5	6.4	9.1	5
35	8.7	8.8	8.5	9.7	6.5	9.4	7
40	9	8.9	8.8	9.8	6.6	9.5	4
45	9.1	9	8.9	9.8	6.8	9.6	3

5.5. Summary

The processes of household decision-making are investigated through analysis of the relationships between goals and strategies, information and production knowledge. A practical method of distinguishing goals and strategies implemented by this study shows that most farmers select goals before setting strategies. Strategies which have proven to be useful by producers have a stronger relationship with household goals (e.g. pesticides and fertilizer). Strategies that influence livestock production are ranked highly on the goals of Selale than Ada farmers. On the other hand, strategies which help boost crop production are ranked highly on goals of Ada than Selale farmers. Farmers who have not received cross-bred cows select non-agricultural activities or strategies to satisfy their goals compared to farmers who own cross-bred cows. Moreover, producers follow sequential or strategic decision-making in ranking goals and strategies.

The sequentiality of decision-making implies that goals and strategies are selected hierarchically. To rank goals and strategies, households have to gather and process information, evaluate the availability of resources and the environment within which their actions are given meaning.

Households gather information from formal and informal sources. Household do not gain equal access to information. Information from extension agents tends to favor peasant associations which are close to cities and service cooperatives. Peasant associations whose officials have stronger relationship with extension agents and are elected at sub-district, district or regional levels have greater access to information compared to peasant associations whose officials are not elected at the sub-district or district levels. Furthermore, producers who are friends with officials of peasant associations or extension agents gain greater access to information than producers who are not. Moreover, test farmers seem to have more access to sources of information than control farmers.

Information gathered is subjected to conscious and unconscious processing. The larger the information set subjected to conscious processing, the greater the quality of knowledge households posses. The cognitive contribution of information is determined by prior experience, knowledge and schooling. The extent to which information is subjected to conscious processing and differential access to information induce differences in production knowledge. Test farmers and politicians received higher scores on problem solving tests than control farmers. As opposed to production knowledge which increases with the age of farmers, higher scores on marketing knowledge are achieved by younger farmers. Households living closer to big cities and market centers gain larger scores on marketing knowledge (e.g. Ada compared to Selale farmers.

The results of causal attributions or explanations of to day-to-day problems indicate that young householders with no schooling provide few causes for or solutions to problems compared to older farmers. Older householders who have attended schools indicated several causes or solutions to production problems relative to answers given by young literate and illiterate householders. Education helps to relate production problems to experience and outside information, thus enabling a comprehensive examination of production problems.

Production knowledge is locale-specific. Households in the Selale area received higher scores in livestock production problems while that of Ada farmers in crop production problems. Production knowledge varies by age and sex.

5.6. Notes

1. Decision trees are flow diagrams that represent actions, events or states and outcomes in a natural or chronological order (see Anderson, et al., 1977:124-128). An aspect is a set of characteristics that represent values along with some fixed qualitative or quantitative dimension (e.g. price, quality) or they can be arbitrary features of the alternatives that don't fit into any simple dimensional structure. For instance, an aspect of deciding to plant a particular crop can be yield, price and knowledge about the crop. These aspects are used as criteria to eliminate some goals and strategies until the decision maker is satisfied with her/his choice.
2. The questionnaire dealing with household decision-making is detailed and specific either to enterprise or region (see Appendix 3b).
3. "Ato" is a local title that is equivalent to "Mr.".
4. "Weizero" is a local equivalent to "Mrs.".
5. "DERG" was the name of the then Provisional Military Government of Ethiopia that replaced the feudal regime in 1974.

6. It should be noted, however, that this method of identification of goals and strategies is subjective because it is based on the sequence of answers given by households and listening to discussions that were taped.

7. The objective of this research is comparative analysis of the processes of decision making and its consequences between farmers who have adopted cross-bred cows (test) and those who have not (control), and between regions with different approaches of technological introduction. Thus, statistical analysis is performed for both groups of farmers in each region. However, discussion focuses only on differences that are significant.

8. Goats are hardy animals. They make a lot of noise, and are continuously on move. Thus, herding goats require more herder-time per day compared to sheep. They browse on leaves or forests (where the snakes are abundant) compared to sheep which feed on grass. Because of the above reasons, they are seen as satan or unholy animals.

9. People who do not live in the village and who are learned or work for an organization are respected. One way in which residents of a village show their respect is to accommodate the guest in a big and clean house. Thus, farmers who own big houses are requested by villagers to host meetings or feasts in their house.

10. Well-to-do farmers may lend grain to poor farmers in June or July when prices reach their peak. They would strike a deal with borrowers to pay in kind at the time of harvest when prices are low, and sell back when prices are high.

11. Opportunistic strategies are those strategies that are intended to increase crop and livestock production above what is required for subsistence. Risk-minimization strategies are strategies chosen by households to ensure subsistence requirements and reduce the effect of technical factors (variability in soil fertility and feed resources), environmental factors (variability in rainfall) and government policies. Long-range strategies are strategies whose effects extend beyond one crop season and possibly over the life time of a household.

12. Dummy variable is a method of capturing the influence of a variable by grouping it into a specific category. For instance, those who chose fertilizer to attain the goal of attaining subsistence requirements or to increase output are categorized into group one. To capture the effect of fertilizer on the level of production of farmers who have used it compared to those who have not used fertilizer, group one is given a value of one. Those who have not used (group two) are assigned a value of 0. In doing so, the impact of fertilizer on the level of output of farmers is determined relative to those who have not used fertilizer.

13. For example: $Y = \alpha_0 + \alpha_n{}^*D^*{}_n + \alpha_t F^*{}_t + u$. Now, the logical estimate of $\alpha_n{}^*$ is $(\alpha_0 - \alpha_n) = k$, $k<0$ if α_0 (the intercept or base) is smaller than α_n (strategy n), and $k>0$ if $\alpha_0 > \alpha_n$.

14. One quintal (qt) is equivalent to 100 kg.

15. The cultural values do not accord status to farmers who are engaged in activities other than farming. The regression result, however, suggests that, for resource poor farmers, obtaining large cash income through trading and off-farm income may help them purchase new technologies or build a house. Subsequently, adoption of technology or building a house may serve as a strategy to gain the status of being a good farmer.

16. Control farmers make use of alternative ways of reducing differences in income between them and those who received cross-bred cows. Thus, control farmers may rank the strategy of sending children highly with the expectation that they may receive higher remittances in the future compared to test farmers.

17. In most cases, the coefficients of securing food and livestock husbandry are statistically more significant than those of building a big house. If securing family survival is obtainable,

then farmers aspire to achieve other goals with confidence.

18. Naturally occurring behavior sequences do not always provide evidence about learning and thinking processes. Thinking is a new combination of previously learned elements, among which problem solving situations are predominant (Bransford and McCarrel, 1983; Cole, 1983).

19. "Black box" in this context refers to a situation in which inappropriate information regarding innovation is rejected by peasants because it can not be given meaning by previous experience or it violates societal value systems. This occurrence is a bad experience or a "black-box" which may result in lags in adoption of technologies.

20. Webster's college dictionary (1991:257) defines cognition as "to become acquainted with, know, act or a process of knowing including both awareness and judgment". The adjective form, cognitive, refers to, capable of being reduced to empirical or factual knowledge.

21. Despite the hypothesis that farmers who hold political office may gain more access to resources, the difference in crop area between farmers who are politicians and non-politicians range between 8 to 10 percent. However, the greatest difference (e.g. 25% for Ada and 40% for Selale) is observed with respect to grazing area. The largest difference between farmers who are politicians and non-politicians is with respect to visits by extension agents (Table 5.8.).

22. Schultz (1975) argued the need to invest in people as a strategy to facilitate adjustment to changes.

23. Households who own plots on a slope of about 10 percent or more face serious problem of soil erosion if they plow in the direction of water flow. Thus, by plowing sideways, they reduce the speed with which the water runs and cause some proportion of the water to permeate the soil.

24. Selale is known for its livestock production. Songs among the Amhara and Oromo areas evoke the slimness and beauty of Selale women. For Ada producers, a good farmer is the one who harvests a lot of grain, not the one who produces a lot of milk (like pastoralists). Ada producers, therefore, do not consider Selalians to be good farmers. Instead, they believe that Selalians should pursue what they are known for, animal production (comparative advantage!).

25. Plant pathologists report that oats carry organisms that cause a disease called "ergotism".

26. It should be noted that there is no one-to-one correspondence between farming experience and knowledge. Nevertheless, experience contributes much more than other sources to the repertoire of indigenous knowledge.

27. In describing the efficiency of trading among the Ethiopians, Pankhurst (1961, 1966) has argued that those who produce grain as their major source of income have a remarkable knowledge of the grain market and sometimes engage in the business. Similar patterns are observed among farmers whose interest is in livestock production.

CHAPTER VI

THE INSTITUTIONAL BASIS FOR DECISION-MAKING

6.1. Introduction

This chapter investigates the nature of institutions which affect household decisions, and their influence on patterns and consequences of household decision-making. Some of the issues regarding institutions are discussed in previous Chapters. Therefore, only selected institutions are examined in this chapter. This chapter will argue that intervention strategies which do not assess the social, economic and natural environment of decision-making are doomed to fail. Institutions, similar to physical resources, should be given appropriate consideration in the design of projects that are intended to increased food production in the peasant sector.

The definition, function and structure of households have been the subject of applied studies in peasant agriculture (see Chapters II and III; Netting, et al., 1984; Wilk, 1991; Barnum and Squire, 1979). These studies could be categorized as follows: those which contend that economic activities shape the structure and function of households, and those which argue that non-economic elements or institutions determine the economic structure of households. Related to the first category, Polanyi pointed out that economic activities determine the functioning of a household only in a capitalist socio-economic system (Polanyi, 1944; Polanyi, et al., 1957). Regarding the second category, Rey (1972), for instance, argued that institutions that control the reproduction of producers shape the patterns of economic and social organization of a society.

Despite these variations, households make use of both the economic and social institutions as an infrastructure for their actions. Producers rely on the social institutions (e.g. self-help groups) and economic institutions (e.g. the market system). These institutions influence the choice of household goals and strategies, and shape the consequences of decision-making. In this study, these institutions are referred as the "environment of decision-making". Each household is faced with choices of actions or decisions. Selection of decisions must be made within the context of the household's needs and goals, the resources available to it, and the decision-making environment or institutions (Barlett, 1980; Ortiz, 1983; Polanyi, et al.,1957).[1]

Chapters III and V indicated that households invest in innovations and other activities provided that their subsistence requirements are secured. Peasants can attain this goal if innovations or investments embody less risk or if the risk involved in using innovations is reduced through government assistance programs. When access to formal risk-spreading institutions is limited, participation in certain informal institutions or social networks is pursued to increase an individuals chain of resources or form a buffer from social and economic risks (Berry, 1980). Therefore, it becomes worthwhile for households to maintain or improve their positions in that group through patterns viewed by outsiders as wasteful. Cooperation among households is necessary to mitigate the effects of social and economic risks. This cooperation is the most important determinant of the stability and integrity of a peasant economy.

When grain prices are low, farmers store and sell when prices are high, thus shifting financial risk to purchasers. Markets provide information that helps reduce uncertainties regarding where, when and how to sell. Farmers may also develop demand for non-farm produced goods. Consequently, these functions create dependency between peasants and markets. Imperfections in the function of markets, however, may result in impoverishing peasants. Credit market imperfections cause peasants to lose and lenders to gain. For instance, traders lend money to peasants. In return, traders purchase grain from peasants at relatively low prices. When land is state-owned, cattle serve as collateral. If there are crop failures, peasants lose their cattle and fall into poverty trap. Policies that facilitate household decision-making through provision of market information, credit, increasing prices of outputs or subsidizing prices of innovations can help reduce agricultural risks and poverty. These kinds of policies, however, introduce another form of reciprocity. Producers obey government institutions, no matter how inappropriate their actions may be, because they have helped them to produce more.

In summary, the range of agricultural options open to farmers is usually determined by the interaction of the natural environment with the larger social environment. Market forces, together with ecological variables, shape the structure of decisions to plant crops (Gladwin, 1980). Furthermore, institutions affect the flexibility of the decision-making framework. For instance, decisions regarding land use may be affected by taxation, regional and national markets, density of population and availability of labor exchange arrangements (Hoben, 1973). The choice of goals and decisions is influenced by the availability of credit and collateral requirements. These influences or dependencies determine the pace of agricultural development of peasant economies (see Binswagner and McIntire, 1987; Binswanger and Rosenzweig, 1986; Binswanger and Rosenzweig, 1984).

Institutions could constrain or facilitate household decision-making.[2] In Ethiopia, these institutions are established by the state (such as the Ministry of Agriculture and Marketing Boards) or villagers (self-help groups). Both types of institutions facilitate the production of goods and services, and serve as a means of extracting part of the produce. Extraction of produce by government institutions is accomplished through the imposition of direct and indirect taxes. Social networks redistribute or extract produce from peasant agriculture through various forms of reciprocity. In addition, these institutions create competition over the use of resources and inequalities. For instance, people gain access to productive resources by making claims on property and productive services. They do so not only through the market but also by drawing on social norms and patterns of authority, obligation and loyalty (Berry, 1984; Chapter V).

In Chapter five, the structural relationships between goals and strategies pursued by households were investigated. The social infrastructure or institutions that made the choice of strategies and goals possible were not examined. This chapter discusses how households make use of local and government institutions as strategies to reduce the adverse effects of government policies and natural disasters, and how institutions influence patterns and outcomes of household decision-making.

6.2. *The Environment of Household Decision-Making in the Ethiopian Highlands*

Household decision-making in the Ethiopian highlands is influenced by: i) neighbors ii) relatives, iii) friends, iv) the village community, v) local authorities, vi) local markets (e.g. merchants, traders), vii) government institutions and viii) development projects. The focus of the research in the Ada and Selale regions is agricultural production. Influences from these elements of the ecosystem of decision-making are discussed with respect to production.[3] Influences from the first four categories are discussed under the heading of "inter-household and micro-macro interface" institutions. The influence of factors in (v) to (viii) are discussed under the category of "macro-integrating forces or institutions".

6.2.1. Inter-Household Institutions: The Coffee Ceremony

It is obligatory for households to create working relationships with neighbors. The saying "bichawin yebela bichawin yimotal", which means "he who eats alone shall die alone", acts as an incentive to establish neighbourliness. Friendship between neighbors in the Ethiopian highlands starts with the celebrated coffee ceremony. Every morning, the household ready to serve coffee sends the youngest child to invite neighbors to the coffee ceremony. Roasted pancakes or cereals ("Yebuna Kurse" which literally means "breakfast of the coffee") are served before the coffee is offered. The coffee is boiled and served three times. Only adults participate in the ceremony while children look after the cattle. Through this ceremony, older children are given a first-hand look at how friendships and neighborships are started, and what kinds of issues are discussed.

At the coffee ceremony every aspect of household life, except very personal matters, are discussed openly. These include the time to start plowing, the number of times a plot should be plowed, and the availability of new crop varieties or livestock feed. Village politics are discussed, such as the usefulness of the activities of local authorities, and ways to reduce the risk of producing less than what is required for survival of the family. Also mentioned is what marketing information and services are available, such as members who are going to the local market, what purchases will they make, whether they can purchase items for a household whose members are not going to the market, and whether they can obtain information about prices of agricultural products and consumer goods. Discussions are conducted regarding labor or oxen-sharing arrangements, and about helping friends and relatives who are experiencing hard times.

The same pattern of coffee ceremony is performed in the morning or evening of the same or next day by the household which was invited. Every ceremony takes one to two hours. The number of households participating in the coffee ceremony ranges from two to five.

A sense of belonging to a village, a community and a specific geographic area begins with this ceremony. Terms such as "Yagerie lij" or "Yewonzie lij" originate with this ceremony. Respectively, these terms mean "my country person" or "a person from the same river". These local social ties extend to the village and district levels. In fact, as the area in which households are situated gets larger, the desire to identify oneself with the region and to consider neighbors or households from the same area as close relatives becomes stronger. Households introduce individuals who come from the same area to people from other regions by saying "wondimie new" or "Ehitie nech", which means "my brother" or "my sister". This does not mean that they are blood brothers/sisters, but that they come from the same region or village.[4] Similar patterns of relationships and ties exist with friends not residing in the same village. The only difference is the frequency of contact and consequently the degree to which they share problems.

Relationships with relatives may be strong or loose depending on how close individuals live to one another. If they live in the same village, relationships tend to be strong. A household with a large number of relatives in a village is given a good social status to the extent that the village is called "Mr. X's hager" implying "Mr. X's country".[5] Relatives are necessary in establishing oneself within a community. In village life, status can be attained through: i) being a religious or political leader, rich and generous and ii) having more relatives and friends. The larger the number of relatives households has, the greater are their chances of obtaining assistance or cooperation in farming, sharing labor, marketing products, and ceremonies such as wedding and burials.

The coffee ceremony plays a significant role in the creation of all forms of production, consumption, marketing, credit, burial and religious institutions among households of the Ethiopian highlands. The ceremony serves as an infrastructure: i) for exchange of information on issues affecting household decisions, ii) to organize reciprocity in the use of labor, oxen and other resources, iii) to pass resolutions regarding activities which may affect their livelihood, including those from local authorities and the state, iv) to establish a sense of identity or belonging and v) to initiate activities of sharing, borrowing and lending.

Neighbourliness that started through the coffee ceremony is the basis for friendship which can be as strong as relationships among blood-related households. Under this kind of social environment, household decisions are strongly influenced by neighbors. The importance of these institutions has not been given an appropriate emphasis in studies of households in the Ethiopian highlands.

The coffee ceremony is an institution where all forms of cooperation begin. At a broader village level, however, relationships are not as strong as those between neighbors. Therefore, social institutions at that level tend to be more specific in their function and structure. In the following pages, social and economic institutions operating at a village level and how they may affect household decision-making are discussed.

6.2.2. "Micro-Macro" Interface Institutions

The term "micro-macro" interface is used to distinguish the micro nature of the coffee ceremony as an inter-household institution and those that extend to a village level. These institutions are seen as forms of cooperation. The micro-macro interface institutions that influence a household's economic and social organization, as well as its decisions, can be grouped into those concerned with: i) production, ii) insurance or credit and iii) distribution. The production-related institutions are reciprocal resource exchange arrangements that include: i) "Debo", ii) "Jigi" and iii) "Wonfel". The insurance groups include: i) "Ekub", ii) "Edir", iii) "Senbetie" and iv) "Mahiber". The distribution related institutions include: i) "Gibzja" and ii) religious holidays observed throughout the month such as "Lideta", "Abo", "Michael", "Mariam", "Giorgis" and "Gabriel" which are respectively the 1st, 5th, 12th, 21st, 23rd, and 29th days of the month. These groups will now be discussed in turn.

6.2.2.1. Production-Related Institutions

"Debo" is a term which probably derives from the word "dabo", which means bread. Farmers involved in cooperative labor use are served with bread and drinks. It may also imply that the crop is going to be used to produce bread.

Debo is a social grouping with no internal by-laws. Often, it is a name assigned to a type of cooperation during plowing. A household head requests friends, relatives, neighbors or members of a village to help in preparing his plots for planting. Several passes are needed before planting. The first plowing takes the longest time and is difficult work. Successive passes and planting have to be completed on time for good harvests. If a household head gets behind in the planting schedule, Debo is called up. The household which organizes Debo is expected to prepare meals for the day. Households which make use of Debo have to reciprocate if and when necessary.

"Jigi" is a term that may have originated from the word "Ejig" which means "very ...". Because several people form Jigi, it may imply very large number of people, thus an indication that the household which called for Jigi should prepare a lot of food and drinks.

Jigi is another institution designed to alleviate bottlenecks in production activity. It refers to a self-help group which cooperates in harvesting crops. Wonfel is the same as Jigi but it may involve only two individuals. When few people participate, reciprocity is almost certain. Thus, it is common for a household to say "Wonfel alebign", implying "I have a commitment to help [in harvesting]".

6.2.2.2. Institutions of Insurance or Credit

Ekub is a credit or saving institution in which weekly or monthly payments of a fixed sum are exchanged for the privilege of receiving a large lump sum at some point in the life of the group. "Ekub" is a term derived from the word "Akebe" which means "to restrict someone from doing something or to prevent something from happening". Ekub is believed to have originated among the Gurage, acquired wider popularity after the Italian invasion when people needed large sums of money for rebuilding homes and purchasing livestock. Ekub is established by all members of low, middle or upper class groups. Its operation is governed by internal by-laws.

Each time they meet, a lottery is drawn to choose the winner of the money collected. People do not join Ekub only to collect money but also to discuss ways of helping each other as friends who are experiencing social or economic crises, and to exchange various kinds of information. If a member needs money urgently, he/she requests officials of the Ekub to give him/her the next draw without resorting to the lottery method.[6]

"Edir" is a term that is derived from the word "Adere" which means "stayed overnight". Edir is the oldest and most widely diffused of the modern institutions in Ethiopia. It is a form of welfare institution. Its functions are primarily to help defray the cost of funerals; to assist families in the event of death, illness, unemployment or imprisonment of their breadwinners; and to help in the case of loss due to fire. This institution is believed to have originated among the Gurage and to have become more widely adopted during Italian invasion, when many people who were killed had no relatives to bury them. This institution is common in cities, fulfilling the vital functions which are performed by kin in the traditional countryside.

Members of Edir contribute money to finance social and investment ventures. Similar to Ekub, their activities are governed by internal by-laws. Every time members meet, decisions are made about problems facing members, their village, church and school. Each member is expected to host members of the Edir at least once a year, depending on how frequently they meet and on the number of members.

"Senbetie" is a term derived from the word "Senbet" which means "holiday" or "a day of saint". Senbetie is a religious association. People build little houses surrounding the church (Senbetie houses), recruit membership and meet in these houses after Sunday mass. The activities of the Senbeties are governed by internal by-laws. A lottery is drawn to determine which members should bring food and drink to the Senbetie house every Sunday. Senbetie feeds the clergy and poor folks. Members of the Senbetie house include those living in and outside the village where the church is located. They talk about issues of concern to the social and economic well-being of their villages and the church. Members of this group cooperate in burial and wedding ceremonies.

"Mahiber" is a term that is derived from the word "Abere" which literally means "to cooperate". The Mahiber consists of a group of people from the same general area, usually of the same sex, who meet once a month on the day devoted to the saint for whom the association is named. Mahiber is a type of formal association which started in Addis Ababa and become popular in the 1960s. Another form of Mahiber, different form the traditional orthodox fraternal association, is also organized on the basis of territorial or tribal bonds and is concerned with social betterment (Levine, 1965; Pankhurst, 1961).

Mahiber operates similar to Edir except that members have a choice of whether or not to contribute money. Their function is guided by internal by-laws. It can be a gathering of people for feasts, to discuss issues important to their community, or to exchange information. Members of Mahiber cooperate in various social activities. Occasionally, a member of these insurance-related institutions can request help from other members to complete his/her farming operations on time.

6.2.2.3. Institutions of Distribution

"Gibzja" literally means "to invite or entertain with food and drinks". Gibzja is mainly a gathering of people who are close to rich households or to political or religious authorities. It can also be a get-together among neighbors, friends or relatives. Food and drinks are served. Invited people chat, sing and bless the house when they leave. Requests for labor, oxen and other resources by people who prepare Gibzja are not refused by households who are invited. If Gibzja is organized by the rich, invited people are not required to reciprocate.[7] However, if it is organized by neighbors, friends and relatives, participants, depending on their economic status, are required to reciprocate. Religious festivities are observed with ceremonies in every household. The rich, however, invite many people. Ceremonies organized for baptizing and wedding children, or for commemorating the dead, have a distributive effect as well as creating ties with people who are called upon when other households face problems.

At baptisms, marriages and funerals, the sponsor provides the food and his neighbors provide the labor for making all the preparations. During this time they are symbolically treated as if they were members of the same household. Although there is a reciprocal set of obligations between neighbors, there is a net flow of food from the rich man's house and a net flow of labor from the poor man's house. None of these inter-household fields of social relations involves ascription as a principle of recruitment (see also Bauer, 1985).

6.2.3. Labor and Micro-Macro Interface Institutions

Micro-macro interface institutions affect household decision-making in various forms. The effect of these institutions can be observed, among others, on decisions regarding allocation of labor. When households are interviewed about the amount of labor input expended on a particular plot, they do not distinguish labor inputs acquired from neighbors and friends from labor supplied by members of the household. They believe that the only reason for acquisition of labor through social-networks is to complete agricultural operations on time. The amount of labor time they spend on crop fields of their friends is reciprocated. Thus, it is equivalent to the time they would have spent on their farm if they were not involved in social-networks. Consequently, it was difficult to ascertain the effect of labor inputs from these groups on a household's resource allocation strategies.

The amount of time spent in insurance and distribution-related institutions is presented in Table 6.1. The result indicates that about half of the year is allocated for participating in social networks. Three major points can be observed from Table 6.1. i) Selale farmers spend less than half the year as working days, ii) households in the Selale region can be seen to be more egalitarian or distributive because they spend several days participating in distribution-related institutions, and iii) the primary interest of Ada farmers is to reduce economic risks through participation in insurance or credit related institutions (Edir and Ekub).

Table 6.1. The Influence of Micro-Macro Interface Institutions on Availability of Labor in Peasant Agriculture

	Number of Days per Year Participated in the Activities of Institutions			
	Selale		Ada	
Institutions	Test	Control	Test	Control
Insurance Related:				
Ekub	12	12	24	24
Edir	8	12	12	18
Senbetie	46	48	48	48
Mahiber	26	22	10	12
Distribution related:				
Gibzja	31	22	12	16
Religious	81	75	60	58
Total	204	191	166	176
Available Working Days per Year	161	174	199	189
Sample size	89	128	26	27

Members of micro-macro interface institutions exchange information, request assistance or enter into reciprocal arrangements in production, consumption, reproduction and marketing. Households' decisions in the Ethiopian highlands are influenced by the inter-household and micro-macro interface institutions. The functions of these institutions are affected by local markets, marketing boards, extension agents, political organizations and development projects. Institutions of such nature integrate households into more complex systems. They are grouped under a sub-title of "macro-integrating" forces. Household decisions, therefore, have to take into account the influence of neighbors, friends, relatives, inter-household and micro-macro interface institutions, and macro-integrating forces.

6.2.4. Macro-Integrating Forces or Institutions

Households are integrated with the larger socio-economic environment through various forces. These include education, markets, political parties, land and price policies, among others.

Schooling increases the chances of obtaining employment outside farming. Thus, sending children to school helps to maintain or reduce the number of people living in a village. Consequently, existing farm size is shared among fewer households. Moreover, schooling helps to increase the contribution of children to decision-making. Children bring with them the product of educational institutions. Furthermore, parents will continue to live on their farm with the help of remittances from schooled children. Links, therefore, are established not only with the state but with the rest of the world through education.

Twenty percent of all household heads in the Selale region have either attended sacred or elementary schools, or participated in the national literacy campaigns. They can usually read and write at the level of grade two or more. An average of three out of six household members have attended schools in the Selale and Ada regions (see Table 3.1., Finnida, 1989).

Through village elites, local markets and government institutions, macro-level institutions influence intra- and inter-household decision-making processes. Each of them play a different role at different stages of integration. The integration of households into a more complex system occurs at the level of the parish, the sub-district, the district, the administrative region, the country and the world system. The linkage or dependency could be as small as through marketing a single product (e.g. a marketing quota) or as great as being a member of a national peasant association (e.g. All Ethiopian Farmers Associations).

Local authorities are contacted by extension agents, tax collectors, politicians, merchants, development project officers and representatives of other institutions. These local authorities complete the link between households and external institutions. An example is the project that provided households of the Selale region with cross-bred cows. The Ministry of Agriculture (MOA) and the FINNIDA project staff communicated with individual farmers through local and district level executives of peasant associations and service cooperatives. Farmers who received cross-bred cows were identified by these officials.[8] Credit repayment is arranged through negotiation between PA officials and MOA. The quality and the amount of milk sold by peasants to the milk collection centers are negotiated through these officials. Individual farmers are reached by institutions such as the grain marketing agency, internal revenue and policy enforcement through local authorities.

Local authorities have to bring benefits to their communities. When integrating-forces negotiate with local authorities, the latter have to compromise in order not to lose credibility in the communities they serve. The society respects and obeys local authorities as long as they abide by their promises. When authorities are squeezed between outside institutions and the traditional value systems, they may lose respect and status (Netting, et al., 1984; Wilk, et al, 1984; Wilk, 1989).[9] Often, however, institutions or politicians make use of the social network systems or micro-macro interface institutions, and organize feasts for relatives and friends to gain support from their communities. Producers have to take into account these influences in their production and marketing decisions.

Respect for authority is a typical value in the Ethiopian highlands. Local authorities, despite working with politicians who are not accepted by peasants, are respected and treated with generosity by their constituencies (see Chapters II and III). Peasants respect this value system to gain access to resources and bend some of the policy guidelines applicable to them.

Through the activities of merchants and traders, local markets facilitate the integration of households into the marketing system. Relationships between households become more impersonal. Village- level negotiations and bargaining among households, that are similar to the market place, becomes the usual practice. This bargaining knowledge is activated within a village when households make purchases or sale of animals and grain, and exchange seed, labor and oxen. Households which make use of market knowledge engage only in balanced reciprocity are thus considered "selfish" or impersonal. This behavior may reduce their possibilities of acquiring labor through social networks during peak cropping season.

The effects of national policies regarding market prices are felt by households through wholesalers, traders and marketing boards who purchase grain or other products. Traders bring imported goods to local markets at relatively cheap prices.[10] Once they establish demand for the product, they raise prices.[11] With imports of non-farm produced goods and exports of agricultural products, households will be tied up with the outside world.

Government institutions organize relief programs when the social networks are incapable of mitigating the risks of crop failure. These positive actions of government institutions, however, are outweighed by policies imposed against the will of producers. As long as producers own the means of production, they resort to social networks, no matter how weak they are, to withdraw from the system. This withdrawal requires changes in household goals and strategies of decision-making. In Ethiopia, withdrawal from the market system and other institutions has been shown to result in reduction of marketable grain supply (see Chapter II).

Increased intervention of governments in household decisions alters resource allocation strategies. For instance, intervention in marketing necessitates: i) planting crops that were not under the quota system but had a higher market value, ii) investing in small ruminants and iii) becoming involved in illegal markets. As a result, the aggregate supply of consumer goods declined and a large amount of food products passed through illegal markets (Lirenso, 1986). Moreover, farmers planted crops which did not maintain soil fertility and moisture (Belete, 1989)

In response to marketing and land use policies that were announced in 1991, a questionnaire was administered by me and my assistants to fifty Selale and twenty Ada farmers. Questions related to changes in crop-area allocation, patterns of marketing, decision regarding when and where to sell. The results are given in Table 6.2.

The following points can be concluded from Table 6.2: i) there is an increase in the area allocated to crops which are widely used as staple (barley) or command high monetary value (teff) and ii) more unrestricted options regarding when and where to sell. The area allocated to crops whose by-product is used for livestock feed (barley and oats) has not changed in the Selale region. The average area for Selale is smaller than Ada because almost an equal amount of land is allocated for grazing and not all crops grown by farmers are listed in the table. They also indicate a significant increase in area allocated to crop in Selale region. It means that with the new free market policy, Selale farmers may reduce grazing area and grow more crops. These are very important changes to producers whose primary objective is securing family survival. It also implies that producers are opportunistic. That is, they switch to crops whose market value is higher (Eisemon and Nyamate, 1988).

Table 6.2. Responses of Households to Changes in Marketing Policy

	Ada Farmers		Selale Farmers	
	Original Crop Plan	New Crop Plan	Original Crop Plan	New Crop Plan
Category	Average hectare per farmer			
Crops to plant				
barley	0.57	0.65	0.61	0.62
wheat	0.36	0.35	0.33	0.58
teff	0.78	0.9	0.28	0.41
bean	0.47	0.48	0.35	0.37
Oats			0.33	0.34
Total	2.18	2.38	1.57	2.32
Where to Sell	AMC/ Consumer Black market	Consumer trader/ /wholesaler	same as Ada	
When to Sell	at harvest	any time	same as Ada	

Source: Compiled from interview results, 1990-1991.

In the livestock sector, access to services such as artificial insemination or bull service, veterinary service and credit are limited. Therefore, the policy of the government to increase milk supply through the introduction of cross-bred cows may not be successful. For instance, farmers may resort to using local bulls for breeding. This will result in higher local breed's gene composition in the off-spring and consequently higher butter content of the milk produced. When this happens, given the low price of fresh milk, producers may find it attractive to process the milk rather than selling it fresh. Consequently, the total milk supply in the region may decline (Tittarelli, 1990; Chapter V).

The discussions under section 6.2. indicate that households are subject to influences from inter-household, micro-macro interface and macro integrating forces or institutions. It is difficult to compare differences in patterns and outcomes of household decision-making that result from variations in influences from external institutions. Differences in household decision-making between regions, and between test and control farmers are discussed below.

6.3. Inter-Household Variations in Resource Allocation Strategies

Social and economic institutions are expected to cause differences in inter and intra-household resource allocation strategies. Discussions regarding these differences are brief since some of the issues pertinent to this topic are discussed in Chapter III and V.

The guiding principles of resource allocation vary within a household, between households, villages, regions and countries. These may include risk spreading, risk minimization, survivalist, disaster avoidance and opportunistic strategies (Marris and Sommerset,1970; Eisemon, et al., 1988).

Household resource allocation strategies primarily focus on securing subsistence requirements. To achieve this goal they employ various combinations of inputs including the inter-household, micro-macro interface and macro-integrating institutions. Securing subsistence is an important goal of households in the Ethiopian highlands because i) of declining farm size, ii) losses in soil fertility and iii) unpredictability in the amount and patterns of rainfall. Households depend on institutions to expand their farm size and borrow money for purchases of fertilizer. However, they do not expect assistance from social networks to reduce the effect of rainfall variability in dryland agriculture.

Table 6.3. shows selected allocation strategies employed by households of Ada and Selale farmers primarily to secure subsistence. Some of the strategies are opportunistic. That is, they are designed to attain increases in production or income over and above what is required for subsistence. The major objective of Ada farmers is to reduce crop failure while that of Selale farmers is equally divided between reducing crop and livestock losses. One striking point is that the percentage of households who would look for government assistance is very low. This implies that even during crises, producers do not expect assistance from the

Table 6.3. Responses to Questions Related to Resource Allocation Strategies (%)

Items	Selale		Ada	
	Test	Control	Test	Control
1. Objective of allocation:				
to reduce crop losses	45	58	93	91
to reduce livestock loss	55	42	5	9
2. Variation of production due to:				
excess/shortage of rain	60	75	84	89
Other reasons	40	25	16	11
3. Worked outside the farm:				
yes	4	5	8	10
no	96	95	92	90
4. If crop fails:				
sell livestock	60	50	36	32
engage in trade	10	14	21	26
trade/wage labor	28	36	42	51.5
request gov't assistance	2	1	1	0.5
5. Criteria for land allocation Between Crop and				
secure subsistence	80	88	91	89
other reasons	20	12	9	11
6. Subsistence Determined by:				
family size, livestock number and Rainfall	90	84	88	91
balancing family and seed needs with livestock	10	16	12	9
6. Potential of a crop is determined:				
by ability to provide for subsistence	86	89	95	96
other reasons	14	11	5	4
7. Fertilizer/pesticides used:				
when rain is dependable	50	70	84	91
when rain is not dependable	20	20	10	9
8. Choices crop technologies:				
improved seed	20	61	51	45
fertilizer and pesticides	65	20	25	21
fertilizer, pesticides and improved seed	15	19	24	34

government. Instead, they resort to other alternatives. Households sell their livestock or engage in trade and wage labor (Table 6.3.).

Poor households would like to supplement their income by working outside the farm. However, societal norms mediated through local institutions act as a constraint on a household's seeking additional income by working outside the farm. Among households of the Ethiopian highlands, norms of the following type are common:

"Kesew Melkam Geberie, Kehile Melkam Firie"

This means that "among all humans, the farmer [is the best], among all crops the seed [the edible] [is the best]".[12] Thus, abandoning farming lowers the status of households in a society. Norms of this nature reduce the individual's ability to gain access to economic resources to reduce the severity of crop losses.

Households which cannot secure subsistence from farming would not receive respect and status from their communities.[13] This, however, does not refer to employment because of educational achievement. Learned people are respected, for they can read, write and communicate with the outside world. If they are remunerated for this achievement, they gain more respect and prestige. However, the majority of people in the study sites of Ada and Selale do not attain a level of education that warrants employment outside the farm.

The study sites of Ada and Selale regions are characterized by mixed farming. Resources, purchased or home produced, have to be allocated between crop and livestock enterprises without affecting the current level of subsistence. Prior to making decisions regarding the allocation of land between crop and livestock enterprises, producers take into account family size, livestock numbers and rainfall (Table 6.3.). Within crop and livestock enterprises, the potential for providing enough for home consumption is the major criterion used in deciding on investments in an enterprise.

Decisions regarding the purchase of inputs are influenced by reliability of rainfall and expectations of assistance from the government, social networks, relatives and friends (Jodha, et al., 1983). Variability in rainfall affects every household, thus possibilities of obtaining assistance from social-networks are limited. Therefore, households have to make conscious decisions. A large percentage of producers invest in fertilizer and pesticides when rainfall is dependable. The percentage of control farmers following this pattern of resource allocation strategy is higher than that of test farmers (Table 6.3.). This implies that one of the goals of control farmers is to increase their income through intensifying crop production when access to cross-bred cows is limited.

Ada has relatively more infrastructure compared to Selale region (see Chapter III). Consequently, the percentage of farmers who would like to work outside farming and want to use crop production increasing inputs are higher in Ada compared to Selale (Table 6.3.)

Cooperation is an essential part of the social fabric of farmers in the Ethiopian highlands. Nevertheless, dependence on other people, including relatives, and institutions may lower status and respect. Thus, households attempt to minimize dependence on social networks or institutions. Strategies to avoid dependence can take several forms, all of which are intended to secure subsistence. For example, all producers plant more than one crop on different plots with different soil characteristics. The reasons for this decision are that each crop has a different: i) growing period, ii) degree of resistance to moisture stress and pests, iii) labor and oxen requirements, and iv) market value. The difference in market value relates to the capacity of the crop to satisfy different consumption requirements. These are important economic and social reasons for planting different crops (see also Chapter V; Table 6.4.). Not only different crops but also different classes of

Table 6.4. Responses to Questions Related to Goals of Resource Allocation Strategies

Items	Selale		Ada	
	Test	Control	Test	Control
1. When rainfall is uncertain:				
Reduce expense and increase saving	40	61	40	55
Sell ruminants/cattle	30	28	10	20
Reduce consumption	20	5	15	10
Sell household valuables	10	6	35	15
2. Criteria for allocation of land:				
Food value	50	47	40	31
Market value	40	53	60	59
3. Shortage of livestock feed, thus:				
Reduce livestock	30	45	80	78
use hay,straw and				
reduce arable land	60	50	10	12
Purchase feed	10	5	10	10
4. Would you abandon farming:				
No	80	85	75	71
Yes	20	15	25	29
5. Preferred employment if forced to leave farming				
Trading	30	35	50	22
Wage labor	10	12	10	8
Trade/wage	60	63	40	70
6. If you obtain more land, You:				
Plow it	50	65	90	85
Grow feed	10	15	0	7
Plow and grow feed	40	20	10	8
7. Plant more than one crop because:				
Pest problems, market				
Value, rainfall, food	80	85	95	98
Others	20	15	5	2
8. Raise more than one class of Livestock				
Traction and milk	40	31	56	50
Traction, milk, meat &				
Transport	50	61	30	44
Others	10	8	14	6

livestock are raised by households. A larger percentage of Selale farmers cited traction, milk, meat and transport as major reasons for raising more than one class of livestock. However, milk and traction are the major reasons for owning more han one class of livestock among farmers of the Ada region.

Avoidance of dependence on friends, relatives, markets and government requires strategic decision-making. This behavior of households is exemplified by the sequential nature of agricultural activities or decisions. Studies in Asia and Africa have shown that household decision-making behavior can be characterized as sequential, even when farmers are affected by drought (Jodha, et al., 1983; Kebede, 1992). Essentially, sequentiality implies that they liquidate assets by the order in which they are necessary to ensure the survival of family members. Among farmers of the Selale and Ada regions, this pattern runs as follows: i) reducing expenses or increase saving, ii) selling the livestock, iii) searching for off-farm work such as trade and wage labor, iv) reducing consumption and v) selling of household valuables (Table 6.4.).

When food production is at risk, major adjustments are made with respect to own resources. For instance, Selale farmers make use of their excess livestock to stabilize household income. However, Ada farmers sell valuables. Ada farmers make use of market prices in most crop production decisions. They are willing to become involved in trading. Additional land is put into crop production in the Ada region, while it may be used for feed and food production in the Selale region (Table 6.4.).

Other forms of avoidance of dependency or strategic decision-making are implementation of adjustment techniques. Jodha and his colleagues (1983) studied adjustment techniques in what they called "self-provisioning" societies. This is accomplished through re-arrangement of resources, management style and making use of social-networks as a last resort. In this regard, producers' decision-making is influenced by the expectation of rainfall. Based on their expectation, adjustments are made with respect to size, number and location of plots (Table 6.5.). In general, when rainfall is uncertain, few plots are planted on low moisture retaining soils (lowland), a large number of plots are planted on high moisture retaining soils (upland), a large number of plots with green or immature crops are harvested for livestock feed rather than letting them dry on the field, weeds are collected from many plots to reduce moisture competition and several weeding and re-plantings are performed to secure subsistence (Table 6.5.). [14]

Resource depletion and rainfall variability force producers to foresee what may happen to their crop and livestock enterprises, and find appropriate remedies. For example, producers anticipate that prospects for adequate supply of livestock feed in Selale and Ada regions are bleak. When environmental degradation threatens survival, households receive very little help from social networks. Access to government assistance to reduce stress on the land as a result of increases in human and livestock population through the provision of fertilizer, improved seed and cross-bred cows is limited. Households, therefore, prefer to improve management of traditional resources to cope with environmentally caused problems. To reduce stress on the land caused by the large number of livestock, households suggest the use of straw, hay, reducing crop area and livestock numbers as solutions.

Table 6.5. Strategic Management or Adjustment Techniques

	Rainfall	Selale		Ada	
		Test	Control	Test	Control
1. low lying plots:	Dependable	5	5	3	3
	Uncertain	3	3	2	3
2. high lying plots:	Dependable	3	3	3	3
	Uncertain	6	4	3	3
3. Harvested green:	Dependable	3	3	3	3
	Uncertain	5	4	2	2
4. No. of plots from Which Weeds are Collected:	Dependable	2	3	4	4
	Uncertain	6	5	2	3
5. Plots from Which mature crops are harvested:	Dependable	5	6	4	4
	Uncertain	3	2	2	3
6. Plots weeded >1:	Dependable	2	3	2	2
	Uncertain	3	5	2	3
7. Plots planted >1:	Dependable	2	3	3	3
	Uncertain	2	3	2	2
Average no. of plots subject to Adjustment when rainfall is:	Dependable	1.57	3.71	3.14	3.14
	Uncertain	4	3.71	2.14	2.71

Until 1991, farm size was frozen to a maximum of 10 hectares per household. With the change in government and policies, opportunities were open for expansion. Given the availability of resources, more than fifty percent of the study farmers would like to grow more crops if they obtained more land. The major contrast between Ada and Selale regions is that Selale producers would like to allocate extra land to grow forage for livestock while most Ada farmers would like to grow crops for food.

Resources owned or purchased by the household are employed to produce grain and livestock outputs. The manner in which resources are allocated are influenced by markets, land tenure, social networks, friends and relatives. Resource allocation strategies that take into account the influence of the above factors determine the amount of crop and livestock output produced. The gross-income from farming in the Ethiopian highlands, market price multiplied by output, approximates the wealth of households (Kebede, 1988).[15] In situation where it is difficult to examine the effect of each factor outside the control of the household, the alternative is to use a proxy. Wealth can be considered an indicator of the effects or consequences of the interaction between resources managed by households and factors outside the control of the farmers.

Schooling increases the chances of off-farm employment. The socialist government was recruiting farmers between the age of eighteen and forty to serve as militia until April 1991. These factors, among others, reduce the amount of labor available to the household. Variation in the worker:consumer ratio, therefore, may serve as a proxy for the influence of factors that may influence the supply of labor.

The result indicates that increases in family size are associated with increases in wealth and in the worker:consumer ratio (Table 6.6a and b.). The increase in these variables is higher for the test compared to the control farmers. The wealth and worker:consumer ratio are relatively low for widows and old couples. In the Ada region, the wealth of widows and old couples is low compared to other family size groups. In Chapter III, it is noted that interpersonal relationships between relatives and friends in the Ada region are weak compared to the Selale region. Discussion with individual farmers in Selale indicate that widows can request help and will receive no less help than wealthy families with higher worker:consumer ratios. Furthermore, children of relatives are asked to live with widows resulting in relatively higher worker:consumer ratio of widows in Selale compared to Ada region. This may explain why the wealth status of widows in the Selale region is closer to the wealth of other family size groups than in the Ada region.

Two cautionary notes ought to be made in reference to Tables 6.6a and b. Firstly, the per capita wealth declines with increases in family size. Thus the smaller family size households may have greater wealth compared to larger family size households. Secondly, with increases in the family size, the worker:consumer ratio may decline because it may take about 15 years before a child enters the working-age group. Consequently, lower worker:consumer ratio may be associated with larger family size households. These issues are not investigated by this study.

Table 6.6a. Changes in Wealth by Family Cycle 1/

	Selale		Ada	
	Test	Control	Test	Control
Family Size	Wealth	Wealth	Wealth	Wealth
2	2180	2271	4568	4419
3	2121	2477	4131	3901
4	2728	2509	3911	3878
5	3360	2702	5132	4141
6	3157	2741	5635	4349
7	3457	2898	5709	4908
Widows 2/				
<0.5	2303	2179	1311	1476
>0.5	3141	3098	2430	2523
Old Couple	2013	2397	1719	2114

1/ Wealth is measured as the sum of the market value of crops, livestock and income from other sources. 2/ > 0.5 and <0.5 refers to worker:consumer ratio.

Table 6.6b. Changes in Worker:Consumer Ratio by Family Cycle

	Selale		Ada	
	Test	Control	Test	Control
	Worker:Consumer Ratio	Worker:Consumer Ratio	Worker:Consumer Ratio	Worker:Consumer Ratio
2	0.66	0.69	0.73	0.78
3	0.67	0.71	0.69	0.81
4	0.93	0.78	0.56	0.79
5	1.01	0.89	1.23	0.84
6	1.04	1.01	1.45	1.43
7	1.5	1.2	1.85	1.53
Old Couple	0.63	0.55	0.31	0.42

6.4. Intra-Household Decision-making

This section examines the patterns of intra-household decision-making processes in an environment where social networks, relatives, friends and macro-integrating forces operate.

Households secure subsistence through joint management of resources under their control (e.g. land and labor). Within a household, responsibilities are assigned by age and sex. This stratification of labor by age and sex corresponds to the timing of farming operation and growth stages of crops and livestock. For instance, crops with a weak root system require a soft top layer or seed bed. The head of the household performs the first two or three plowings because they require more physical strength and hence cannot be performed by younger children. As the seed bed gets softer, the head moves to another plot or engages in other activities while children perform the remaining passes. Children are not allowed to weed crops which are susceptible to trampling especially after attaining knee height. However, they engage in harvesting pulse and threshing of most crops. Women engage in weeding and threshing of most crops, and harvesting of only pulse crops. Adults tend to look after working animals, milking and pregnant cows while children are assigned to ruminants, dry cows and other classes of cattle. The latter group of animals require less care and management than the former. Thus, planting different crops and raising different classes of livestock make use of the household labor pool. Household members, depending on their age or physical strength and sex, are assigned responsibilities. This pattern of allocating responsibility helps members to ascertain status in a village and inculcates a sense of achievement.

Most household heads have not attended secular education, even though their children have. Their input as members of the family may produce variations in the outcome of daily activities of households (Korten, 1972). Children of "good farmers" are expected to show superior knowledge of farming compared to children of other farmers. This will take the form of early and intensive training by the head of the household and acquisition of information (Eisemon, et al., 1988; Warren, et al., 1989). Production knowledge of children reflects that of their parents. When a child demonstrates his knowledge, he is labeled as "lik inde abatu" or "yekelie lij". It implies "just like his father" or "Mr. X's son" . Similar metaphors are used for girls except that their experience and knowledge is related to their mothers.

Social networks are the most common relation through which household decision-making is influenced. This influence comes mainly from friends, neighbors and relatives. The strength of social ties between households (not related by blood) varies from exchange of greetings to being considered part of the family. Consequently, participating in family affairs involves more than close relatives. The proverb "Mewoled Quanqa new" (to be born from the same parents is just a saying) serves as evidence that the terminologies which identify blood related groups (e.g., brother and sister) are not bounded. Inter-household decision-making is dominated through sharing of labor. To make use of labor, from a person fed by another household, is to enter into an obligation to help secure the survival of each other's family. The number of relatives of a household living close by heavily influences decisions regarding crop and livestock production, including adoption of new innovations. Farmers of the Ethiopian highland are characterized by highly individualized decision-making, occasionally influenced by close relatives and friends. Individual household heads are the major decision makers regarding where to live and how, where, how much, what and when to cultivate. Thus, in studying peasants in various societies it is important to see how strong the relationship amongst extended relatives, friends and neighbors is. The degree of relationship with relatives ranges from poor to average for Ada while it is relatively strong for Selale. Villagization has brought major changes, psychologically and socially, such as impersonal relationships between neighbors. A household in the Selale and Ada regions respectively has on average 10 and 7 relatives within an area under three PAs (i.e., 2400 ha).[16] If a household has insufficient labor to accomplish its duties on time, it may adopt relatives, friends or poor households' children, look for traditional mutual help groups, or request help from relatives.

Delegation of responsibilities implies that members of the household contribute to each kind of decision-making. Intra-household decisions regarding crop and livestock production are influenced by social networks and macro-integrating forces, relatives and friends. The degree of influence from within the family, relatives and friends with respect to decisions regarding crop and livestock enterprises, classified as high, medium, low or joint, are discussed below.

6.4.1. Crop Farming Decisions

The contribution of household members, relatives and friends to crop and livestock production and marketing are presented in Table 6.7. In general, crop production decisions are mainly the responsibility of the husband. The input of the wife is medium in both study regions.

The contribution of children to crop production decisions is low in the Selale region but high in the Ada region. Relatives contribute more to this decision in the Selale than in the Ada region. The strength of relationships among relatives is higher in the Selale than in the Ada region (Chapter II and III). As a result, relatives contribute more to decisions regarding crop production. Parents tend to teach children knowledge of the enterprise in which they foresee good prospect. Thus, Ada farmers encourage participation of children in crop farming. Selale farmers involve their children mainly in livestock production decisions.

As a result of the villagization program, few relatives reside in the same village (see Chapter III). This has increased intimacy or frequency of contact among neighbors or friends. Traditional mutual help groups such as "Debo", "Wonfel", "Ekub", "Edir" and "Mahiber" are centered around non-relatives. Thus, economic and social cooperation is intensified between friends or colleagues. This may be the reason for the high input of friends in household decision-making regarding crop production. Most producers are members of different local mutual help groups. While help from these groups comes as a result of the internal by-laws or reciprocal arrangements, it does not guarantee sufficient coverage of risks. Thus, decisions regarding the purchase of inputs are influenced more by blood-related groups.[17]

Use of income from the sale of crops in Ada and purchase of inputs in Selale are made jointly (by husband and wife). Disposition of crop output and the manner in which household use income from the sale of crops are important to their survival. Thus, the contributions from relatives, friends and even children are low. It means that the husband and wife are responsible for making sure that their family obtains its food requirement.

6.4.2. Livestock Production Decisions

Decisions regarding whether to sell fresh or processed milk is the responsibility of the wife if a household's milk supply is from local cows. If the milk production is obtained from cross-bred cows, however, the decision on how much milk to sell is made jointly by the husband and wife (Table 6.7.; Chapter V).

Decisions which endanger a household's survival as a result of huge financial investment and increases in the work load of the family members tend to be made jointly. For instance, cross-bred cows require a lot more time for feeding, caring, watering and milking compared to local cows. This will increase the workload of the wife or daughters. At the same time, the large amount of money invested to acquire these cows forces the head of the household to be cautious of other decisions which require financial expenditures. Thus, both the husband and the wife share the decision to purchase cross-bred cows.

Table 6.7. Contribution of Individuals in Household Decision-Making

	Decision participants[a/]									
Decision Activity	Husband		Wife		Children		Relatives		Friends	
	A	S	A	S	A	S	A	S	A	S
1. Crop Production (when, how and what to plant, weed and harvest)	H	H	M	H	H	L	L	M	L	M
2. Input Use (how much, when and how to apply)	H	J	M	J	M	L	M	M	L	L
3. Purchased Inputs (which input to purchase, how much and for what crop)	H	J	M	J	L	L	M	M	L	M
4. Disposition (how much to sell, where, to whom, When, and which crop to sell)	J	M	J	M	M	L	L	L	L	L
5. Use of Income from the Sale of crop:	J	M	J	M	L	L	L	L	L	L
6. What class of Livestock to raise:	J	J	J	H	L	H	L	M	L	M
7. How much feed to grow or purchase:	H	H	M	H	M	M	L	M	L	L
8. Use of A.I. and veterinary service:	H	H	L	M	M	M	L	M	L	M
9. How much milk to sell:										
Cross-bred	J	J	J	J	L	L	L	L	L	L
Local breed	L	M	H	H	L	L	L	L	L	L
10.Use of income from:										
Sale of livestock	J	H	J	H	L	M	L	M	L	M
Sale of livestock products	L	M	H	H	L	L	L	L	L	L

a/ Where: A= Ada, S= Selale, H= high, M=medium, L=low, J=joint

Households consider several factors to make decisions on whether to sell fresh or processed milk. These factors include: i) distance to milk collection centers, ii) the time it requires to produce one "kubaya" (local unit equivalent to 0.5 kg) of butter, iii) the need for immediate cash, and iv) the season of the year (see Tittarelli, 1990; Chapter V).

Besides decisions regarding production, marketing, and the use of income from the sale of crop and livestock products, households make decisions such as to who should be sent to school, where and when should they be sent; who should approach local social networks to request cooperation; who should go to the local market and purchase household necessities; and how should they react to interventions. Each of these decisions integrate the household into a larger institution or socio-economic organization.

Parents are responsible for decisions regarding whether to send children to school or not. Changing the components of human capital within a family, and planning the future of their children are critical decisions for the whole family. The wife, husband and children contribute substantially to this decision compared to relatives and friends.

6.5. Summary

The institutional basis of decision-making has been investigated. Peasants of the Selale and Ada regions are not different from peasants in other LDCs in the sense that they also operate in an environment that includes the inter-household, micro-macro and macro-integrating institutions. However, the manner in which institutions are organized and their influence on household decision-making is different.

Institutions require contributions or sharing of labor, crop output, tools, livestock and livestock by-products and production risk. Consequently, these processes of sharing or contributions influence household goals, resource allocation strategies and patterns of decision-making. They influence households differently depending on family size, number of relatives, participation in political or religious activities and markets.

The most important resources affected by institutions are labor, land, production knowledge, and livestock. This influence may affect allocation strategies. The ultimate effect of changes in strategies is reflected on crop and livestock output as well as on the efficiency of production.

Institutions were analyzed at three levels. These include the micro institutions (e.g., the coffee ceremony), the micro-macro interface (e.g., social networks) and macro-integrating forces (e.g., markets). The coffee ceremony is an institution where all forms of cooperation begin. At a village level, however, relationships are not as strong as those between neighbors. Therefore, social institutions tend to be more specific in their function and structure.

The micro-macro interface institutions that influence household's economic and social organization as well as their decisions concern: i) production, ii) insurance or credit and ii) distribution. Micro-macro interface institutions affect household decision-making in various forms. Labor is the most important factor of production affected by these institutions. The result for Selale and Ada regions indicates that about half of the year is allocated for participating in social networks. This effect is more pronounced in Selale than in Ada region. Households in the Selale region spend more days participating or working in distributive institutions, those in of Ada in insurance or credit institutions. Thus, social networks and relationships among farmers are stronger in the Selale than in Ada. The primary risk-minimizing strategies of households in Ada concern economic management while those of the Selale consist of using social institutions.

The functions of inter-household and micro-macro interface institutions are affected by local markets, marketing boards, extension agents, political organizations and development projects. Institutions of such nature are, for the purpose of this study, called "macro-integrating" forces. One example of the effect of macro-integrating forces is agricultural policy. The response of Selale and Ada farmers to the elimination of forced procurement and low market prices of grains, and to the possibility of expanding farm size, indicates that they will increase the area allocated to crops which are widely used as staples (barley) or command high monetary value (teff).

The choice of goals is influenced by the degree of assistance household received from institutions and physical resources. The major objective of Ada farmers is to reduce crop failure while that of Selale farmers is equally divided between reducing crop and livestock losses. Resource depletion and rainfall variability force producers to improve management of traditional resources rather than request help from social networks or the government.

Worker:consumer ratio and wealth determine the structure, function and stability of households. These parameters vary according to family cycle. The result of this research indicates that increases in family size are associated with increases in wealth and worker:consumer ratios. The increase in these variables is higher for the test compared to the control farmers. The wealth and worker:consumer ratio are relatively low for widows and old couples in Ada compared to Selale regions. This is attributed to the existence of stronger social relationships in Selale than in the Ada regions.

Households secure subsistence through joint management of resources under their control (e.g., land and labor). Within a household, responsibilities are assigned by age and sex. This stratification of labor by age and sex corresponds to the timing of farming operations, and to the growth stages of crops and livestock. The contribution of children to crop production decisions is low in the Selale region but high in the Ada region.

The villagization program in Ethiopia has increased economic and social cooperation between friends or colleagues. Furthermore, the contribution of friends and neighbors is confined to areas that encounter labor or oxen shortages. Specifically, their contribution is high in crop production (e.g., Ada region) and in livestock production (e.g., Selale region).

Intervention strategies which do not consider the manner in which these institutions affect household decisions are doomed to failure. Chapters 7 and 8 provide empirical evidence regarding factors that influence risk taking behavior, technology adoption, production decisions, and production risk, in light of the institutional framework examined in this chapter.

6.6. Notes

1. Berry (1980) and Cancian (1980) have shown that the decision making environment includes decisions made by others. Seemingly irrational decisions to spend resources to participate in the social environment are logically rational when seen in their total context (see also Sahlins, 1965).
2. Institutions can be considered as social strategies or collective actions when individuals make use of them to justify their actions or request assistance.
3. Production, in its broadest definition, includes all the processes that take place from planting a crop until the grain or output reaches the final consumer. These processes transform the primary commodity or increase the quality of the product, and are considered part of the production process. Marketing, as a process of transforming the primary commodity, can be considered as a production process.
4. The association of brotherhood or sisterhood, with friends or neighbors sometimes comes in conflict with another societal value system. For example, people say "sew kezemedu ahiya kamedu", meaning "humans with relatives and donkeys with ashes". This means relationships between those which are blood related is better than relationship between the non-blood related. This proverb is evoked in times of conflict, especially if one household's request for help is denied by another.
5. It must be noted that "hager" which literally means country is contextual. It can be a village, a parish, district or the totality of all administrative regions. Among the Ethiopian highlanders, "Hager" may be equated to identity.
6. Those who win the lottery will be excluded from the next draw.
7. Furthermore, rich households lend seed, money and oxen but may request less than what they give or none at all. Those who benefit from the rich are committed to reciprocating in any form, including voting for them for political, Edir, Ekub and other leadership roles that rich farmers want.
8. PA officials helped to identify farmers who were able to construct sheds for cows and had received training in the management of cross-bred cows, among other criteria (see Chapter III).
9. The success of development strategies in helping producers is dependent on the interaction of public and political institutions. Where political institutions (e.g. PAs) take the commanding position over social networks (e.g. Mahiber), the mechanisms of integrating the household into a larger system cannot be sustained. Under this situation, conflict of interest between the state, local political institutions, and peasants is inevitable.
10. In most parts of Ethiopia, there was no restriction on penetration of local markets by traders. This was not the case between 1978 to 1990 because of government policy (see Chapter II). Since 1990, however, the restriction on traders was lifted.
11. The effect of competition from local traders on traders from major markets is not strong because local and major market traders collaborate on ways of maximizing their profit or start joint venture.
12. Furthermore, households which are engaged in craft production, lower caste groups, such as iron-smiths are considered to be social outcasts and less than human.
13. Thus, there is a conflict in the value system of peasants in the Ethiopian highlands. On one hand, working outside the farm accords lower status and on the other inability to secure subsistence lowers status.
14. The lowland plots are dominated by soils which have low moisture retaining capacity

(e.g. sandy soils), the upland by black soils which retain moisture.

15. Wealth is measured by the value of livestock and livestock products, crop output and income from other sources.

16. Relatives may cooperate in various activities of the household. While the degree with which they cooperate and influence decisions may differ, it is the obligation of a relative not to let other relatives fall into poverty or starvation.

17. This finding is related to the proverb "sew kezemedu ahiya kamedu", meaning "humans with relatives and donkeys with ashes". The people most trusted when households face problems are not friends or neighbors but relatives. Yet, another proverb, "meweled Quanqa new" which means "it is not to be blood-related that matters as it is just a saying" contradicts the above proverb. Thus, households may have stronger relationship with non-blood related people than with their relatives. In either case, these norms complicate intra and inter-household decision-making processes.

CHAPTER VII

TECHNOLOGY ADOPTION AND RISK

7.1. Introduction

Decision-making is greatly influenced by skills that households possess. Chapter V has shown that goals, strategies, information and skills differ by age, sex and region, among others. Chapter VI discussed differences in resource allocation strategies because of influences from micro and macro institutions through which households operate. The factors examined in these chapters and other elements of the socio-economic environment influence decisions regarding adoption of innovations and the extent to which households are willing to take risks. This chapter investigates patterns of and consequences of household decisions regarding adoption of technologies and risk-taking behavior in the Selale and Ada regions.

Agricultural technology is suggested as a strategy to increase food production in LDCs (Chapter I and II). However, mistakes in the generation and implementation of technologies greatly contributed to failures of agricultural projects. These may include inappropriate policies, unavailability of preconditions, incompatibility of technologies with the socio-economic characteristics of adopters, among others. Furthermore, adoption decisions are influenced by the risk-taking behavior of households.

Risk is a function of several socio-economic factors. The degree to which these factors influence the behavior of producers varies. In some cases, risk aversion is argued to have little effect on adoption decisions (Cancian, 1980). Mason and Halter (1980) argued that early adopters are, in fact, risk-averse farmers. Early adopters have acquired greater knowledge and skills from which they learned to achieve success by adopting new practices that reduce the risks of farming (Mason and Halter, 1980). Earlier adopters are generally shown to have more knowledge of innovations than late adopters, and the diffusion of innovations is viewed as a process by which risks in farming are reduced (Mason, et al., 1980).

Innovations are not necessarily risky because of their newness (Morrison, et al., 1974). Rather, incomplete knowledge and uncertain outcomes associated with some innovations may lead farmers to view the new practices as risky. Risk-averse farmers are not willing to adopt until adequate information are available. As knowledge becomes more diffused and outcomes more certain, the risks might be expected to be reduced (Mason, et al., 1980).

Several interpretations of what risk is and its role in household decision-making are provided in the literature regarding peasants' resource allocation strategies (See Roumasset, et al., 1979; Barlett, 1980). Some argue that peasants operate in a risky environment, thus risk-taking behavior is a function of various factors in peasant agriculture. Actions peasants undertake may follow strategies of ensuring them against adverse events. Others contend that peasants are inherently risk-averse. Both groups have attempted to classify peasants into those who are risk neutral, moderately risk-averse and extremely risk-averse.

This chapter examines the preconditions for adoption and the factors influencing adoption decisions in the Selale and Ada regions of Ethiopia. Information from actual production decisions is used to construct measures that characterize the risk-taking behavior of households. The impact of risk-taking behavior on technology adoption is assessed.

7.2. Patterns of Technology Adoption

Prior to performing statistical analysis, frequency distributions of patterns of adoption of innovations for the Ada and Selale regions are calculated. The results are presented in Table 7.1.

The results show that at least eighty-five percent of households in both regions adopt fertilizer as their first choice, among other production technologies. Most Ada producers adopt innovations in the following order: fertilizer, pesticides, improved seed, and cross-bred cows. Selale producers adopt fertilizer, pesticides, cross-bred cows, and improved crop varieties in that order. These patterns of adoption indicate that the primary goal of producers in both regions is to produce adequate grain through the adoption of crop production technologies.

Table 7.1. Patterns of Adoption of Innovations by Region

	First	Second	Third	Fourth
	percentages of farmers			
Ada Region				
Fertilizer	90	30	15	16
Pesticides	5	60	31	21
Improved seed	4	6	48	23
Cross-bred cows	1	4	8	38
Selale Region				
Fertilizer	85	15	10	13
Pesticides	9	74	28	21
Improved seed	1	2	5	28
Cross-bred cows	5	9	56	48

Source: Calculated from data gathered from the study sites, 1990/91.

The performance of improved seed varieties is not promising in the Selale area. Therefore, producers adopt this innovation only if it does not reduce gain from adoption of other innovations. These results confirm that the package approach to the provision of technologies is not a feasible strategy in the grain surplus regions of the Ethiopian highlands. In fact, sequentiality of decisions regarding goals and strategies is a universal characteristic of subsistence farmers in

LDCs (see Chapter V and VI; Kebede, 1992; CADU reports, 1975; Lele, 1976; Molnar, et al., 1983; Barlett, 1980).

Success in adoption of agricultural innovations in the LDCs is linked to the existence of preconditions or "modernizing" conditions (Molnar, et al., 1983; Hayami and Ruttan, 1985). For instance, adoption of fertilizer requires conditions such as accessibility to distribution centers, availability of credit, subsidies and favorable output prices (Table 7.2). When production constraints are binding, however, preconditions become irrelevant. Producers of Ada and Selale regions have realized that the fertility status of their farm lands is declining. Adoption of fertilizer is a feasible strategy to reduce the risk of not producing enough grain to satisfy family food requirements. Households may adopt innovations as the only solutions to ensure subsistence requirements, regardless of the existence of preconditions. Furthermore, producers in the Selale and Ada regions adopt improved crop varieties if increases in production from these varieties is twice that of local cultivars and if they can sell their grain at fair market prices. With respect to livestock production, producers indicate availability of feed and veterinary services as the two important preconditions necessary for the adoption of cross-bred cows.

Adoption of innovations necessitates shifts in allocation of financial and physical resources. For instance, adoption of cross-bred cows requires growing feed on the farm to reduce milk production costs. Given the restriction on average land holdings per household, growing feed on the farm reduces land allocated to crop production. More than sixty percent of Selale farmers chose to reduce the area allotted to barley and oats. These two crops occupy the largest percentage of area per household. The by-product of these crops is a good source of livestock feed. However, the market value of these crops is low when compared with crops such as wheat, teff and beans. In the Ada area, fifty percent of the test and thirty percent of the control farmers reduced the area devoted to barley.

The objective of introducing "more productive" animals or cross-bred cows is to reduce the less productive native breeds of livestock. Eighty percent of all the study farmers intend to reduce the number of local cattle if they can obtain cross-bred cows instead.[1]

Adoption of cross-bred cows or crop production technology involves decisions regarding what use can be made of the outcomes of new innovations (e.g., higher crop or milk yield). Seventy percent of the Selale farmers prefer to use calves for husbandry or traction purposes. One third of producers in the Ada region intend to use calves for traction and husbandry as their first choice, and for fattening as their second choice. Ada producers are equally divided between using calves for traction and husbandry, fattening or both.

In the Ethiopian highlands, elites are the first to adopt innovations (Ayana, 1985; Waktola, 1980). The elites or early adopters are characterized by, among other factors, having greater resources or wealth, exposure to sources of new agricultural practices, belonging to formal community organizations, and holding political offices in these organizations (Alemayehu, 1975). However, the results

Table 7.2. Preconditions for Adoption of Innovations

	Selale		Ada	
Categories	Test	Control	Test	Control
	Percentage of producers			
1. Adopt cross-bred cows if:				
a. Availability of Feed & veterinary serv.	45	68	85	87
b. Fair Price	39	51	71	77
c. Grow Feed	90	81	40	20
2. Adopt crop technology if:				
a. Increase in production is the same as local cultivars	10	15	12	9
b. ≥ twice the local cult.	81	79	96	94
c. Price is good	78	85	89	92
d. Adequate distribution Centres	42	32	65	75
e. Credit/subsidies	78	69	81	78
3. If grow forage, then reduce area under:				
a. Barley & oat	74	61	50	30
b. Wheat	31	29		
c. Guaya	30	33		
4. Type of feed purchased:				
a. Wheat bran	15	11	35	23
b. Noug cake	5	14	5	6
c. a+b	80	75	60	71
5. If adopt cross-bred				
a. Reduce livestock no.	80	61	98	87
b. Increase livestock no.	6	11	-	-
c. No change in Livestock Number	14	28	2	13
6. Use of calves from Cross-Bred Cows for:				
a. Husbandry and traction	68	76	35	46
b. Fattening	15	22	32	29
c. a+b	17	2	33	25
Sample Size	89	128	26	27

from adoption patterns of elites cannot be used to design strategies of technological intervention for the majority of smallholders. In fact, diffusion strategies catered towards elites increase inequality in net farm income among peasants (see for example Feder, et al., 1982; Jamison and Lau, 1982).

Extension contacts and information from other sources influence adoption of new technologies (Feder, et al., 1985). The degree to which smallholders process information related to new production techniques varies depending on their production knowledge. This knowledge is shaped by social experience (Warren, et al., 1988). In their study of Kenyan farmers, Eisemon et al., (1988) demonstrated that producers can accurately process production information about new innovations provided that it is related to their day-to-day activities and given meaning by their social experience (see also Warren, et al., 1988). Ethiopian farmers not only have extensive farming experience but they also modify the innovations over time in a way different from what is recommended by extension agents (Toborn, 1978; Waktola, 1980; Tecle, 1975). However, studies in the adoption of innovation in Ethiopia have not examined the influence of production knowledge on adoption decisions.

If producers follow the safety first rule of decision-making, then inputs that have proved to produce successful results are adopted first. Subsequent adoption decisions depend on several factors such as wealth, number of relatives and goals of the households. It follows that there are no patterns of adoption of complex or simultaneous technologies but rather a sequential adoption pattern (Feder, et al. 1982; Jamison and Lau, 1982; section 7.1.).

Accessibility or supply of inputs and improved agricultural management practices were found to influence the adoption of innovations in the Central Ethiopian Highlands (Kiros, 1976). Michie (1986) argues that because unequal access to technologies developed outside a system has become a common phenomenon, locally available technology using locally available resources at low cost should be the goal of agricultural development. This approach can also help to reduce the dependence of peasants on national and international markets for imported technologies.

7.3. Empirical results of Adoption Decisions

Decisions regarding adoption of cross-bred cows, fertilizer, pesticides and improved seed technologies are analyzed using a linear probability or logistic model (Chapter IV). From a practical and statistical point of view, it is impossible to analyze all factors that influence decisions to adopt innovations. Based on correlation analysis, therefore, a limited number of factors are selected to examine adoption decisions. Spearman's correlation analysis is performed for several variables. If two variables have a correlation coefficient of 0.6 or more, only one of them is selected for inclusion in the logistic regression. Inclusion of highly correlated variables in a statistical decision analysis not only reduces the degrees of freedom but also results in lack of solution (Maddala, 1983).

To investigate adoption decisions in the Ada and Selale regions of Ethiopia, cross-bred cows, fertilizer, pesticides and improved seed varieties are selected.

Chapter II has shown that the two regions are different with respect to selected variables, thus adoption decisions are investigated by region. Several variables were included in the logit regressions. However, it was not possible to attain statistical convergence. In lieu of parsimony and convergence, a maximum of 14 variables were selected to investigate decisions regarding each production technology. The independent variables were family size, education (years of schooling), awareness, experience (years), number of relatives, purchase price of technologies, productivity, production and marketing knowledge, wealth, household expenses excluding expenses for technologies, off-farm income, number of technologies adopted, input prices, veterinary services, feed area (hectares) and farm size (hectares).[2]

Production and marketing knowledge has been assessed through problem solving tests (Chapter V). Wealth is measured as the sum of the market values of livestock and crop outputs, without accounting for home consumption and seed requirements. Expenses include the sum of money spent to purchase consumer goods, taxes and contributions, school fees and clothing. Off-farm income includes money earned from wage labor by working on plots of other farmers, for government agencies or in cities.

Producers were asked to rate the productivity of each technology compared to local breeds or crop cultivars. These ratings were coded as high, average and low. Respectively, they were assigned values of 3, 2 and 1. Input prices include feed, insemination and other costs related to the management of cross-bred cows. Purchase prices apply only to crop production technologies because prices of cross-bred cows are fixed or constant and producers consider the cost of managing the cows more than the purchase prices.[3]

Awareness is calculated as a geometric average or composite value as follows:

$$\text{Awareness} = [A_1 + A_2 + A_3]/3$$

Where A_1 is the number of visits a farmer makes to the nearby city and A_2 is the number of days household members visit local markets. Both variables are calculated on a yearly basis and divided by the highest number of visits in the sample. Owning a radio (A_3) is given a value of 1 or 0. A farmer who owns a radio is assumed to listen to news from outside the vicinity, thus is expected to have a 100 percent exposure to outside information compared to farmers who do not own radio.[4] Although owning a radio is recorded as a discrete variable and city visits as a continuous variable, the need to capture influences from these sources and reduce the number of variables in the logit regression necessitated the construction of this index. This way the values of A1, A2 and A3 are constrained to lie between 0 and 1. The anticipated or hypothesized effect of variables is obtained from results of previous research on adoption of technologies, consultation with agencies working in the study sites and discussion with farmers in the study sites (Table 7.3.).

Table 7.3. Anticipated Effects of Socio-Economic Variables on Adoption Decisions

	Adoption of							
	Cross-Bred Cows		Fertilizer		Pesticides		Improved Seed	
Variables	Selale	Ada	Selale	Ada	Selale	Ada	Selale	Ada
Family Size	+	-	+	+	+	+	-	+
Education	+	+	+	+	+	+	+	+
Awareness	+	+	+	+	+	+	-	+
Experience	+	-	+	+	+	+	-	+
No. of relatives	+	-	+	-	+	-	+	+
Purchase price	-	-	-	-	-	-	-	-
Productivity	+	+	+	+	+	+	+	+
Knowledge:								
Production	+	+	+	+	+	+	-	+
Marketing	-	+	-	+	-	+	-	+
Wealth	+	+	+	+	+	+	+	+
Expense	-	-	-	-	-	-	-	-
Off farm income	+	+	+	+	+	+	+	+
Number of technologies adopted	-	-	+	+	+	+	+	+
Input price	-	-						
Veterinary service	-	-						
Farm /Farm size	+	+	-	+	+	+	-	+

7.3.1. Adoption of Cross-Bred Cows[5]

The results of adoption of cross-bred cows and fertilizer technologies are presented in Table 7.4. Education, awareness, experience, production knowledge and off-farm income positively and significantly influence adoption of cross-bred cows in the Selale and Ada areas (Table 7.4.). Marketing knowledge is excluded from the model of adoption of cross-bred cows because prices of cows and milk are fixed. Therefore, limited use can be made of this knowledge. However, prices of crop technologies and crop outputs vary. Thus, knowledge of grain marketing affects adoption of crop production technologies.

The influence of wealth on adoption of cross-bred cows is positive and significant in the Selale region but negative and significant in the Ada region. The largest percentage of the wealth of Selale producers is embodied in livestock while that of Ada in crop (see Chapter III). Thus, wealth exerts a positive and significant

Table 7.4. Results of Logit Regression of Adoption Decisions[1/]

	Adoption of			
	Cross-Bred Cows		Fertilizer	
	Selale	Ada	Selale	Ada
Intercept	1.092	2.454	2.078	3.033
	-1.303	-1.6823	(3.679)*	(4.466)*
Family Size	-0.062	0.04	0.626	0.744
	-0.0197	-0.0565	(2.4523)*	-0.4797
Education	0.228	0.555	0.173	0.091
	(2.535)*	(2.908)*	(2.470)*	(1.971)#
Awareness	0.502	0.722	-0.195	0.192
	(3.65)*	(3.1748)*	-0.2561	-0.0125
Experience	0.362	0.476	-0.279	0.965
	(2.682)*	(3.23)*	(2.237)*	(4.260)*
No. of relatives	-0.058	-0.077	0.269	0.268
	(1.978)#	-1.0165	(2.183)#	(2.141)#
Purchase price			-0.443	-0.310
			-1.2593	(3.137)*
Productivity	0.25	0.048	0.602	0.415
	(1.961)#	-0.9401	(1.989)#	(2.030)#
Production knowledge	0.674	0.471	0.624	0.563
	(3.959)*	(2.516)*	-1.2336	(3.735)*
Marketing knowledge			-0.830	-0.021
			(2.635)*	-0.0002
Wealth	0.191	-0.244	0.77	0.134
	(2.025)#	(2.304)#	(2.067)#	(2.002)#
Expense	-0.880	0.847	-0.769	-0.917
	(2.524)#	-1.0708	(2.356)#	(1.963)#
Off-farm income	0.22	0.044	0.064	-0.110
	(3.538)*	(1.946)#	(1.967)#	-0.3589
No. of technologies adopted [3/]	-0.433	-0.172	0.191	0.425
	(-2.360)#	(2.504)*	-0.4576	-0.2315
Input price	-0.308	-0.191		
	(-2.098)#	-1.0663		
Veterinary service	0.852	0.011		
	(4.878)*	-0.9475		
Feed Area	0.795	0.957		
	(2.147)#	-0.4118		
Farm Size	0.121	0.431	-1.752	(4.35)*
-2 Log L[3/]	98.789	56.71	21.189	12.888
	(0.041)*	(0.072)#	(0.069)#	-0.4565

1/ Values in Parenthesis are Wald's Chi-square. * and # indicate statistical significance at 1 & 5 percent respectively. 2/ This refers to fertilizer, pesticides and improved seed for the cross-bred cows model, and pesticides, improved seed and cross-bred cows for the fertilizer model. 3/ -2 Log L is a criterion for assessing model fit and test the joint effect of explanatory variables included in the model.

influence on adoption of livestock related technologies. In the Ada region, however, the prospects for livestock production are limited. Producers do not want to risk their family survival by investing in livestock technologies. Thus, wealth exerts a negative influence on adoption of cross-bred cows in the Ada region.

Producers engage in wage employment if they: i) encounter difficulty in securing food from farming, ii) have potential that can best be utilized outside farming (e.g., education), or iii) cannot achieve their goals, other than securing food, by relying only on farming. Wage employment helps to raise money that can be used to purchase new inputs. Despite its implication for lower social status of households, off-farm income positively influences adoption decisions (see Chapters V & VI).

The wealthier a producer, the less interest he will show in livestock husbandry, especially if the preconditions for livestock husbandry are not available. This is particularly true for the Ada region. Group discussions with farmers of both study sites reveal that the absence of complementary inputs such as feed and veterinary services are the major constraints or risk factors in livestock production. These constraints are more binding in the Ada than in the Selale region. Thus, feed and veterinary services exert a positive but non-significant influence on adoption of cross-bred cows in the Ada region.

The wealthier a producer, the less interest he will show in livestock husbandry, especially if the preconditions for livestock husbandry are not available. This is particularly true for the Ada region. Group discussions with farmers of both study sites reveal that the absence of complementary inputs such as feed and veterinary services are the major constraints or risk factors in livestock production. These constraints are more binding in the Ada than in the Selale region. Thus, feed and veterinary services exert a positive but non-significant influence on adoption of cross-bred cows in the Ada region.

The amount of household expenditure significantly and negatively influences adoption of cross-bred cows in the Selale area. Family size and number of relatives are negatively correlated with adoption of cross-bred cows. Livestock and crop production are competing and complementary enterprises. Both enterprises compete for the limited financial and physical resources of households. Income earned from one enterprise helps boost production in the other enterprise. Prior to adoption of cross-bred cows, producers evaluate prices of inputs required for better management of cross-bred cows. The higher the prices of inputs, the lower the probability of adoption of cross-bred cows in both study sites.

Among the social variables, education, experience, production knowledge and awareness significantly influence adoption of cross-bred cows. Several studies have documented that households that are actively involved in social networks, especially those with a large number of relatives, are better insured against unforeseen risks of failures or financial losses compared to households who are less involved in social networks and have few relatives (Barlett, 1980). Nevertheless, there is a limit to the insurance that relatives or networks can provide against risks of crop or livestock losses. Cross-bred cows or purchases of livestock are risky investments. Households may not be completely buffered from financial risks associated with adoption of cross-bred cows. Therefore, they have to take full responsibility for the consequences of their investment decision. Another interpretation for the negative effect of family size and relatives on adoption of cross-bred cows is that if households invest in expensive innovations, they may not have sufficient financial or physical resources to participate in social networks. They may not be in a position to help relatives and feed their families. Therefore, increases in family size and relatives may negatively influence decisions regarding adoption of cross-bred cows.

Technical variables such as productivity of cows, veterinary services and availability of land to grow feed or grazing area, positively and significantly influence adoption of cross-bred cows in the Selale area. The number of crop production technologies adopted negatively and significantly influences adoption of cross-bred cows in both study sites.

7.3.2. Adoption of Fertilizer

Adoption of fertilizer is positively influenced by economic factors (wealth and off-farm income), social factors (family size and relatives) and technical factors (farm size, productivity). Awareness, purchase price, market knowledge and household expenses are negatively correlated with the adoption of fertilizer (see below).

Increases in agricultural output can be expected if fertilizer is complemented by other crop production-increasing inputs such as pesticides, herbicides and improved crop cultivars. Crop lands that are affected by water-logging condition and susceptible to erosion reduce the yield increasing effect of fertilizer. Application of less than the recommended amount of fertilizer per hectare and inadequate rainfall reduce the impact of fertilizer. Producers obtain these inputs either on cash or credit. Their ability to pay is dependent on the amount of grain production attained and market price. In situations where the market price is variable (free market case) or controlled by the state (until 1990), awareness of factors influencing the impact of fertilizer on crop yield, purchase price and general market knowledge may reduce the probability of adoption of fertilizer. The impact of these factors on adoption of crop production technologies is higher in the Selale than in Ada region. This due to the fact that Ada has relatively better infrastructure (e.g., roads and access to major markets) than Selale.

Ada producers have experimented with fertilizer for several years compared to farmers in the Selale region. Thus, their knowledge about the importance of fertilizer is expected to be higher than Selale producers. This may be the reason for the positive and significant effect not only of production knowledge but also complementary inputs such as land on the decision to adopt fertilizer in the Ada region.

7.3.3. Adoption of Pesticides

The results of adoption of pesticides and improved seed varieties are presented in Table 7.5. Adoption of pesticides is positively influenced by economic factors (wealth and off-farm income), social factors (education and production knowledge) and technical factors (farm size) in both Selale and Ada regions. Similar to fertilizer, Ada farmers have several years of exposure to crop production innovations compared to Selale farmers. Producers in the Ada region have the know-how about the advantages and disadvantages of using pesticides. Consequently, family size, experience, number of relatives, and yield increasing ability of the technology (i.e., productivity) positively and significantly influence adoption of pesticides in the Ada region. Purchase price, marketing knowledge and household expenses negatively influence decisions to adopt pesticides.[6]

7.3.4. Adoption of Improved Crop varieties

Adoption of improved crop varieties is positively and significantly influenced by awareness, relatives, and wealth in both study sites. Improved crop varieties were recently introduced in the Selale region. Their performance on fields of early adopters was not promising. Thus, adoption of this innovation may endanger securing subsistence. Family size, experience, the yield increasing impact of the technology (productivity), production knowledge, and number of technologies adopted reduces the probability of adoption of improved seed in the Selale region.

Research conducted in the Ada region shows that Ada farmers have access to most crop production technologies and that they show a keen interest in adopting improved crop varieties (Debrezeit Agricultural Experiment Station report, 1975, 1980, 1985). Experience and access to complementary inputs may have contributed to the positive and significant effect of family size, experience, productivity, production knowledge, and number of technologies adopted on decisions to adopt improved seed varieties in the Ada region.

The impact of technologies, when they are incorporated in the adoption equations as *separate* regressors, on the probability of adoption of the technology in question produced results that were comparable to those obtained when the *number* of technologies adopted was used as a regressor (Kebede, 1993). The former approach reduces the degrees of freedom. Therefore, the number of technologies adopted prior to adoption of the technology in question was used as a regressor to examine the sequentiality of adoption decisions. In all adoption equations, the number of technologies adopted positively and significantly influenced the probability of adoption of the technology in question. It means that producers followed sequential adoption decisions.

Table 7.5. Results of Logit Regression of Adoption Decisions[1/]

	Adoption of			
	Pesticides		Improved Seed	
Variables	Selale	Ada	Selale	Ada
Variable	Selale	Ada	Selale	Ada
Intercept	1.428	4.613	1.578	3.044
	-1.432	(5.714)*	-0.1089	(2.738)*
Family size	0.116	1.816	-0.267	0.038
	-0.089	(2.435)#	-0.466	(1.990)#
Education	0.603	0.065	0.037	0.244
	(5.439)*	(5.213)*	-0.0205	-0.132
Awareness	0.063	0.43	0.581	0.988
	-0.0286	-0.0425	(2.306)#	(3.129)*
Experience	0.019	0.779	-0.344	0.419
	-0.0025	(4.570)*	-0.8486	(1.972)#
Relatives	0.033	0.85	0.47	0.502
	-0.0342	(2.051)#	(2.839)*	(2.584)*
Purchase price	0.24	-0.816	0.092	-0.976
	-1.0643	(3.004)*	-0.0409	-0.7575
Productivity	0.14	0.654	-0.018	-0.415
	-0.2637	(3.824)*	-0.0038	-0.1434
Production	0.715	0.945	-0.523	3.248
knowledge	(3.616)*	(4.000)*	-0.6767	(2.450)#
Marketing	0.301	-0.918	0.216	-0.822
knowledge	-0.374	(2.370)#	-0.4648	-1.313
Wealth	0.526	0.122	0.779	0.441
	(2.513)*	(2.02)#	(3.027)*	(2.417)#
Expense	-0.463	-0.115	-0.064	0.45
	(2.072)#	(2.411)#	-1.0204	-0.2269
Off-farm income	0.156	0.219	0.069	0.213
	(2.988)*	(1.987)#	-0.7605	(1.992)#
No. of technologies	0.37	0.713	-0.184	0.269
adopted [2/]	(1.997)#	(2.248)#	(1.988)#	(1.968)#
Farm size	0.214	0.531	0.212	0.191
	(1.972)#	-1.971	-1.752	(2.723)*
-2 Log L[3/]	20.943	28.731	7.358	14.63
	(0.074)#	(0.007)#	-0.8828	-0.331

1/ Values in Parenthesis are Wald's Chi-square.

2/ This refer to fertilizer, cross-bred cows and improved seed for the pesticides model, and pesticides, fertilizer and cross-bred cows for the improved seed model.

* and # indicate statistical significance at 1 & 5 percent respectively.

3/ see footnote in Table 7.4.

7.3.5. Summary

Education of household head, wealth, number of relatives, productivity of technologies, off-farm income, extension contacts, feed area and/or farm size exert positive impacts on adoption of most of the innovations in both study sites. The impact of education, experience, and production knowledge on decisions to adopt innovations is not only positive but larger than other variables. Compared to other innovations, Adoption of improved crop varieties in both study sites and adoption of cross-bred cows in the Ada region are negatively influenced by several variables. Improved crop varieties face serious competition from local (traditional) cultivars with respect to yield and resistance to disease. Introduction of livestock production technology is not feasible in environments where "modernizing conditions" and potentials are lacking. For instance, Ada region lacks adequate livestock feed and milk collection centers and A.I. services. Thus, most variables exert a negative influence on adoption of cross-bred cows in the Ada region.

Various studies have argued that provision of technologies as a package raises the income or well-being of producers compared to provision of a single technology (Feder, et al., 1982; Waktola, 1980; CADU, 1973, 1974). Producers characterized by "disaster avoidance" or "safety-first" behavior, however, ranks goals and strategies hierarchically or according to the order of their importance. This section has shown that the number of technologies adopted prior to making decisions to adopt other innovations are positively associated with the adoption of fertilizer and pesticides but negatively with cross-bred cows and improved crop varieties. Fertilizers and pesticides have proven to be helpful in increasing crop production. Producers have a better chance of producing sufficient grain for family requirements if they adopt these innovations. Therefore, most variables exert a positive and significant influence on adoption of fertilizer and pesticides.

The number of relatives and input prices negatively influence adoption of cross-bred cows but positively influence the rest of the innovations. Purchase price of cross-bred cows and prices of inputs required for better management of these cows are high compared to other innovations. Therefore, these factors reduce the probability of adoption of such lump-sum innovations.

Tests for zero statistical significance of all variables indicate that all variables contribute significantly to the adoption of cross-bred cows, fertilizer and pesticide technologies in both study sites. This implies that all the variables included in these adoption models warrant attention in the design of agricultural policy for the peasant sector. Given the limited financial and physical resources of governments of LDCs, priority should be given to those variables which are statistically significant determinants of adoption decisions.

7.4. Predicted Probabilities of Adoption Decisions

Predicted probabilities are obtained by inserting parameters obtained from estimation into the logit equation specified in Chapter IV. Given the weights associated with the independent variables (beta's), calculation of predicted probabilities helps to calculate the probability with which a given farmer will adopt a technology. The predicted probabilities of adoption decisions are calculated for cross-bred cows, fertilizer, pesticides and improved crop varieties. This study hypothesizes that decisions to adopt a technology are influenced by prior number of technologies adopted. Furthermore, the number of technologies adopted and the probability with which they are adopted are influenced by the number of adopters.

Figures 7.1 and 7.2 depict the predicted probability of adoption of one, two, three or four production technologies. The probability of adoption of cross-bred cows declines sharply among producers who have already adopted three and four technologies for the Selale and Ada regions respectively. The probability of adoption of fertilizer and pesticides is higher than improved seed. However, the probability of adoption of improved seed increases with increases in the number of technologies adopted. The above results imply that the probability of adoption of cross-bred cows is higher among producers who have adopted few or none of the crop production innovations.[7] It means that producers do not want to add more risk by adopting several innovations. Producers take the risk of huge financial investment by adopting cross-bred cows. Thus, they would either adopt none or a few crop production innovations.

The relationship between the predicted probability of adoption and the number of adopting farmers is presented in Figures 7.3 and 7.4 for the Selale and Ada regions respectively. The result shows that with increases in the number of farmers who adopt innovations: i) the probability of adoption of cross-bred cows increases in both sites, ii) the probability of adoption of pesticides and fertilizer increases slightly, and iii) the probability of adoption of improved crop varieties decline or the probability with which most farmers adopt improved crop varieties is low. The difference in the probability of adoption of fertilizer and pesticides is very small among Ada farmers compared to Selale farmers. The probability of adoption of cross-bred cows increases sharply among farmers who have adopted four four technologies in the Ada region. It may imply that farmers who have adopted a large number of crop production innovations may gain a large amount of outputs. Thus, they may afford to take the risk of purchasing cross-bred cows.[8] The increase in the probability of adoption of cross-bred cows is small among Selale farmers relative to Ada farmers. Most Selale farmers have a large number of livestock. If they think that preconditions are lacking, they may decide not to adopt cross-bred cows. Selale farmers show a higher probability of adoption of cross-bred cows compared to Ada farmers. Thus, the marginal increase in the probability of adoption with increases in access to preconditions for the adoption of cross-bred cows may be small among Selale farmers compared to those in Ada.

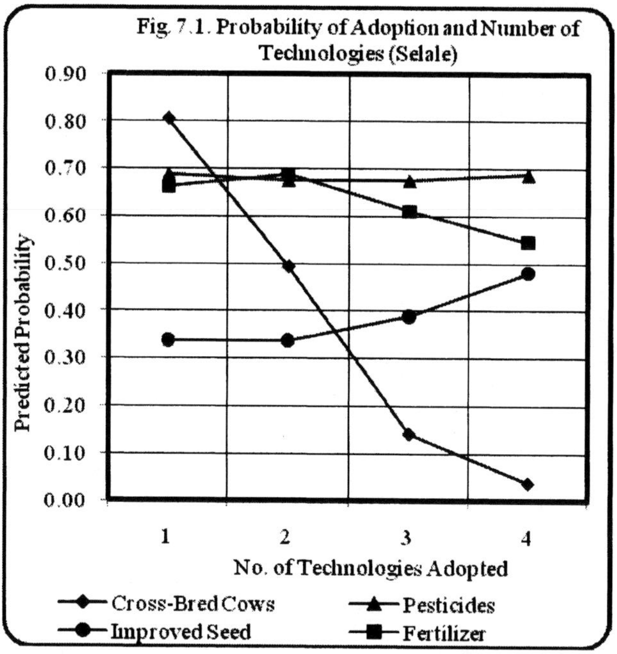
Fig. 7.1. Probability of Adoption and Number of Technologies (Selale)
Predicted Probability
0.90
0.80
0.70
0.60
0.50
0.40
0.30
0.20
0.10
0.00
1
2
3
4
No. of Technologies Adopted
Cross-Bred Cows
Pesticides
Improved Seed
Fertilizer

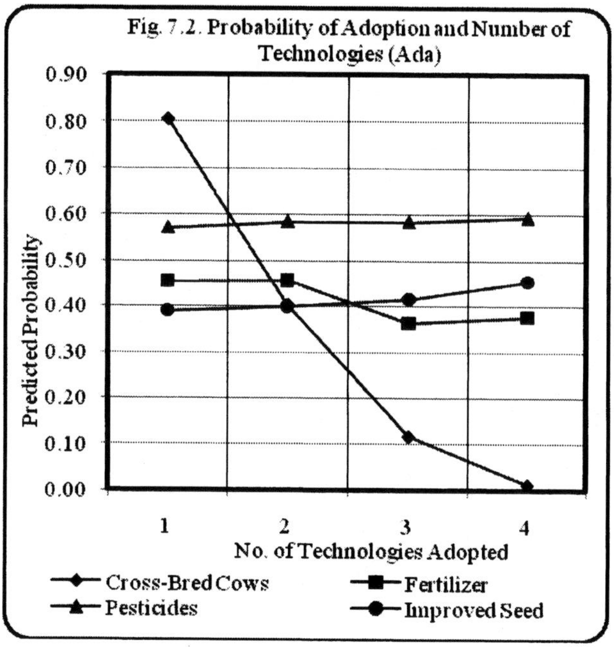
Fig. 7.2. Probability of Adoption and Number of Technologies (Ada)
Predicted Probability
0.90
0.80
0.70
0.60
0.50
0.40
0.30
0.20
0.10
0.00
1
2
3
4
No. of Technologies Adopted
Cross-Bred Cows
Fertilizer
Pesticides
Improved Seed

Fig. 7.3. Probability of Adoption and Number of Adopting Farmers (Selale)

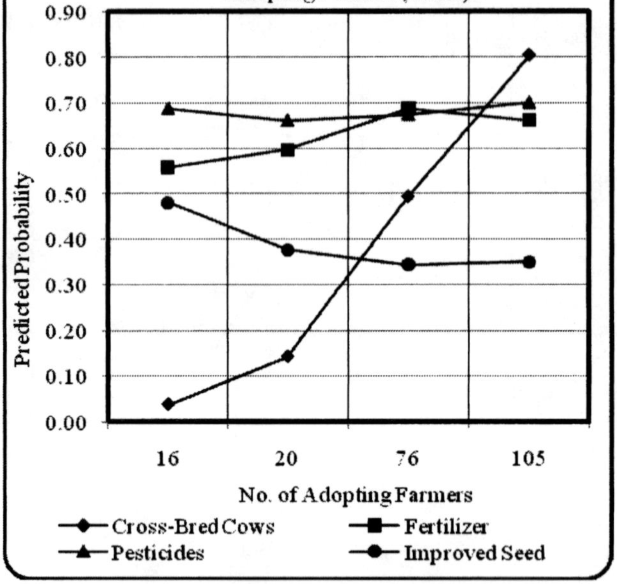

Fig. 7.4. Probability of Adoption and Number of Adopting Farmers (Ada)

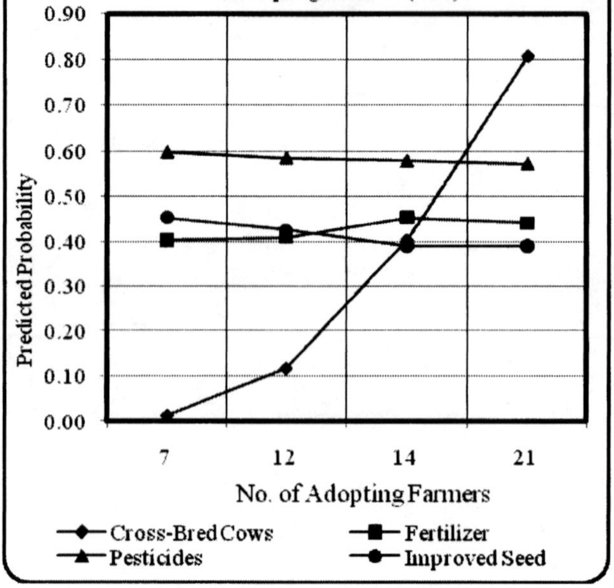

One method of assessing the number of farmers who have adopted specific innovations is the use of a frequency distribution. Table 7.6 presents the percentage of test and control farmers grouped by ranges and averages of probability classes. These probabilities are obtained from predictions of the estimated logistic regressions. The probability of adoption of cross-bred cows for the control group is calculated on the basis of their *willingness* to adopt rather than on the number of farmers who have actually adopted the new technologies. Eighty percent of *test* farmers in both sites show less than 0.6 probability of adoption of cross-bred cows. However, seventy percent of Ada farmers show an average probability of 0.3 while the same percentage of farmers in the Selale region have a probability of adoption 0.6 (Table 7.6). Seventy percent of *control* farmers in the Selale region would like to adopt cross-bred cows with a probability of 0.7 or more. The respective figure for the Ada region is fourteen percent. This implies that not only do households who have adopted cross-bred cows show low probability but there is also a small percentage of Ada farmers who are willing to adopt with a probability of at least 0.7. The probability of adoption of cross-bred cows for most control farmers in the Selale region is greater than 0.6. However, the opposite is true for Ada farmers. It implies that preconditions for adoption of cross-bred cows such as availability of grazing area and milk collection centre are lacking in the Ada region.

A sharp contrast is observed with respect to adoption of crop production technologies between Selale and Ada farmers. Sixty or more percent of Ada farmers adopt fertilizer with a probability of 0.8 while only 20-30 percent of Selale farmers adopt fertilizer with the same probability. Similar patterns are observed for adoption of improved crop varieties but the differences between regions are less than those observed for adoption of fertilizer. A relatively high percentage of farmers adopt pesticides with a probability of 0.7 or more. These results indicate that most Selale producers would rather adopt pesticides than other innovations. The results also confirm that most households would like to adopt crop production technologies whose effects on crop outputs are more certain before adopting technologies which either involve huge financial outlay (e.g., cross-bred cows) or whose outcomes are not certain (e.g., improved seed).

Various studies have concluded that wealth significantly influences decisions regarding adoption of innovations (Feder, et al., 1985; Molnar, et al., 1983). Contrary to the assumed risk avoidance behavior of small farmers, Cancian (1980) has shown that poor farmers are willing to take risk or innovate under uncertainty because they have less to lose. The rich farmers are more likely to innovate later, when uncertainty is less. Thus, uncertainty avoidance on the part of rich farmers can be reinterpreted as profit maximization because they can afford to let poorer farmers test out innovations. The statistical significance of wealth on adoption decisions confirms the hypothesis that it increases the probability of adoption of technologies (Tables 7.4 and 7.5). However, the influence of wealth on adoption in a regression analysis may not provide clear-cut evidence.

Table 7.6. Percentage of Producers and Probability of Adoption

Technologies\Ranges		Selale	Ada	Selale	Ada	Total	Total
Cross-bred cows	Mid Point	Test	Test	Control	Control	Selale	Ada
0.00-0.5	0.25	10.2	73.1	14.1	41.1	15	41.5
.51-0.6	0.555	71.8	7.7	10.9	28.3	39.7	15.6
0.61-0.7	0.655	0	7.7	7	10.4	7.8	12.5
0.71-0.8	0.755	9	11.5	15.7	8.5	12.9	9.1
0.81-0.9	0.855	7.9	0	25.7	2.3	18.5	10.3
0.91-1.00	0.955	1.1	0	26.6	3.4	16.1	8
Fertilizer							
0.00-0.5	0.25	29.1	10.8	18	13	28.4	11.7
.51-0.6	0.555	19.1	9.2	14.8	14.8	16.6	17
0.61-0.7	0.655	21.4	10	26.6	14.8	24.4	7.5
0.71-0.8	0.755	16.8	70	19.5	57.4	28.5	63.8
0.81-0.9	0.855	1.4	0	20.3	0	10.7	0
0.91-1.00	0.955	2.2	0	0.8	0	1.4	0
Improved seed							
0.00-0.5	0.25	36.7	13.1	20.6	15.9	23.7	14.5
.51-0.6	0.555	26.9	18.4	37.2	17.1	27	17.8
0.61-0.7	0.655	18.1	30.8	19.8	33.3	25.1	32
0.71-0.8	0.755	10.2	27.7	12	23.7	15.4	35.7
0.81-0.9	0.855	7.9	0	8.6	0	8.3	0
0.91-1.00	0.955	0	0	0.8	0	0.5	0
Pesticides							
0.00-0.5	0.25	4.4	2.3	7.7	6.3	7.3	4.3
.51-0.6	0.555	15.6	27.7	17.3	30	13.7	23.8
0.61-0.7	0.655	30	40	50	33.7	30	31.9
0.71-0.8	0.755	19.1	23	20	20	40	20
0.81-0.9	0.855	30.9	4	5	1	9	10
0.91-1.00	0.955	0	0	0	0	0	0

The relationship between wealth and the number of technologies adopted by region is presented in Figure 7.5. The result indicates that in both study sites, wealth increases with increases in the number of technologies adopted. In addition, the increases in wealth (vertical difference of bars corresponding to the number of technologies adopted) declines with increases in the number of technologies adopted for Selale region. However, with increases in the number of technologies adopted, the difference in wealth increases among farmers of the Ada region.

Producers gain information and knowledge about innovations from extension agents, city dwellers, the market places, radio, newspapers and magazines. This

Fig. 7.5. Wealth ('000 Ethiopian Birr) & Number of Technologies Adopted

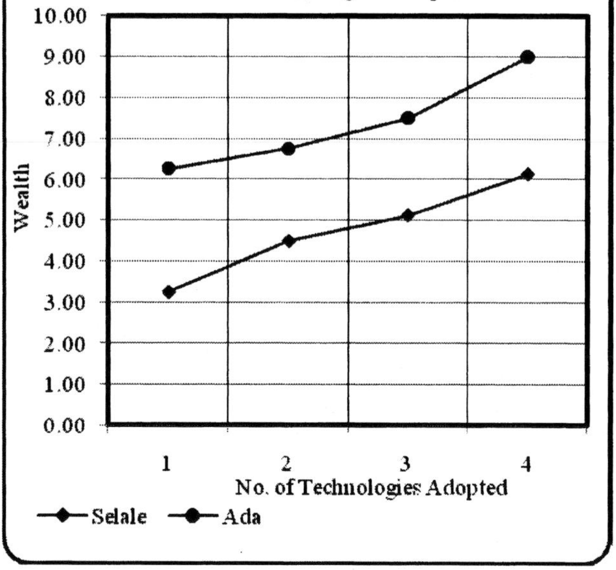

study hypothesized that the greater the amount of information subjected to conscious processing (hence more cognitive contribution), the better is the quality of decisions regarding adoption of innovations. The composite index of awareness about innovations or agricultural practices is plotted against the number of technologies adopted.

The number of technologies adopted increases with increases in the index of awareness. The difference in the value of this index declines among farmers of Selale region while it increases among farmers of Ada region (Fig. 7.6).

Neo-Chayanovians argue that with increases in the number of household members capable of working relative to those who are not (high worker:consumer ratio), households are better able to secure family survival (Durrenberger, 1984).[9] The present study hypothesizes that households with a high worker:consumer ratio can cultivate larger farms and adopt innovations more readily than households with a low worker:consumer ratio. The relationship between the worker:consumer ratio and the number of innovation adopted are presented in Figure 7.7. The result indicates that with increases in the worker:consumer ratio, the number of technologies adopted increases. However, the difference in worker:consumer ratio among farmers who adopted three or more innovations is less than those who adopted one or two technologies.

Fig. 7.6. Index of Awareness and Number of Technologies Adopted

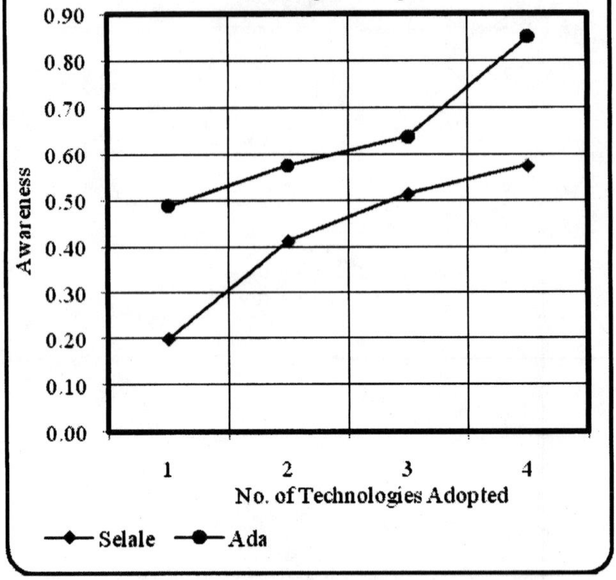

Fig. 7.7. Worker:Consumer Ratio and Number of Technologies Adopted

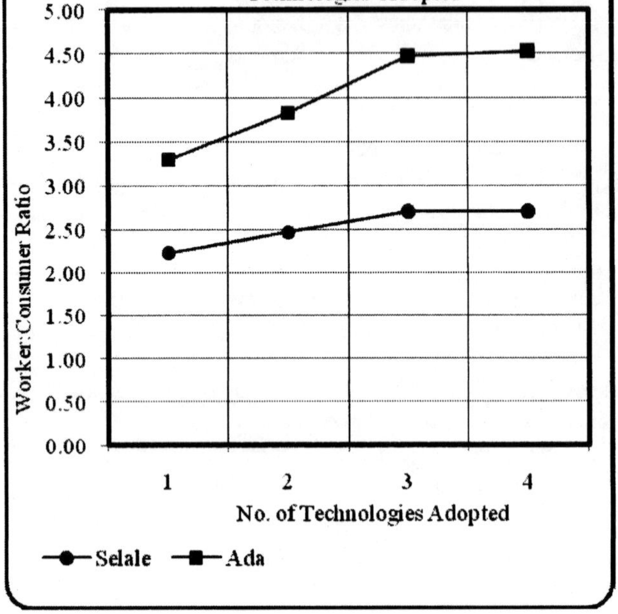

The number of relatives may exert significant influence on decisions regarding adoption of innovations. If innovations result in poor production, households with more relatives may receive more help (financial or physical) than those with few relatives. The strength of social relationship among relatives and friends is stronger in the Selale region compared to the Ada region. The result from Figure 7.8 indicates that with increases in the number of relatives, the number of technologies adopted increase in Selale. The larger the number of relatives, the more likely that households share their resources. A smaller number of relatives means fewer obligations to share resources. Thus, households can afford to adopt more technologies. This may be the reason for negative relationship between the number of relatives and the number of technologies adopted in the Ada region. That is, an increase in the number of relatives is associated with a decline in the number of technologies adopted (See Chapter III and VI, Figure 7.8.).

Fig. 7.8. Number of Relatives and Number of Technologies Adopted

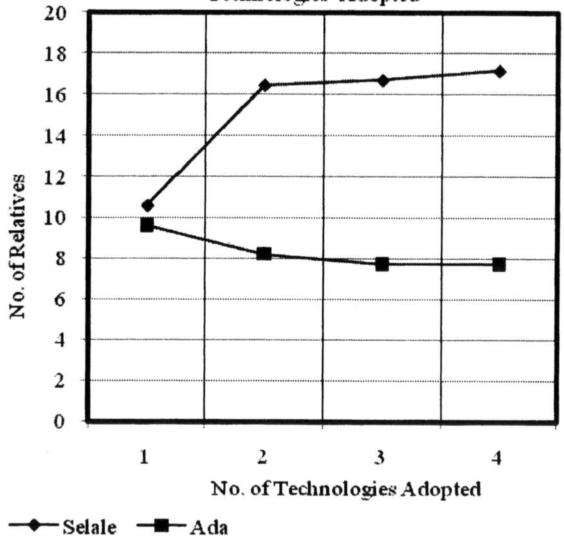

7.5. Measures of Risk Attitudes

The concept of risky choice has dominated the literature of decision analysis since Somer (1954). Since then, various methods of examining decisions under uncertainty have been developed (see Anderson, et al., 1977; Roumasset, 1976, Roumasset, et al., 1979; Barlett, 1980; Barry, 1984). Most of these methods, however, are not adequate to represent the decision-making processes of producers in LDCs. These methods are based on interviews and experiments of what a decision-maker will do if certain conditions are present or absent. They are characterized as hypothetical and consequently may not reflect actual decisions. The result obtained from these kinds of models may be of little value to design policies that can mitigate production risks of peasants.

Peasants have various choices of goals and strategies of decision-making. According to Schluter and Mount (1976), knowledge about strategies of households' decision-making is important in the formation of agricultural policy in LDCs. They found that farmers allocate land to different crops by striking some balance between the competing criteria of increasing income and decreasing risk measured by income variability. Thus, not only risk-taking behavior but also factors that shape this behavior are equally important in the design of intervention strategies.

Policies can help household decision-making by mitigating farm level risks through provision of production augmenting services or inputs (Dillon, 1986). However, in many LDCs, governments show unbounded capacity of inappropriate intervention into almost every sphere of economic activity (McNamara, 1985). The government is seen as an invader. When peasants do not receive physical or financial support to reduce the effect of crop or livestock risk, they make use of experience and village-level social and economic institutions. Measures of risk-aversion in these kinds of systems can be determined from decisions made in crop and livestock production.

Behavioral characteristics such as risk-taking attitudes are shaped by several socio-economic and technical factors. Technologies bear economic and social risks. Failure of a technology to provide enough for the family may result in indebtedness and lower social status. The higher the probability of these risks, the less acceptable the technology is to producers. To design programs to mitigate these risks, it is important to identify factors that influence risk-taking attitudes of decision makers. Instead of calculating measures of risk-taking behavior from utility functions, this study determines approximate values of the degree of risk-aversion from actual crop and livestock production activities of households.

In a mixed farming system, the risk of falling below the subsistence requirement (sensitivity index) is the result of failures in crop or livestock production (Chapter IV). Nevertheless, the degree to which producers are sensitive to the risk associated with these enterprises may differ. For instance, households with no working oxen may exchange their labor for oxen to produce for subsistence. If households have oxen but not an adequate supply of labor, they exchange their oxen for labor. Therefore, it is logical to examine the degree of risk-taking behavior by enterprise rather than by an aggregate measure.

In examining the livestock diversification index, two important considerations need to be addressed. The first one is with respect to environmental suitability or regional advantages for livestock production. Selale farmers prefer to increase their stock of different classes of cattle (e.g., oxen, cows) rather than diversifying into different types or species livestock (e.g., sheep, goat, draft, and cattle). For instance, farmers of the Selale region own a larger proportion of cows in their stock compared to Ada farmers (Tittarelli, 1990; Finnida, 1988; Gryseels, et al., 1983). Ada farmers, on the other hand, prefer to increase the percentage of oxen in their stock (Gryseels, et al., 1983, 1985). Secondly, a larger proportion of household's wealth is embodied in cattle than in small ruminants or draft animals. Therefore, in calculating the livestock diversification index, cows, oxen, steer, heifer and calves are considered as a separate class similar to sheep, goats or draft animals.

7.5.1. Risk Sensitivity Index

A risk sensitivity index, is the amount of income necessary to avoid falling below some threshold or disaster level. This threshold is different between households. The higher the value of the disaster income level the more risk-averse a household is, and vice versa.

Assuming that a household cares about its net income (without deducting home consumption and remunerating family labor), a risk sensitivity index is calculated per hectare of crop land and per cow. This index measures the income required per unit of land or cow to avoid the necessity of the farm family selling some part of its non-liquid assets.[10] For simplicity, the computed values of risk sensitivity indices are scaled such that the indices range between 0 and 1. The results are presented in Table 7.7.

Table 7.7. Percentage of Producers by Classes of Crop and Livestock Sensitivity Indices (Crop indices and livestock indices)

	Selale Test	Ada Test	Selale Control	Ada Control	Total Selale	Total Ada
Degree of risk-aversion grouped by crop indices[1/]						
Risk-neutral	0.00	0.00	0.00	0.00	0.00	0.00
Low risk-averse	36.50	76.90	46.90	85.20	31.30	81.10
Moderate risk-averse	73.50	19.30	43.10	14.80	68.70	17.00
Extreme risk-averse	0.00	3.80	0.00	0.00	0.00	1.90
Degree of risk aversion grouped by livestock indices[1/]						
Risk-neutral	0.00	0.00	0.00	0.00	0.00	0.00
Low risk-averse	86.50	35.40	33.90	35.60	82.00	36.10
Moderate risk-averse	16.50	54.60	56.10	64.40	17.70	64.90
Extreme risk-averse	0.00	0.00	0.00	0.00	0.00	0.00
N	88	26	127	27	216	52

1/ Respectively, risk-neutral, low, medium and high risk-aversion groups correspond to risk-sensitivity or disaster level intervals of 0-0.01, 0.011-0.3, 0.31-0.6 and 0.61-1.

Following classical approaches, producers are grouped by their degree of risk-aversion into four intervals. These intervals, 0-0.01, 0.011-0.3, 0.31-0.6 and 0.61-1 represent risk-neutral, low, medium and high risk-averting farmers respectively.[11]

The result for crop production enterprise shows that most producers in the Ada area fall in the low risk-aversion group. The risk-averse behavior of most Selale farmers can be seen as moderate relative to Ada farmers. The result for livestock production indicates that most producers of the Selale region fall in the low risk-aversion group while those of Ada exhibit medium to high degree of risk-aversion. The implication of these findings is that the risk-taking behavior of producers takes into account farming experience and resources available within the farming system. Ada farmers are more knowledgeable of crop production and invest more in crop than in livestock production. In other words, Ada producers have extensive land- and location-specific experience (Ekanyake, et al., 1989). The same conclusions can be drawn for livestock production in the Selale region.[12] The low risk-averse behavior of producers in Ada and Selale regions with respect to crop and livestock production respectively confirms these arguments.

7.5.2. Diversification Indices

A further characterization of the risk-taking behavior of households is obtained from the crop and livestock diversification indices. Crop diversification or planting several crop varieties generally increases with risk-aversion (see Jodha, et al., 1983). However, this pattern of resource allocation may not necessarily indicate risk-averse behavior of households (Norman, 1978, 1980).[13] Producers plant several crops to satisfy the family food requirements, to match farming operations with the different growing time requirements of crops and to maximize the use of labor and other resources (Ayalew, 1986; Chapter V and VI).

The crop diversification index ranges between 0 and 1. A zero diversification index implies that households plant only a single crop.[14] With increases in the number of crops planted, the index increases, reaching a maximum of one. The results of the crop diversification index are presented in Table 7.8. Respectively, sixty-three and twenty-two percent of the study framers in the Selale and Ada regions have a crop diversification index of at least 0.6. This implies that most producers in the Ada region are characterized as more risk-averters than Selale farmers. However, the same finding may have a different interpretation. Households in the Ada region are more knowledgeable about crop production. Their cropping pattern may be designed not only to minimize losses that may occur from planting a single crop but also to maximize income per hectare by planting crops with different water and soil requirements, and market prices.

With respect to the livestock production index, most producers are either moderately or extremely risk-averse. It means that they own more than one class of cattle and/or livestock. For producers who have the resources and expertise to raise livestock, owning more classes of livestock may not imply a risk-averse behavior.

Table 7.8. Percentage of Producers by Crop (Cropdiv) and Livestock Diversification Indices (Lsdiv)

	Selale Control	Ada Test	Selale Control	Ada Control	Total Selale	Total Ada
Degree of risk-aversion grouped by Crop diversification indices1/						
Risk-neutral	0	0	7.8	0	0	0
Low risk-aversion	38	26.9	27.2	17	37.5	32.1
Moderate risk-aversion	32.5	18.2	25.1	23.6	39.8	25.5
Extreme risk-aversion	20.5	54.9	39.9	59.4	22.7	62.7
Degree of risk-aversion grouped by Livestock diversification indices1/						
Risk-neutral	0	0	0	0	0	0
Low risk-aversion	6.4	34.6	27.3	44.4	17	44.2
Moderate risk-aversion	30.1	46.8	22.1	43.8	29.3	39.7
Extreme risk-aversion	63.5	18.6	50.6	11.8	53.7	16.1
N	88	26	127	27	216	52

1/ Respectively, risk-neutral, low, medium and high risk-aversion groups correspond to diversification index intervals of 0-0.01, 0.011-0.3, 0.31-0.6 and 0.61-1.

This pattern of resource ownership can be considered as a means of creating insurance against economic and social risks or a strategy of risk-spreading. It may also be a strategy to attain goals such as being a good farmer, utilizing resources efficiently and maximizing income from livestock.

Risk-averseness with respect to livestock production, under conditions where the potentials for expansion of this enterprise are binding, may be reflected by a large percentage of farmers who own one or two classes of livestock (e.g., Ada). Selale farmers own more classes of livestock when compared with Ada farmers. Fifty or more percent of Selale producers are extremely sensitive to risk (risk-averters). Group discussions revealed that most Selale producers have at least a cow, an ox, a calf and a heifer/steer. However, an average Ada producer owns a cow and an ox only. As a result, the diversification index for Selale farmers is greater than the index for Ada farmers. The fact that the largest proportion of the wealth of Selale producers is embodied in livestock may induce households of this region to be more sensitive to fluctuations in livestock number or output (i.e., they become risk-averters).

Graphical representation of both kinds of risk show that with increases in the number of technologies adopted both kinds of risks tend to increase (Figures 7.9. and 7.10). It means that intervention strategies can succeed only if they provide a

Fig. 7.9. Diversification and Number of Technologies Adopted

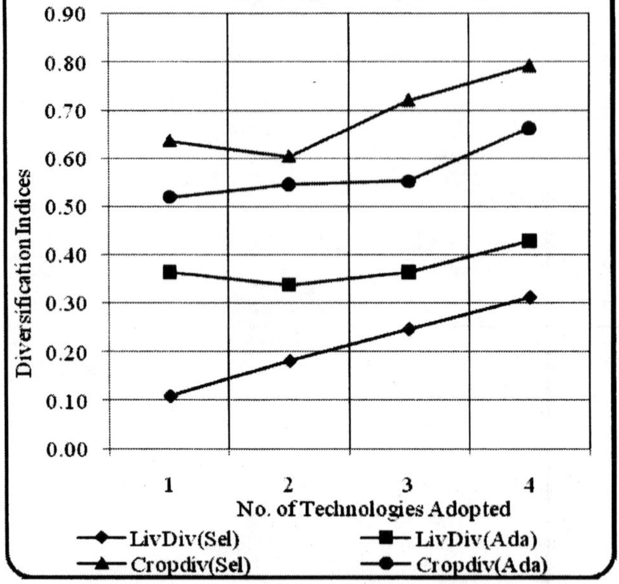

Fig. 7.10. Sensitivity Indices and Number of Technologies Adopted

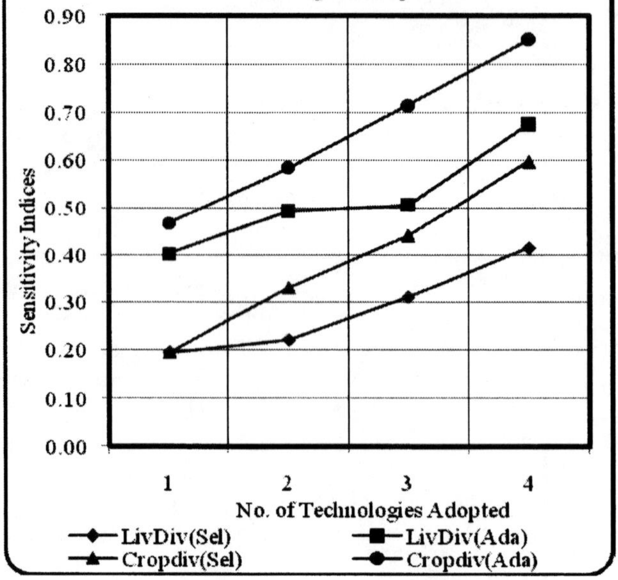

means to reduce livestock production risks or if there are “modernizing conditions”.

7.5.3. Risk and Social, Economic and Technical Factors

Examining the risk-taking behavior of households with the help of such measures as sensitivity or diversification indices is part of the process of acquiring information that may help to design policies to reduce risks associated with farming. Identifying variables that influence this behavior of households is the second part or process. The second part examines the relationship between the indices of risk-taking behavior and socio-economic and technical factors in a regression framework. Preliminary findings indicate that the effects of variables included in the regression models do not significantly change whether the dependent variable is the sensitivity or diversification index. The results of regressing the sensitivity index on selected socio-economic and technical variables are presented in Table 7.9.

The larger the family size, the more the amount of production needed to satisfy the family food requirements. Family size, expenditure and the number of technologies adopted increase the burden of household decisions in securing subsistence, thus increasing a households' degree of risk-aversion. Production and marketing knowledge, awareness, experience and schooling help to better understand problems and opportunities associated with crop or livestock enterprises. The better households understand the impact of various factors on crop production, the better able they are to design strategies to reduce the risk of not securing their subsistence requirements. Thus, factors that enable producers to comprehend new information and ideas and improve management styles reduce the risk-averse behavior of households.

The more relatives households have, the more likely that they will receive support if there are failures in crop or livestock production, or if they want to invest in new technologies. This assertion holds true only for the Ada region. Chapters II, III, V and VII have indicated that relationship between households is stronger in the Selale region than the Ada region. Households in the Selale region have greater responsibility to help relatives. Thus, the more relatives that Selale producers have, the more risk-averse they will be. In the Ada region, weaker relationship among relatives implies that households can invest in innovations without considering sharing their resources with relatives. Furthermore, the average number of relatives per household in the Ada region is less than that in the Selale region. Thus, relatives may have a risk-decreasing effect in the Ada region. An alternative interpretation of this result is that relatives serve as insurance, therefore exert a risk-reducing effect. However, observations of the relationship among households during this study period (1990/91) in the Ada region have not proven the validity of this assertion.[15]

Table 7.9. The Impact of Socio-Economic Variables on Crop Sensitivity Indices[1/]

			Selale		Ada	
	All Selale Farmers	All Ada Farmers	Control	Test	Control	Test
Intercept	0.162	-0.041	0.176	0.148	-0.565	0.11
	(4.196)*	(-0.188)	(3.007)*	(2.665)*	(-0.331	-0.65
Family size	0.501	0.103	0.003	-0.012	0.214	0.018
	(2.174)#	(2.357)#	-1.199	(-1.166)	(2.91)*	(2.317)#
Schooling	-0.001	-0.04	-0.007	0.009	-0.069	-0.025
	(-0.123	(-0.951)	(-0.566)	-0.923	(-0.62)	(-0.935)
Wealth	-0.121	-0.209	-0.107	-0.205	-0.104	0.091
	(-4.108)*	(-2.337)#	(-2.652)*	(-0.041)	(-1.986)#	-0.917
Experience	-0.281	-0.102	-0.304	-0.361	-0.323	-0.221
	(-1.964)#	(-2.015)#	(-2.603)#	(-2.204)#	(-2.497)#	(-1.437)
Awareness	-0.026	-0.169	-0.061	0.012	-0.391	-0.088
	(-2.406)#	(-2.426)#	(-2.81)*	-0.827	(-1.883)	(-0.939)
Relatives	0.401	-0.313	0.241	-0.171	-0.515	-0.002
	(2.929*	(2.636)*	(2.558)*	(-2.61)*	(-1.84)	(-0.492)
Expense	0.119	0.001	0.276	0.472	0.012	0.102
	(4.8)*	-0.198	(3.663)*	(3.231)*	-0.966	-1.119
Marketing knowledge	-0.001	-0.013	0.001	-0.005	-0.156	-0.41
	(-0.432	(-0.953)	-0.208	(-1.294	(-2.549)*	(-2.25)#
Crop production knowledge	-0.401	-0.605	-0.103	-0.211	-0.622	-0.511
	(-3.604)*	(-1.297)	(-2.692)*	(-2.31)#	(-2.704)*	(-2.964)*
Off-farm income	-0.012	-0.201	-0.019	-0.047	-0.051	0.17
	(-4.57)*	(-2.073)#	(-2.414)#	(-3.664)*	(-0.666)	-0.571
No of technologies adopted	0.301	0.414	0.403	0.276	0.17	0.039
	(2.102)#	(2.49)#	(2.303)#	(2.522)#	(1.978)#	-0.893
Soil characteristics	-0.408	-0.454	-0.317	-0.003	-0.199	-0.066
	(-1.972)#	(-1.952)#	(-1.999)#	(-0.396)	(-1.986)#	(-3.09)*
Crop Area	-0.512	-0.415	-0.219	-0.209	-0.434	-0.312
	(-4.694)*	(-2.117)#	(-3.029)*	(-3.102)*	(-2.52)#	(-2.545)#
Plot Characteristics	0.026	0.144	0.439	0.535	0.007	0.168
	-1.415	-1.403	(2.447)#	(2.283)#	-0.026	-2.332
N	216	52	88	127	25	26
R2	0.52	0.68	0.56	0.57	0.79	0.89

1/ Values in parenthesis are t-statistics

* and # indicate statistical significance at 1 & 5 percent respectively.

Economic variables such as wealth and off-farm income provide households with a leverage against undesirable events or outcomes. Wealthy and high income earners need less money for investment to avoid falling below the minimum subsistence requirements. The larger the crop area, the less the need to intensify production to secure subsistence. Hence, income needed per unit of crop land is less among producers with a relatively large crop area. The more expense households incur, the less money is available for investment. The results suggest negative (risk-reducing) effects of wealth, farm size, off-farm income, and knowledge variables, and positive (risk-increasing) effects of household expenses and family size.

Technical variables such as soil and plot characteristics are sources of risk in crop production.[16] Fertile soils not only require less oxen input but also retain moisture and contain more nutrients, hence are more productive. Fertilizer and seed inputs are less exposed to soil erosion on fertile soils. Therefore, soil type influences success in using crop production inputs. The more fertile a plot is, the less the investment required per unit of land. The more difficult a plot is to work, the more physical investment (oxen and labor inputs) required to produce for subsistence. The results suggest that the higher the value of soil characteristics, the less the sensitivity of households to fluctuations in income or production. The higher the value of plot characteristics (e.g., difficult to work), the more sensitive households will be to fluctuations (more risk-averse).

The more technologies adopted, the more the amount of income needed to keep the family from falling below subsistence requirements. Similar to expenses, this variable is positively associated with the risk sensitivity index.

The direction and level of significance of social, economic and technical variables on livestock sensitivity indices follow similar patterns to the model in which the dependent variable is the crop sensitivity index (Table 7.10.). Crop production knowledge and farm size in the crop sensitivity index model are replaced by livestock production knowledge and feed area. Wealth and off-farm income are expected to have a risk-reducing effect. Wealth in this study is equivalent to gross farm income. Most households receive only a small amount of off-farm income because of the absence of employment opportunities outside agriculture.

Investments in livestock production are expensive. Households who have invested in livestock technologies exhaust most of their gross farm and off-farm income. Thus, they may require a larger amount of income to protect their families from falling below subsistence requirements. That seems to be the reason for the positive (risk-increasing) effect of wealth and off-farm income on the livestock sensitivity index.

The results in Tables 7.8 and 7.9 indicate that production knowledge exerts a consistent risk-reducing effect in both study regions. Crop and livestock production knowledge helps to reduce the amount of income invested per unit of land or per cow through better management techniques. The magnitude of their impact varies

Table 7.10. The Impact of Socio-Economic Variables on Livestock Sensitivity Indices[1/]

			Selale		Ada	
Variables	All Selale Farmers	All Ada Farmers	Control	Test	Control	Test
Intercept	1.7	1.9	1.96	1.1	4.612	0.972
	(3.286)*	-1.292	(2.105)#	(2.051)#	-1.411	(2.132)#
Family size	0.023	0.0496	0.033	0.09	0.4	-0.188
	-0.194	-1.154	-0.177	(0.051	-1.43	(-0.296)
Schooling	-0.480	-0.130	-0.21	-0.820	-0.136	-0.547
	(-2.792)*	(-0.474)	(-2.205)#	(-2.918)*	(-1.025)	(-1.96)#
Wealth	0.182	0.312	-0.451	-0.012	-0.721	-0.605
	(3.084)*	(3.536)*	(-2.225)#	(-1.965)#	(-3.087)*	(-3.712)*
Experience	-0.9	0.121	-0.552	-0.16	-0.651	-0.168
	(-2.615)*	(1.946)#	(-3.28)*	(-1.993)#	(-2.824)*	(-0.90)
Awareness	-0.38	-0.026	-0.168	-0.093	-0.056	-0.376
	(-2.326)#	(-0.314)	(-2.244)#	(-1.97)#	(-0.789)	(-1.983)#
Relatives	-0.57	-0.063	-0.135	-0.496	-0.258	0.279
	(-1.962)#	(-2.194)#	(-1.985)#	(-1.995)#	(-2.092)#	(2.838)*
Expense	0.6	0.4	0.6	0.5	0.221	0.8
	(3.503)*	-0.857	(2.523)#	(2.014)#	-1.16	-0.823
Marketing knowledge	0.08	-0.621	-0.422	0.51	-0.383	-0.881
	-1.41	(-2.645)*	(-1.533)	-0.306	(-1.96)#	(-2.936)*
Off-farm income	0.101	-0.061	0.7	0.313	0.412	0.04
	(5.114)*	(-1.322)	(2.7)*	(4.253)*	(4.138)*	-1.185
No of technologies adopted	0.611	0.023	0.029	0.08	0.425	0.431
	(1.986)#	-0.501	-1.34	-1.757	(1.969)#	(2.907)*
Livestock production	-0.564	-0.172	-0.689	-0.462	-0.0691	-0.189
	(-2.258)#	(-1.774)	(-3.532)*	(-2.387)#	(-1.087)	(-1.667)
Feed area	-0.501	-0.43	-0.52	-0.081	0.125	0.48
	(-2.59)*	(-1.986)#	(-2.734)*	(-2.204)#	-1.917	-1.923
N	216	52	88	127	25	26
R^2	0.55	0.69	0.57	0.59	0.65	0.63

1/ Values in parenthesis are t-statistics.

* and # indicate statistical significance at 1 and 5 percent respectively.

between regions. The impact of crop production knowledge is larger in the Ada region, while that of livestock production knowledge is larger in the Selale region.

Chapters II through VI have indicated that the potential for, and experience in, crop production is extensive in the Ada region compared to the Selale region. Conversely, the potential for and experience in livestock production in the Selale region is greater than in the Ada region. Attributes of a good farmer in the Ada and Selale regions are closely linked to crop and livestock production respectively.

Thus, Ada farmers attain higher scores in crop production and Selale farmers receive higher scores in livestock production problems. This inter-regional difference in production knowledge seems to be the reason for differential impact of knowledge on measures of risk-taking behavior. The measure of goodness of fit or adequacy of the model is not high. Nevertheless, this is expected for cross-sectional data of this sort.

7.6. Risk and Adoption of Technologies

Classifying farmers into different risk-taking groups and evaluating factors influencing this behavior of households are important to assess what impact risk may have on adoption decisions. Before evaluating the impact of risk on adoption decisions, correlation analysis is performed to detect problems of multicollinearity.[17] As a rule of thumb, one of any pair of variables with a correlation coefficient of 0.6 and over is excluded from the analysis. The results of logit regression are presented in Table 7.11.

Most variables exert a positive and significant effect on adoption of all technologies, except on improved crop varieties. Family size and household expense exert significant and negative impacts on adoption of most of the technologies. Knowledge variables (education, awareness, experience and production knowledge) exert positive and significant impacts on adoption decisions.

The performance of improved crop cultivars managed by farmers is poor especially in regions which have less access to complementary inputs such as fertilizer and pesticides (e.g., Selale). Farm size as an input to crop production shows negative association with adoption of improved crop cultivars in the Selale region. Natural pasture or grazing lands in the Ada region are marginalized because of over grazing and erosion compared to Selale (Gryseels, et al., 1983; Tagoe, 1983). Thus, feed area is negatively associated with adoption of cross-bred cows in the Ada region. Risk sensitivity index exerts a negative and significant effect on adoption of most of the technologies. Similar results are observed when the diversification index is used in the analysis.

Despite the controversies regarding the interpretation of risk-taking behavior, this chapter has demonstrated that risk-taking behavior calculated from actual production decisions of households negatively influences adoption decisions. This is true not only because of the newness of innovations but also because of environmental or resource constraints that inhibit peasants from attaining the maximum potential of innovations.

Risks from adoption of innovations can be insured by social networks, by an individual's socio-economic status, and by the government. In dryland agriculture, all producers are subjected to a common source of risk, namely rainfall variability. Under these farming conditions, the first two sources of insurance may not buffer

Table 7.11. Risk and Adoption of Technologies[1/]

Variables	Cross bred Cows		Fertilizer	
	2.122	0.954	Ada	Ada
Intercept	(4.803)*	(2.987)#	4.182	3.413
	-0.462	-0.14	(5.772)*	(4.986)*
Family size	-0.129	-0.236	-0.096	0.92
	0.328	1.055	-1.082	(2.731)#
Education	-1.425	(4.340)*	0.213	0.021
	0.001	-0.102	-1.572	-0.417
Awareness	-0.09	-0.111	-0.203	0.609
	0.231	-0.029	-1.426	(2.919)#
Experience	(2.981)#	-0.43	0.399	0.871
	0.572	-0.392	(2.941)#	(3.214)#
Relatives	(2.721)#	(4.815)*	0.609	0.508
	0.174	0.231	(4.002)*	(5.201)*
Production Knowl-	(4.099)*	-1.031	0.322	-0.29
edge	-0.474	-0.571	(6.103)*	-1.109
Risk	(2.859)#	(4.163)*	-0.594	-0.509
	0.408	-0.024	(2.833)#	(2.735)#
Wealth	(2.995)#	-1.311	0.95	0.216
	-0.781	-0.268	(4.983)*	(3.321)#
Expense	(6.902)*	-1.427	-0.331	-0.731
	0.177	0.255	(2.949)#	(2.994)#
Off-farm	-1.342	(2.972)#	0.129	0.927
	0.991	-0.467	-1.999	(4.989)*
Farm/Feed Size	(6.457)*	(2.958)#	0.241	0.724
size	217	52	(3.001)#	(2.91)#
N	2.122	0.954	217	52

1/ Values in parenthesis are Wald's Chi-square.

* and # indicate statistical significance at 1 and 5 percent respectively.

2/ Most of these variables significantly influence measures of risk-taking behavior (Table 7.10). Naturally, inclusion of these variables and measures of risk-taking behavior in a single model would cause collinearity and lack of convergence. That is why correlation analysis is carried out to exclude some of the variables that are highly correlated with sensitivity or diversification indices. For instance, wealth, marketing knowledge and number of technologies adopted were highly correlated with risk sensitivity index. Therefore, these variables were excluded from the above mode.

Table 7.11. Risk and Adoption of Technologies (Continued)

Variables	Improved Seed		Pesticides	
	Selale	Ada	Selale	Ada
Intercept	2.428	1.923	1.201	3.942
	(5.437)*	(4.342)*	1.1131	(3.501)*
Family size	-0.326	-0.096	-0.197	-0.429
	-1.899	(3.093)#	-1.036	(3.041)#
Education	0.529	0.211	0.029	0.422
	(3.012)#	(2.945)#	-0.249	(1.992)#
Awareness	-0.197	0.138	-0.211	0.821
	-1.728	-0.209	-1.209	(3.401)#
Experience	0.001	0.591	0.217	0.631
	-0.039	(2.724)#	-0.932	(2.991)#
Relatives	0.004	0.09	0.31	-0.201
	-1.009	-1.004	(2.997)#	-1.074
Production Knowl-	0.035	0.021	0.481	0.398
	(2.97)#	-1.679	(2.948)#	(3.757)#
Risk	-0.601	-0.415	-0.401	-0.648
	(3.01)#	(2.945)#	(2.949)#	(4.653)*
Wealth	0.219	0.241	0.925	0.092
	(2.95)#	(4.91)*	(4.904)*	-1.751
Expense	-0.209	-0.921	-0.167	-0.329
	-1.781	(4.39)*	(3.092)#	-0.97
Off-farm	0.227	0.401	-0.124	0.001
	(3.82)#	(3.023)#	-1.3201	-0.077
Farm/Feed Size	-0.182	0.647	-0.019	-0.459
size	-1.891	(6.783)*	-1.009	(6.997)*
N	217	52	217	52

1/ Values in parenthesis are Wald's Chi-square.
* and # indicate statistical significance at 1 and 5 percent respectively.

peasants sufficiently from risks. Therefore, governmental or non-governmental assistance should accompany innovations to mitigate the effects of risks, thereby increasing food production (see Chapters V and VI).

7.7. Summary

This chapter investigated the impact of social, economic and technical factors on decisions to adopt new technologies. It examined the risk-taking behavior of households and how this behavior may affect adoption decision. Empirical results from adoption studies indicate that schooling, experience, extension contacts, wealth, off-farm income, crop/grazing area, the number of relatives, the worker:consumer ratio, and productivity of technologies exert positive and significant effects on the adoption of most technologies in both study sites.

The worker:consumer ratio is hypothesized to exert a significant influence on technologies that require large labor inputs. The result of this study suggests that the impact of this variable relates more to the objective of securing adequate food grain than to the specific labor requirements of new technologies. That is, its impact is higher on food crops than on cross-bred cows that require greater labor input per head compared to local breeds of cows. The impacts of production knowledge and experience on adoption decisions are not only positive but larger than most variables. This is particularly true on adoption of innovations which households believe will have more certain outcomes (e.g., fertilizer and pesticides) than on adoption of innovations with uncertain outcomes (e.g., cross-bred cows and improved seed).

In regions where modernizing conditions or preconditions are not satisfied, the introduction of innovations may be met with more failure rather than success. Consequently, adoption of cross-bred cows in the Ada region and improved crop varieties in both regions are negatively influenced by several variables when compared with fertilizer and pesticides. Sequentiality of adoption holds for technologies which are either proven to be successful or do not involve huge financial investments. Typical examples are fertilizer and pesticides as opposed to improved crop varieties and cross-bred cows.

With respect to prospects for the diffusion of innovations, the result of this research indicate that with increases in the number of farmers adopting innovations, the probability of adoption of cross-bred cows, pesticides and fertilizer increases. Moreover, the findings of this study indicate that the probability of adoption of cross-bred cows increases sharply among farmers who have adopted most of the crop production technologies.

The research results from the Selale and Ada regions of Ethiopia indicate that with increases in wealth, not only do the number of technologies adopted increase but also the distribution of that wealth among households becomes more uniform. Awareness of new innovations helps to increase adoption. The results of this research indicate that with increases in the index of awareness the number of technologies adopted increases.

A larger worker:consumer ratio increases the chances of securing the family food requirements. Increases in this ratio are positively associated with increases in the number of technologies adopted.

As opposed to the usual assertion that relatives serve as an insurance for households which have adopted innovations, more relatives are associated with a decline in the number of technologies adopted in the Ada region. However, in the Selale region increases in the number of technologies adopted are associated with increases in the number of relatives. This is probably due to the stronger relationship and cooperation among relatives in the Selale region compared to the Ada region.

Risk-taking behavior of households in the Selale and Ada regions is examined using information on actual production decisions. Crop and livestock risk sensitivity and diversification indices are calculated. Households are categorized into four groups: risk-neutral, low, moderate and extreme risk-averters. The results of sensitivity and diversification indices indicate that risk-taking behavior varies by the concentration of farmers in specific enterprises and by the regional possibilities that exist for either crop or livestock production. With respect to crop sensitivity indices, most Ada farmers are in the low risk-aversion group while those of Selale in moderate to high risk-aversion group. The opposite holds true for livestock sensitivity indices. Producers are willing to take risk in enterprises where they have location- and enterprise-specific experience. This experience is reinforced by availability of resources for expansion of a particular type of farming or enterprise (Ekanyake, et al., 1989). Thus, Ada and Selale farmers may be willing to take more risk related to crop and livestock production respectively.

With respect to crop diversification indices, most producers in the Ada region are relatively more risk-averse compared to Selale farmers. Most study farmers are either moderate or extreme risk-averters when the index of classification is livestock diversification. Selale farmers have extensive experience and resources for livestock production. They are hypothesized to take more risks in livestock production when compared with crop production. However, when each sub-class of cattle (e.g., oxen, cows, heifer, steer and calf) is considered as a class, Selale farmers exhibit a higher index, hence are more risk averters than Ada farmers. With increases in the number of technologies adopted both kinds of risks tend to increase.

The process of characterizing risk-taking behavior does not necessarily specify the factors which shape this form of household behavior. The results of this study indicate that production knowledge exerts a consistent effect of increasing the risk-taking or reducing the risk-averse behavior of households in both regions. The impact of crop production knowledge is high among the Ada farmers while that of livestock production knowledge is higher among the Selale farmers.

Risk-averse behavior of households reduces the probability or chances of adopting technologies in both Selale and Ada regions. Risk-averse behavior can be reduced through insurance from social-networks, wealth of households and the government. However, farmers of the Ethiopian highlands are commonly exposed to the vagaries of crop and livestock risks caused by variability in amount and distribution of rainfall. Consequently, governmental or non-governmental assistance that accompanies innovations mitigate the effects of risk, increase the probability of adoption of technologies and may contribute to increases in food production.

7.8. Notes

1. Control farmers were asked about their willingness to reduce local breeds of cattle if they received cross-bred cows. Test farmers own cross-bred cows. However, they have not reduced their local stock. Thus, they were asked about their plan to reduce local breeds of cattle.

2. Relatives include brothers, sisters, uncles and aunts of the husband and wife, and their spouses; and members of households who are intermarried with relatives of the head or wife of another household. They are included in the study if they live within an area under three PAs (i.e. 2400 ha). Group discussion with farmers revealed that relatives living within this geographic limit cooperate on a regular basis.

3. Unlike production-increasing inputs such as fertilizer, investment in livestock production technology is risky because it can not be adopted in a sequence of small amounts. This is a typical characteristic of indivisible or lump-sum technologies. Thus, the rate of adoption of livestock production technologies is low (Jamison and Lau, 1982; Feder, et al., 1985).

4. In fact, group discussion with farmers has revealed that farmers who do not own a radio have less knowledge about factors outside the control of households that influence agricultural production. Although the index involves biases because of the manner in which the variables are aggregated, it is expected to provide some measure about the influence of outside information which would not have been studied otherwise.

5. The dependent variable in the logistic regression is the logarithm of the odds that a particular choice (i.e. adoption of the technology in question) will be made. Thus, the interpretation of the results must be done with care. A one percent increase in the percentage of `x' variable (continuous) will lead to an increase of `β' (estimated coefficient of x) in the logarithm of the odds that a farmer will adopt the `y' technology. The effect of a dummy variable `D' on the probability of adoption of the y technology, however, approximately equals: estimated coefficient of D (Γ) H (the odds ratio) or Γ H (ln [Pi(N|Xi)/Pi(T|Xi)]) (see Chapter IV; Appendix 6a; Pindyck and Rubinfeld, 1981).

6. For the pattern of the impact of purchase price and marketing knowledge on adoption decisions, see discussion under section 7.3.2.

7. This result does not necessarily contradict the sequential pattern of adoption decisions discussed earlier. Farmers of both regions indicated that they will adopt cross-bred cows after they adopted fertilizer and pesticides. This means that the probability of adoption of cross-bred cows may increase with increases in the number of technologies adopted. A possible explanation for the decline in the probability adoption of cross-bred cows with increases in the number of technologies adopted may be that adoption of fertilizer and pesticides have not contributed to increases in crop production. Thus, producers do not want to jeopardize family food requirements by investing in an expensive technology such as adoption of cross-bred cows.

8. This should not be confused with an earlier result that with increases in the number of technologies adopted, the probability of adoption of cross-bred cows declines. The argument made here is with respect to the number of farmers who may adopt rather than the probability with which they adopt cross-bred cows.

9. Chayanov's original thesis was that a low consumer:worker ratio increases effort to feed

more mouths.

10. According to Kminsky (1979), we can talk about risky conditions when we are referring to events or outcomes subject to a probability distribution which is assumed to be known. Thus, risk cannot be defined as falling below a subsistence level or not. Rather, a farmer is operating under risky conditions and trying to acquire insurance against undesirable events by directing the decision-making process towards the goal of not falling below subsistence. Instead of examining the role of risk in agriculture, therefore, household studies should address the role of risk aversion in agriculture (Kminsky, 1979:65-66).

11. In a more simplistic way and for the purpose of examining the risk-taking behavior of households, risk-neutrality refers to a situation where producers are not sensitive to fluctuations in farm income or production. If producers show little, average or a high degree of sensitivity compared to the risk neutral situation, they can be considered respectively low, moderate or extreme degree of risk-averters. This definition is subjective, thus cannot be logically deduced from empirical findings as to what is a low, medium or high degree of risk aversion.

12. Thus, farmers take risks regarding enterprises they know more about, and have more investments in.

13. Information from actual decisions may show two aspects of farmer's characteristics: the degree of risk-taking behavior and "rationality" of decisions. Farmers who are efficient, producing a given crop or milk with minimum outlay of inputs, may show extreme risk-averseness. Therefore, characterization of farmer's behavior has to be made with great care.

14. It may not, however, be possible to find a farmer with an index of one (growing a single crop) in the Ethiopian highlands (see Chapter IV for the formula).

15. The number of relatives has a risk decreasing effect among test farmers while it increases sensitivity of control farmers to fluctuations in income. It may be that the control farmers have extended families which are relatively poor. Therefore, the burden of resource sharing falls heavier among these farmers. The opposite may be true among the test farmers.

16. The index of plot characteristics is measured as follows: Plot characteristics= (S1 + S2 + S3)/3, where S1=1 if most fields are located in hilly areas and 0 otherwise, S2=1 if the soil type poses problems for plowing and 0 other wise, and S3=1 if more than 50% of the plot is covered with stones, and 0 otherwise.

17. Multicollinearity is a situation where at least two independent (cause) variables are linearly related (e.g., one becoming a product or sum of the other). When this holds, the estimated results will not be reliable or statistically valid.

CHAPTER VIII

PRODUCTION STRUCTURE, EFFICIENCY AND RISK IN PEASANT AGRICULTURE

8.1. Introduction

Most activities of households are directed at securing subsistence food requirements. Therefore, production decision are very crucial to the survival of households. Their decisions involve "rational" combination of traditional and new technologies. This chapter investigates factors that influence the amount and efficiency of production, and production risk. Furthermore, the impact of selective adoption of technologies on the efficiency of production are investigated.

Production function analysis is a method of analyzing the consequences of household decision-making regarding crop and livestock production. It has been used extensively by economists, sociologists, political scientists and economic anthropologists. Variations in this method of analysis occur, however, with respect to the specification of decision inputs. Attempts are made to include non-economic variables that are assumed to influence access to resources, choices of enterprise location and management (see Ortiz, 1980; Gladwin, 1979). In this study, variables which directly influence the level and efficiency of production are selected. The derived effect of socio-cultural factors are assumed to be reflected by the magnitude of the impact of these variables.

8.1.2. Empirical Results

Various parametric functional forms are available to estimate a response surface relating inputs to outputs. These forms are dependent on specific distributional assumptions about variations not explained by variables included in a regression model. Non-parametric econometrics offers an alternative specification of response functions. However, a statistical package for estimation of non-parametric econometrics has not yet been developed (Vinod, et al., 1986; Ullah, 1988). As a result, a Cobb-Douglas production function is chosen for reasons discussed in Chapter IV.

To minimize biases from aggregation, estimation is carried out for six major crops on a per plot basis. These include ***Eragrostis Abysinica*** (teff), wheat, barley, horse beans, field peas and oats. Cereals (which include teff, barley, wheat, and oats) make up the bulk of a household's daily food intake. Thus, an aggregate function is also estimated for cereals.

Prior to performing production function analysis for each crop and by study groups, tests for structural changes or differences in crop and milk response functions are carried out. The premise behind this analysis is that if the functions are statistically significant, it means that the effects of inputs on the amount of outputs produced by the groups compared are different. It may also imply that the efficiency with which households of the groups compared produce outputs is different. Tests for differences in the structure of production are carried out to compare: i) Selale and Ada regions, ii) all test and control farmers, and iii) test and control farmers within the Selale and Ada regions. The results are presented in Table 8.1.

Table 8.1. Structural Change Test of Response Functions for Crops and Milk

Category	Teff	Barley	Bean	Wheat	Peas	Oats	Cereal	Milk
By Region (Selale Vs Ada)[1]	-8.34*	2.84**	-4.37*	-7.40*	-8.96*		-8.96*	6.58*
All Test Vs All Control[2]	-3.17*	4.34*	2.39**	3.89*	2.13**		4.33*	6.53*
Selale[3]	-2.76**	5.81*	-4.08*	-5.99*	-9.80*	4.08*	3.58*	7.69*
Ada[4]	4.58*	3.14*	3.39*	-8.62*	8.56*		7.18*	4.79*

1/ Positive values indicate that Ada farmers have larger production than Selale farmers.
2/ Positive values indicate that test farmers produce more than control farmers.
3/ and 4/ Respectively indicate that test farmers produce more than control farmers when the estimated values are positive. Respectively, test and control farmers refer to households who have adopted cross-bred cows and those who have not.
* and # indicate statistical significance at 1 & 5 percent respectively.

The results show that production of barley, peas and milk by Selale producers is larger and statistically significant than that of Ada farmers. Test farmers of Selale show positive and statistically different production structure for barley and milk compared to control farmers in the same region. Test farmers of the Ada region exhibit a positive shift in the production structure of teff, barley, beans, peas, cereal and milk relative to the control farmers of the same region. The results also show that test farmers consistently exhibit a positive shift in the production of milk relative to control groups for both regions. Production of most crops and milk by test farmers has shown a positive shift compared to control farmers. Furthermore, the response functions for crops exhibit positive shifts among Ada producers while that of milk production among Selale producers. These results suggest that there are differences in the impacts of inputs on outputs and that the structure of crops and milk production differ across regions. Therefore, analysis of production decisions should proceed by region as well as by test and control farmers.

8.2. Inter-Region Analysis of Crop Production Decisions

The results of statistical analysis of factors of production on the level of outputs by region are presented in Table 8.2. Plot size exerts a positive and significant effect on barley, beans, wheat and cereals (Table 8.2.). Its effect on teff is negative. The yield per unit area for teff is lower than other cereals. However, the market value of the output per unit area is much higher than other crops. It requires a larger number of oxen inputs to prepare the seed bed, and large labor inputs to weed several times, harvest and thresh on time. The yield of teff is low compared to other cereals. Breeding for higher yields is at an infant stage. Because of the high market price, farmers spend a lot of labor and financial resources on the production of teff. However, teff yields does not increase proportionately with increases in farm size. Consequently, farm size may be negatively associated with the production of teff.

The number of oxen inputs exerts negative and significant impacts on teff grown in the Selale region and beans cultivated in both study sites.[1] Oxen inputs, however, are positively and significantly associated with the production of most crops. The seeding rate shows a positive and significant influence on most crops. Similar patterns of effects and degree of significance are observed for the labor inputs.

The effects of pesticides and fertilizer are not consistent across crops. In general, their effects are statistically significant only on barley and wheat grown in the Selale and Ada regions respectively. The larger the value of plot characteristic the more difficult it is to prepare a plot for planting.[2] Thus, plot characteristics exerts a negative and significant effect on most crops except on barley and beans.

Barley is a crop that can be grown both in the short and main(long) rainy seasons. It requires a relatively short growing period. It is a crop which poor households use to fill deficits in their food consumption requirements throughout the year. For better results, households frequently manure this crop more than other crops. Green beans serve as a source of food when household's food stocks decline. In many cases, these crops are planted on plots closer to the farmstead. The plots on which they are planted are better situated, fertile, free of stones and located on plains. Thus, plot characteristics can positively influence outputs of these crops.

Black soil is treated as a separate variable because its effect on crop yield differs in the Selale and Ada regions. Its effect is constrained by a water-logging condition in the Ada region and by soil erosion in the Selale region. Thus soil exerts a positive and significant effect only on teff and peas cultivated by Ada farmers.[3]

Experience is positively and significantly associated with most crops grown in the Ada and Selale regions. The effect of extension education on most crops is not statistically significant. Schooling is positively and significantly associated with production of most crops. Similar to experience, production knowledge shows positive and significant effect on all crops.

Table 8.2. Estimates of Average production Functions by Region[1/]

	Teff		Barley		Beans	
	Selale	Ada	Selale	Ada	Selale	Ada
Intercept	-0.616	-0.563	0.629	1.411	0.895	-0.855
	(-2.103)#	-0.487	(2.116)#	(2.201)*	-4.789	(-1.562)
Plot Size	-0.091	-0.156	0.124	0.569	0.306	0.362
	(-1.704)	(-2.63)*	(1.987)#	(1.969)#	-1.982	(2.56)*
Oxen	-0.169	-0.202	0.346	0.234	-0.107	-0.21
	(-1.817)	(-1.597)	(5.684)*	(2.27)*	(-3.233)	(-1.95)#
Seed Rate	0.419	0.675	0.314	0.325	0.162	-0.131
	(2.152)#	(3.195)*	(2.191)#	(3.052)*	-2.935	(-1.406)
Labor	0.506	0.47	0.58	0.459	0.168	0.154
	(2.267)*	(3.443)*	(2.075)#	(2.431)*	-1.143	-1.72
Fertilizer	0.115	0.308	0.322	0.464		0.401
	-1.79	(2.253)*	(1.963)#	(2.39)*		-1.077
Pesticide	0.028	0.003	0.383	0.02		0.204
	-0.269	-1.003	(1.998)#	-0.206		-1.047
Plot Characteristic	-0.371	-0.11	0.194	0.092	0.347	0.322
	(-3.001)*	(-0.513)	-1.73	-0.497	(4.142)*	(2.11)#
Black Soil	0.217	0.334	-0.131	-0.043	-0.026	-0.028
	(3.364)*	(2.7)*	(-2.278)*	(-0.422)	(-0.619)	(-0.326)
Extension Education	0.114	0.095	0.41	0.019	0.011	0.415
	-1.952	-1.498	(2.45)*	-1.174	-0.622	(2.315)*
Production knowledge	0.379	0.454	0.407	0.438	0.312	0.241
	(1.951)#	(2.99)*	(2.097)#	(2.86)*	(2.247)*	(1.95)#
Schooling	0.329	0.431	0.462	0.521	0.263	0.304
	-1.221	(1.99)#	(2.107)#	(2.622)*	(1.95)#	(1.99)#
N	216	52	216	52	216	52
R2	0.72	0.734	0.607	0.786	0.806	0.803

1/ Values in parenthesis are t-statistics.

* and # indicate statistical significance at 1 and 5 percent respectively.

Table 8.2. Estimates of Average production Functions by Region (continued)

	Wheat		Peas		Cereal	
	Selale	Ada	Selale	Ada	Selale	Ada
Intercept	0.014	1.561	-0.332	-0.825	1.999	1.767
	-0.028	(2.785)*	(-1.195)	(-1.326)	-2.498	(2.863)*
Plot Size	0.657	0.551	0.157	0.3	0.579	0.68
	(2.182)*	(3.905)*	(2.413)*	(1.949)#	(1.97)#	(2.832)*
Oxen	0.165	0.153	0.002	-0.057	0.135	0.288
	(2.422)*	(1.964)#	(1.978)#	(-0.38)	-1.747	(1.973)#
Seed Rate	0.293	0.015	-0.044	0.264	0.186	0.187
	(2.261)*	-1.94	(-0.80)	(2.375)*	-2.217	(2.48)*
Labor	0.539	-0.443	-0.02	-0.088	0.437	0.546
	(2.416)*	(1.986)#	(-0.313)	(-0.828)	(1.965)#	(2.745)*
Fertilizer	0.547		0.416	0.62		
	(1.966)#		-1.84	(2.09)*		
Pesticide	0.752		0.024	0.032		
	(2.129)*		-0.81	-0.436		
Plot Characteristic	-0.083	-0.389	-0.088	-0.335	-0.077	-0.142
	(-0.651)	(-1.977)#	(-0.854)	(-2.012)#	(-0.72)	(-0.495)
Black Soil	-0.022	-0.096	0.046	0.206	-0.076	0.313
	(-0.293)	(-0.895)	-0.753	(2.05)#	(-1.35)	(1.98)#
Extension Education	0.012	0.185	0.013	0.03	0.024	0.262
	-0.399	(1.953)#	-0.518	-0.578	-1.043	(2.69)*
Production knowledge	0.49	0.946	0.769	0.319	0.531	0.444
	(1.97)#	(3.857)*	(1.965)#	(2.603)*	(1.97)#	(2.59)*
Schooling	0.325	0.521	0.299	0.437	0.836	0.374
	(1.98)#	(2.123)#	(1.947)#	(2.007)#	(2.06)#	(2.03)#
N	216	52	216	52	216	52
R2	0.576	0.89	0.754	0.838	0.54	0.553

* and # indicate statistical significance at 1 and 5 percent respectively.

8.3. Intra-Region Analysis of Crop Production Decisions

Most of the statistical results regarding the influence of inputs on the level of production of test and control farmers in each study sites are similar to the inter-regional analysis of production decisions (see Appendix 10a&b). Explanations of the statistical analysis are presented below.

The effect of plot size on the level of production of most crops is statistically significant and positive except on barley cultivated in both areas. Oxen and labor inputs show similar influence on production of most crops. Seeding rate and labor affect wheat and peas grown in both areas negatively but are not statistically significant.

Fertilizer exerts a positive and significant effect on crops cultivated by Ada farmers as do pesticides on crops cultivated by Selale farmers. Fertilizer has been used in the Ada area longer than in the Selale region. Not only did the use of fertilizer started recently in the Selale area, but most of the plots are steep and expose fertilizer granules to soil erosion. Moreover, crops in the Selale region are damaged by rodents. Application of pesticides mixed with grain has reduced the effect of rodents. This measure contributed to increased yields of crops in the Selale area.

In the Ada region competition arises between cheap labor migrating from crop deficit regions and the increasingly expensive and less effective herbicides. Households in the Ada region have to make a choice between hiring labor and purchasing herbicides to reduce the effects of weed infestations. It is not easy to get access to knapsack sprayers to apply herbicides or pesticides. Furthermore, herbicides or pesticides have the side effects of being a health hazard. Labor from Ada region is expensive compared to that of migrant workers.[4] Migrant laborers come from a radius of 500 kilometers and are mainly from regions that are grain deficit or do not specialize in grain production. Thus, Ada farmers may prefer to hire labor for weeding rather than purchasing pesticides or herbicides. These factors may have contributed either negatively or insignificantly to the effect of pesticides on crop production in the Ada region. Out of six group discussions held with farmers of Ada region, half of them considered hiring labor, while the rest purchased pesticides or herbicides. It can be seen that the results from group discussions are not conclusive. It is, however, possible to infer that households compare costs of repeated application of herbicides with that of hand-weeding using hired labor.

Slope and other characteristics of a plot influence teff cultivated in both regions negatively. However, the effect of plot characteristics on other crops is positive and in most cases significant. Black soil has the effect of reducing barley and peas production in the Selale area while its influence on other crops is positive and significant. Black soil retains moisture. Plots planted with barley tend to be fertile. Fertile and high moisture retaining plots cause barley plants to be weak and fall because of wind drift and decay. Peas require relatively infertile and coarse seedbed. Thus, black soil may cause excessive growth and less yield of peas. Experience and production knowledge exert positive and significant effects on all crops. Schooling exerts a similar effect but the magnitude of its impact is relatively smaller.

Secular education in Ethiopia emphasizes "scientific" methods of production. It does not address actual problems in peasant agriculture. Alternatives to improve production and productivity do not examine issues such as family labor, limited capital, farming experience and production knowledge. It is not surprising to observe negative or insignificant effects of schooling and extension education on the amount of crop production (see Eisemon, et al., 1988).[5]

The statistical results presented in Tables 8.2. and Appendix 10a indicate that farm size and production knowledge consistently exert positive impacts on the level of production in both regions. Comparisons of all test with all control farmers by crop type indicate that most variables exert significant and positive effects on most crops (Appendix 10b). Furthermore, the patterns of influences of inputs on the level of outputs are similar to that presented in Tables 8.2. and Appendix 10a.

8.4. Analysis of Milk Production Decisions

The results of statistical analysis of inputs that influence the amount of milk produced by each region, and by test and control farmers are presented in Table 8.3. Production knowledge, grazing area, the amount of concentrates fed and experience exert significant and positive effects on the amount of milk produced by Selale producers. The number of cows, labor, use of atela (by-product of local brewing) and veterinary services exert significant and positive effects on the amount of milk produced in both areas.

When farmers are grouped by test and control, regardless of the region, production knowledge, grazing area, veterinary services, atela use and experience positively and significantly influence the level of milk production of test farmers (Table 8.3.). Moreover, the number of cows, grazing area and labor input are positively and significantly associated with milk production of both test and control farmers.

Within each region, production knowledge, atela use, veterinary services, concentrates and experience positively and significantly influence milk production of test farmers. In the Ada area, stubble feeding, production knowledge, number of cows and experience positively and significantly influence milk yield of test farmers. In chapter VII, it is pointed out that grazing lands are destroyed because of over-grazing or erosion. The quality of roughage in the Ada region is lower than that in the Selale region. Test farmers in the Ada region receive access to livestock feed from ILCA and EPID (a project under MOA). Thus, they may not use their grazing lands as the only source of feed. Consequently, the size of grazing plots and roughage negatively and significantly influence milk yield of the control farmers in the Ada region.

Table 8.3. Estimates of Average Production Function by Test and Control Farmers[1/]

	Groups of farmers			
Variables	All Selale Farmers	All Ada Farmers	All Test Farmers	All Control Farmers
Intercept	4.462	-3.699	4.389	5.315
	(4.39)*	(-1.88)	(2.623)*	(4.521)*
Grazing area	0.42	0.295	0.427	0.211
	(2.21)#	-1.24	(2.212)#	(1.951)#
Stubble	0.175	0.041	0.191	0.438
	-1.91	-0.183	(1.98)#	(2.501)*
Production Knowledge	0.765	-0.156	0.588	0.277
	(3.43)*	(-1.944)	(1.96)#	-1.94
No. of cows	0.439	0.265	0.329	0.263
	(3.18)*	(3.167)*	(1.967)#	(2.663)*
Labor	0.385	0.251	0.124	0.134
	(2.48)*	(2.344)*	(2.234)#	(2.157)#
Atela	0.131	0.062	0.136	0.05
	-1.75	-1.305	-1.73	-1.18
Veterinary Services	0.111	0.259	0.273	0.07
	-1.17	-1.21	-1.4	-0.54
Roughages	0.243	-0.219	0.007	-0.112
	(2.37)*	(-2.23)#	-0.07	(-1.318)
Concentrates	0.196	0.117	0.021	-0.013
	-1.04	-1.614	-1.364	(-0.289)
Extension Education	0.019	0.041	0.026	0.077
	-1.14	-1.26	-1.07	-1.2
Schooling (years)	0.347	0.232	0.477	0.097
	(1.98)#	(1.968)#	(2.09)#	-1.34
N	216	52	114	154
R2	0.512	0.582	0.543	0.382

1/ Values in parenthesis are t-statistics.
* and # indicate statistical significance at 1 and 5 percent respectively.

Table 8.3. Estimates of Average Production Function by Test and Control Farmers[1] (continued)

	Selale Region		Ada Region	
Variables	Test Farmers	Control Farmers	Test Farmers	Control Farmers
Intercept	5.796	5.305	-2.71	2.961
	(3.357)*	(3.86)*	(-1.97)	(4.637)*
Grazing area	0.51	0.519	0.324	0.282
	(2.754)*	(2.15)#	-1.91	(2.201)#
Stubble	0.212	0.182	0.322	0.437
	(2.09)#	(1.99)#	(2.52)*	(2.209)#
Production Knowledge	0.717	0.171	0.475	-0.108
	(2.39)*	-1.68	(1.96)#	(-1.467)
No. of cows	0.447	0.249	0.176	0.28
	(2.869)*	(3.09)*	(2.36)*	-0.81
Labor	0.464	0.528	0.448	0.086
	(2.36)*	(2.49)*	-1.718	-0.596
Atela	0.134	0.044	0.219	0.287
	-1.45	-0.959	-1.62	(1.78)#
Veterinary Services	0.255	0.035	0.297	0.049
	-1.91	-1.26	-1.94	-0.424
Roughages	0.257	0.286	0.243	0.161
	(1.981)*	(1.96)#	-1.59	-1.543
Concentrates	0.193	0.035	0.279	-0.005
	-1.29	-0.7	(1.96)#	(-0.012)
Extension Education	0.221	0.156	0.17	0.112
	-1.24	-1.79	(1.97)#	-1.354
Schooling (years)	0.533	0.249	0.235	0.319
	(2.346)*	(1.99)#	(2.01)#	(1.948)#
N	88	127	25	26
R2	0.535	0.34	0.812	0.711

1/ Values in parenthesis are t-statistics.

* and # indicate statistical significance at 1 and 5 percent respectively.

8.5. Adoption of Mixes of Technologies and Production Decisions

This study hypothesizes that producers who adopt different mixes of technologies are relatively more efficient. In other words, inputs will have greater impacts on the level of production of producers who have adopted several innovations compared to those who have adopted one or none. Study farmers are grouped by the number of production technologies adopted. These groups are those who have used: i) fertilizer and pesticides, ii) fertilizer and improved seed, iii) improved seed and pesticides, iv) improved seed, fertilizer and pesticides and v) none (control). The results of statistical analysis of the effects of these combinations of crop technologies on the amount of cereals produced are presented in Table 8.4.

Table 8.4. Estimates of the Impact of Inputs for Different Mixes of Crop Technologies

	Fertilizer & Pesticides	Fertilizer & Seed	Seed & Pesticides	Fertilizer, Seed & Pesticides	Control (crop)
Intercept	1.236	2.001	-1.014	-1.389	5.546
	(2.154)#	(2.221)#	(-1.09)	(-1.453)	-4.521
Plot size	0.245	0.341	0.531	0.176	0.211
	(2.984)*	(2.771)*	(2.98)*	(1.962)#	(1.97)#
Oxen/co	0.541	0.245	0.221	0.191	0.438
ws	(2.967)*	(2.104)#	(1.948)#	(1.96)#	-1.601
Seed rate		0.265	0.256	-0.119	
		(2.03)#	(2.741)*	(-1.884)	
Labor	0.401	0.404	0.312	0.108	0.423
	(2.94)*	(2.19)#	(1.967)#	(1.968)#	-2.003
Fertilizer	0.215	0.277		0.278	
	(3.21)*	(2.96)*		(2.117)#	
Pesticides	0.421		-0.177	0.448	
	(2.778)*		(-2.116)#	(2.797)*	
Plot	-0.142	-0.211	-0.448	-0.273	-0.163
characte-	(-1.665)	(-1.97)#	(-2.45)*	(-2.98)*	(-1.91)
Black soil	0.181	0.114	0.122	-0.107	-0.315
	-1.711	-1.102	-1.93	(-1.09)	(-1.219)
Extension	0.191	-0.217	0.178	0.119	-0.207
education	-1.819	(-1.196)	(1.994)#	(1.97)#	(-1.791)
Produc-	0.828	0.633	0.561	0.515	0.466
tion	(2.357)*	(3.17)*	(2.29)#	(3.09)*	(1.98)#
Schooling	0.51	0.451	0.324	0.417	0.297
	(2.00)#	(1.996)#	(2.168)#	(2.701)*	(1.97)#
N	88	43	41	23	49
R2	0.87	0.71	0.93	0.81	0.78

* and # indicate statistical significance at 1 and 5 percent respectively. Values in parenthesis are t-statistics.

The results indicate that: i) inputs such as plot size, oxen, experience, production knowledge and schooling positively and significantly influence production of cereals under all combinations of crop technologies, ii) physical inputs and knowledge exert large and significant positive impacts on production when farmers use a combination of fertilizer and pesticides, iii) technical factors such as plot characteristics negatively influence production, iv) several variables seem to have significant and large impacts on the amount of production when the production structure of households is examined using this type of grouping instead of grouping by test and control farmers (see Tables 8.2. and 8.3, Appendix 10a and 10b). It means, the structure of production is greatly influenced by the combination of technologies adopted by households than by region or use of one technology as a criteria to group farmers.

8.6. Measures of Production Efficiency

8.6.1. Introduction

Several researchers have studied production efficiency of peasants in LDCs. Most of these studies can be seen as attempts to characterize the performance of peasants using neoclassical method of production analysis (e.g., Ellis, 1988). These studies assume ideal socio-economic environment such as appropriate marketing policies, access to technologies and other inputs, and adequate infrastructure. The results of these studies portray peasants' behavior in several ways. For some the behavior of peasants is static, largely because they are not innovative or efficient. For others, they are efficient because the environment in which they are operating is static (Wharton, et al., 1969; Mellor, 1969). Yet, there are several studies which convincingly argue that peasants are efficient, that their goals, strategies and decisions are logical and rational given the constraints and choices they face.[6]

In the following section, the most recent advances in methods of efficiency analysis are used to examine household production in the Central Ethiopian highlands. It is hoped that examining efficiency or inefficiency of households will help to characterize peasants whose decision-making processes are investigated in previous chapters. Factors that contribute to variation in measures of production efficiency are examined. Furthermore, the impact of adoption of different mixes of innovation on production efficiency is investigated.

8.6.2. Results of Data Envelopment Analysis (DEA)

Chapter IV suggests that one method of examining production efficiency is the use of a linear fractional program that compares differences in efficiency across farmers for all enterprise combinations (i.e., a measure of aggregate efficiency). Aggregate measures of efficiency such as DEA rely on very few assumptions. It has the ability to incorporate both crop and livestock enterprises in a single model. In so doing, it is possible to compare all farmers relative to those whose efficiency measures 1.00 (i.e., efficient). Details on the construction of the model are provided in the appendix. The results of DEA are presented in Table 8.5.

Chapters II, III, V and VI have described that Selale farmers have large grazing land, cattle and livestock production knowledge compared to Ada farmers. On the other hand, farmers in Ada have more fertile soil, have greater access to crop production inputs, markets, roads and gained greater scores on crop production solving problems compared to farmers in the Selale region.

Table 8.5. Percentage of Producers Grouped by Efficiency Scores Obtained from Data Envelopment Analysis

Efficiency Ranges	Efficiency Mid-point	Selale Test Farmers	Ada Test Farmers	Selale Control Farmers	Ada Control Farmers	All Selale Farmers	All Ada Farmers
0.00-0.5	0.255	2.2	0	6.1	0.4	7.6	1.8
0.51-0.6	0.555	2.3	5.5	6.3	4.4	4.6	9.4
0.61-0.7	0.655	10.5	6	11.2	3.7	9.2	1.9
0.71-0.8	0.755	17	15.3	12.5	11.1	11.2	10.1
0.81-0.9	0.855	13.4	13.7	18.8	18.5	16.6	11.3
0.91-1.00	0.955	55.6	69.5	45.1	61.9	50.8	65.5

The results of DEA indicate that Ada farmers are more efficient than Selale farmers. Fifty-six percent of the test farmers in Selale area achieved an efficiency score of 0.9 (90%) or more out of 1.0 (100). Seventy percent of Ada test farmers attained a similar efficiency score. Similar efficiency scores are observed when all farmers of the Selale region are compared with all farmers of the Ada region. The analysis also reveals that test farmers are more efficient than control farmers.

DEA is helpful to analyze the efficiency of the whole farm. At the same time it may hide other aspects of the production system, especially when different enterprises are managed by households. Therefore, this method is complemented with analysis using stochastic frontier production functions.

8.6.3. Results of Frontier Production Function Analysis

Statistical analysis using stochastic frontier production function is performed for cereal, milk and the major crops for which average production function analysis is carried out (section 8.2). As the findings for individual crops is very similar to that of cereals, only results from cereal and milk production efficiencies are presented (Table 8.6.). The results of efficiency scores for individual crops and for each farmer are presented in the appendix.

Table 8.6. indicates that: i) Ada farmers exhibit relatively higher cereal production efficiency compared to Selale producers, ii) test farmers of both study sites show higher efficiency scores when compared with control farmers and iii) most Selale farmers are efficient in milk production relative to Ada producers.

Table 8.6. Percentage of Producers Grouped by Efficiency Scores Obtained from Stochastic Frontier Production Function Analysis

Efficiency Ranges	Efficiency Mid-point	Selale Test Farmers	Ada Test Farmers	Selale Control Farmers	Ada Control Farmers	All Selale Farmers	All Ada Farmers
Crop Production Efficiency							
0.00-0.5	0.255	1.1	1.8	2	2.7	0.5	3.8
0.51-0.6	0.555	2.3	2	7.8	1.5	1.3	0.0
0.61-0.7	0.655	6.7	3.9	13.5	9.2	15	6.9
0.71-0.8	0.755	28.9	5.4	22.5	11.1	10.6	10.3
0.81-0.9	0.855	30.9	40.3	28.6	40.5	46.6	40.4
0.91-1.00	0.955	30.1	46.6	25.6	43	26	41.8
MILK Production Efficiency							
0.00-0.5	0.255	2.4	4	16.5	20	4.8	9
0.51-0.6	0.555	3.4	5.8	10.6	19.1	14.1	7.5
0.61-0.7	0.655	10	23.9	14.9	7.4	12.9	12.7
0.71-0.8	0.755	20.5	25.4	21.1	28.2	28	30.1
0.81-0.9	0.855	30.1	20.6	20.5	14.8	30.8	20.5
0.91-1.00	0.955	38.3	20.3	14.1	10.5	37.1	20.2

8.6.4. Explaining Efficiency

Variations in production efficiency arise from various factors. Efficiency scores are bound between 0 and 1. In statistical terminology, variables that are restricted to lie in a specific range are called truncated. Thus, logit regressions are performed to identify variables which may contribute to variations in efficiency scores (Maddala, 1983). Several variables are included in the regression analysis. Because of collinearity among the regressors, only ten variables are selected for analysis. The results suggest that most of the variables significantly influence production efficiency (Table 8.7.). The magnitude of the impacts from knowledge related variables, especially production knowledge and education, are very high relative to other variables.

Table 8.7. Impacts of Selected Variables on Measures of Efficiency[1/]

	Selale region			Ada region		
Variables	DEA	Crop	Milk	DEA	Crop	Milk
Intercept	1.015	2.768	1.799	2.102	3.412	1.786
	(2.454)*	(3.945)*	(2.98)*	(1.963)#	(2.53)*	(3.98)*
Depenexp	0.151	0.279	0.756	0.211	0.522	0.178
	-1.784	(1.981)#	(1.97)#	(2.022)#	(1.981)#	(2.77)*
Indepexp	0.351	0.341	0.942	0.493	0.839	0.532
	(1.987)#	(2.014)#	(3.184)*	(2.98)*	(2.051)#	(3.11)*
Education	0.755	0.396	0.577	0.689	0.733	0.571
	(2.376)*	(3.57)*	(1.994)#	(2.973)*	(3.09)*	(2.01)#
Extension education	0.211	0.439	0.215	0.711	0.563	0.561
	(1.99)#	(2.09)#	(2.257)#	(1.981)#	(3.03)*	(1.97)#
Worker:cons.ratio	0.384	0.485	0.351	0.344	0.611	0.215
	(1.97)#	(1.98)#	(1.994)#	(2.124)#	(2.057)#	-1.77
Region	0.454	-0.213	0.669	0.011	0.789	-0.699
	-1.981	#(-2.45)*	(2.394)*	-1.544	(2.23)#	(-1.98)#
Production knowledge	0.894	0.677	0.945	0.881	0.987	0.612
	(2.312)#	(1.981)#	(3.452)*	(2.101)#	(3.002)*	(1.96)#
one Technology	0.214	0.576	0.874	0.355	0.791	0.157
	(2.15)#	(2.49)*	(3.12)*	(1.984)#	(3.017)*	-1.91
Two Technologies	0.329	0.602	0.984	0.631	0.834	0.411
	(1.98)#	(2.96)*	(3.81)*	(2.874)*	(2.96)*	(1.97)#
Three Technologies	0.201	-0.301	0.914	0.611	0.935	0.041
	(1.91)#	(1.96)#	(2.15)#	(2.12)#	(3.91)*	-1.71
2 Log L	35.129	42.809	29.954	21.02	45.48	31.72
	(0.051)*	(0.026)*	(0.07)*	(0.09)*	(0.02)*	(0.07)*

1/ Values in parenthesis are Wald's Chi-square statistics.
* and # indicate statistical significance at 1 and 5 percent respectively.

Region is included as dummy (0-1) variable to capture differences (including modernizing conditions) between Ada and Selale. It contributes negatively to cereal production efficiency in the Selale and to milk production efficiency in the Ada region, and reflects comparative advantage, location and enterprise-specific experiences. In other words, region favors livestock and crop production in the Selale and Ada regions respectively.

In peasant economies, children are brought up learning methods of raising livestock and growing crops from childhood (Sperling, 1987). Specifically, they are taught the practices of production and managing enterprises which parents believe profitable. Thus, the number of years that children spend under the control of their parents (depenexp) exerts a relatively greater impact on milk efficiency scores in the Selale region, and on crop efficiency scores in the Ada region. The impact of experience gained while working for parents is strengthened by the number post-marriage or independent years of farming experience (indepexp). In general, the model performed well and most variables significantly influence variations in production efficiency.

The result also indicates that different combinations of innovations affect production efficiency of households differently. Despite differences in the combination of innovations, most of the combinations of technologies affect efficiency scores positively and significantly. A combination of two innovations shows a consistently large, positive and significant effect on all measures of production efficiency in both study sites. It seems that with current policies and production constraints, households can attain maximum efficiency with adoption of two innovations.[7]

8.6.5. Technology, Risk and Production Efficiency

Discussions in previous sections and Chapter VII indicate that new technologies embody risk. This risk can be social, economic or technical. Decision-makers rely on information of recently experienced events and anticipate the future prior to selecting actions (Ortiz, 1980). These actions are influenced by the degree of risk-taking behavior. This behavior may constrain or enhance decision-making ability related to new innovations, mixes of locally available inputs, management styles and the amount of production and the efficiency with which it is produced. Figures relating adoption, risk and efficiency are presented below.

Crop and livestock production efficiency, risk and adoption of innovations are plotted in Figures 8.1. and 8.2 for Ada and Selale regions respectively. The results indicate that: i) the probability with which test farmers adopted cross-bred cows is low compared to the probability with which control farmers are willing to adopt, ii) livestock diversification indices are higher among control farmers while livestock sensitivity indices are higher among test farmers, and iii) production efficiency scores are higher among test farmers.[8] One important observation is that the number of classes of cattle and/or livestock decline among test farmers. This indicates that households who adopt cross-bred cows reduce either classes of livest-

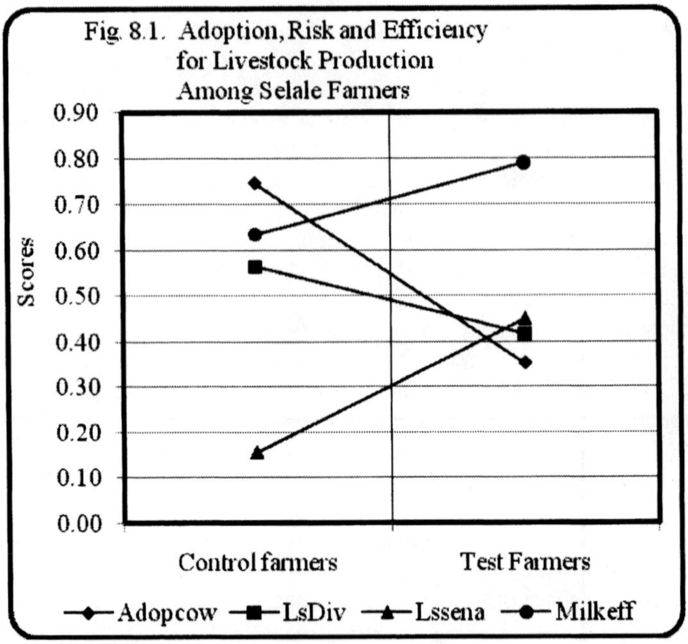

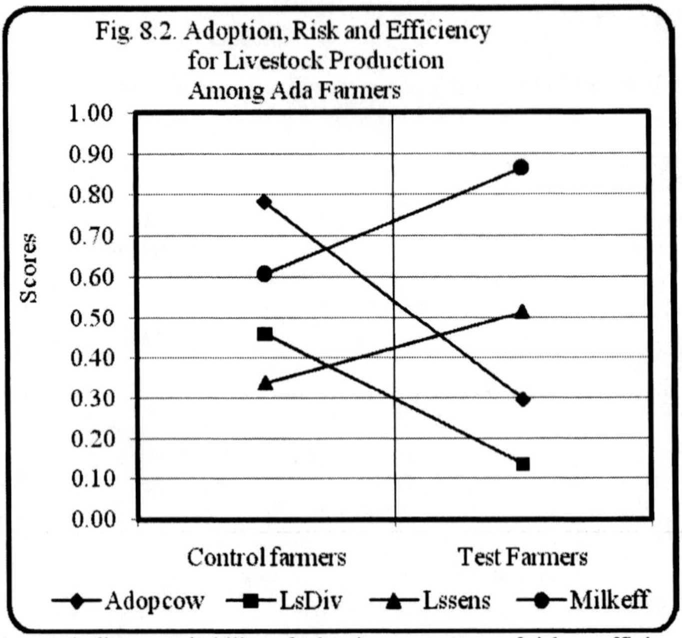

Note: Scores indicate probability of adoption, measures of risk or efficiency

ock or cattle to gain the maximum advantage from this technology. Compared to households who purchased crop production technologies, households who invest large sums of money in livestock production technology require a substantial amount of money to avoid falling below substance requirement. Thus, livestock sensitivity indices are higher among test farmers.

Figures 8.3 and 8.4. present relationships between adoption of crop technology, risk and crop production efficiency. The results indicate that the probability of adoption of crop technology and diversification indices do not show substantial differences between test and control farmers of either region. The major difference between the study sites is that crop production efficiency between test and control farmers in the Selale region are approximately equal, but not in the Ada region.

It is argued that innovations in either crop or livestock production would provide additional resources to increase the productivity of another enterprise (Gryseels, et al., 1985, 1986, 1988). Introduction of livestock production technology should provide extra financial support to purchase inputs for crop production. While this seems to hold true in the Selale region, the assumed trickle-down effect of adoption of livestock technology may not induce differences in the amount or efficiency of crop production between test and control farmers in the Ada region. That is, it was anticipated that test farmers would have significantly higher efficiency score in crop production. In general, the probability of adoption, crop diversification and sensitivity indices are slightly higher among the test farmers in both regions

From Figures 8.1, 8.2, 8.3 and 8.4, it can be concluded that there is not an exact correspondence between risk and technology adoption, technology adoption and efficiency, and risk and efficiency scores. Introduction of innovations increase the risk-taking behavior as well as production efficiency. This means that households are rational in their decision-making. It may well be argued that the innate ability of producers to implement appropriate decision-making strategies may be the reason for this rationality. However, the effectiveness of this rationality may be obstructed by government policies. Thus, development strategies should design ways of utilizing knowledge of households to attain sustainable increases in food production.

8.7. Results of Production Risk Analysis

Peasant agriculture is exposed to several sources of risk: economic risks such as availability of inputs and price variability; technical risks such as uncertain outcomes of innovations; and environmental risks such as excess or shortage of rain. It is not possible to include all factors which are expected to influence variability in production because of statistical problems associated with too many variables and a small number of observations (Koutsoyiannis, 1981; Johnston, 1984).

Fig. 8.3. Adoption, Risk and Efficiency for Crop Production Among Selale Farmers

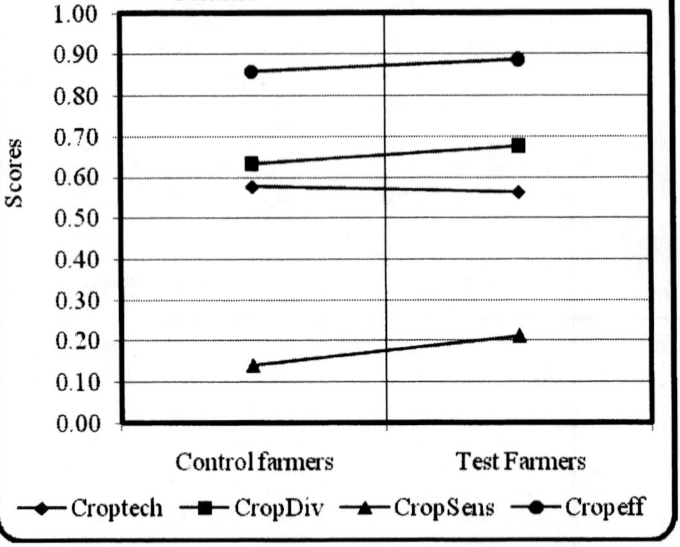

Fig. 8.4. Adoption, Risk and Efficiency for Crop Production Among Ada Farmers

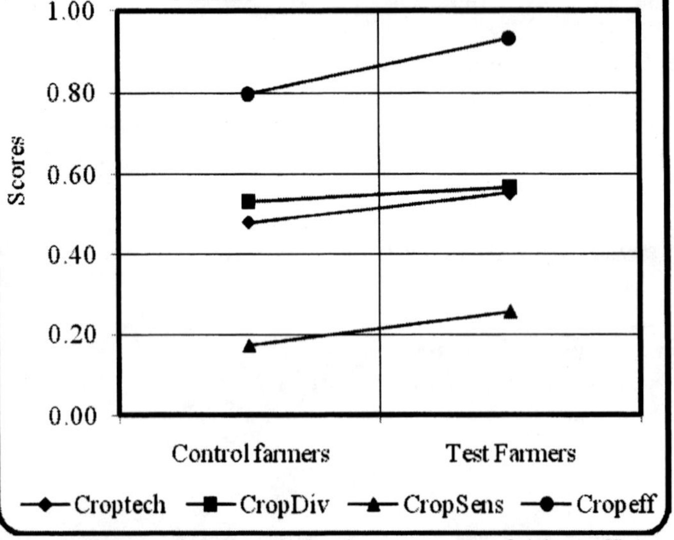

Note: Scores indicate probability of adoption, measures of risk or efficiency

Several methods are available to examine production risk. These could be mathematical such as programming or production functions analysis (transcendental or variable-parameter functions) (Maddala, 1977, 1983; Hazel, et al., 1986). However, there is no consensus as to which method of analysis is suitable under specific circumstances. Neither is there agreement on how to examine production risk with cross-sectional data.

In the study of production risk in Selale and Ada regions, deviations of individual yields from the average are regressed on variables that are anticipated to influence variability in crop yields. Deviations reflect the degree to which actual yields are dispersed, thus are expected to reveal actual production risk. The variables selected to examine production risk and criteria used for coding are given in Table 8.8.

Analysis is carried out for cereal and milk production by region, and by test and control farmers for each region. The results indicate that rainfall and price variability, soil type, improved seed, steepness of a plot and uncertain land tenure are negatively and significantly associated with yield deviations (Table 8.9). This means that these variables increase variability or production risk relative to the impacts of essential inputs.[9] On the other hand, whether or not a plot is located in a valley or flat land (Selale only), use of fertilizer, pesticides, schooling, number of crops planted, experience and production knowledge are positively associated with deviations in crop outputs. It means these variables contribute less to variability in production compared to essential inputs such as land and oxen.

Most variables in the milk production risk model exert a risk- reducing effect (Table 8.10). Rainfall variability, region, growing feed and availability of concentrates increases variability of milk production.[10] From Tables 8.9 and 8.10, it can be concluded that knowledge related variables (schooling and production knowledge) exert a consistently large risk-reducing effect while variability in rainfall and/or price and insecure tenure exert a risk-increasing effect. These variables can be influenced through appropriate intervention strategies or policies. Policies targeted at variables that reduce production risk, especially knowledge related, price and rainfall, can contribute to stabilization of production. If production is stable, then introduction of innovation may be acceptable at a relatively higher rate, and the efficiency with which crops and milk are produced may increase.

8.8. Summary

Production decisions and risk are examined in this chapter. Analysis of production involves evaluating the impact of inputs on the level of outputs and the efficiency with which crop and milk are produced.

Examination of the impact of inputs on the level of outputs of Selale and Ada farmers suggest that physical factors such as land, labor, oxen and seed rate exert positive and significant impacts on the amount of production attained. Management related variables such as experience, schooling and crop production knowledge

Table 8.8.: Variables used in production Risk Analysis and Criteria of Coding

Extension education	≥12, then =1, 0 otherwise[1/]
Rainfall and/or price variability	Favorable=1, 0 otherwise
Soil type	Suitable=1 0 otherwise
Steep Plots	If $ 50 % of plots are steep=1, 0
Plots in Valley or flat land	If □ 50% of plots are located in a valley or flat land=1, 0 otherwise
Number of crops planted	If plant ≥4 crops=1,0 otherwise[2/]
Fertilizer, pesticide and improved seed	If used=1, 0 otherwise
Experience	If ≥25 years=1, 0 otherwise[3/]
Production Knowledge	If scored ≥0.5=1, 0 otherwise
Schooling	If ≥ two years=1 and 0 otherwise[4/]
Atela	If used=1 0 otherwise
Stall Feeding	If used=1 0 otherwise
Grow feed	If grow=1 0 otherwise
Availability of concentrate	If used=1 0 otherwise
Veterinary service	If used=1 0 otherwise
Cross-bred cows	If adopted=1 0 otherwise
Region	For crop production, Ada=1 and Selale=0, for livestock production Selale=1 and Ada=0[5/]
Land tenure	not secure=1 secure=0

1/ The average number of extension contacts is 12 times per year.
2/ The average number of crops planted is four.
3/ The average farming experience is 25 years.
4/ The average number of school-years for the study sites is two.
5/ Selale is suitable for livestock and Ada for Crop production.

Table 8.9. Estimates of Causes of Variability in Crop Production[1/]

	All Selale Farmers	All Ada Farmers	All Test Farmers	All Control Farmers
Intercept	1.287	0.998	2.621	0.549
	-1.821	-0.624	(2.296)*	-0.816
Extension education	-0.022	(0.085	(0.011	-0.016
	(-0.202)	-0.647	-0.068	(-0.21)
Rainfall Variability	-0.586	(-0.7112	-0.7959	-0.8102
	(-2.905)*	(-3.445)*	(-2.833)*	(-3.715)*
Price Variability	. -0.4039	-0.549	-0.9761	-0.5322
	(-2.873)*	(-3.987)*	(-2.755)*	(-2.735)*
Soil type	-0.4389	0.2038	-0.5798	-0.6361
	(-3.114)*	(2.206)#	(-3.507)*	(-1.963)#
Steep plots	-0.2162	-0.5399	-0.2444	-0.2615
	(-0.702)	(-0.688)	(-0.425)	(-0.924)
Plots in valley or flat lands	-0.138	-0.667	0.973	0.644
	(-1.632)	(-2.992)*	(1.991)#	(2.709)*
Land tenure	-0.217	-0.788	-0.638	-0.964
	(-2.828)*	(-3.381)*	(-4.372)*	(-2.536)*
Fertilizer	0.506	0.4717	-0.3627	-0.2648
	(1.964)#	-0.611	(-0.719)	(-0.925)
Pesticides	0.0081	0.6608	-0.7803	-0.2658
	-0.027	(2.171)#	(-1.478)	(-0.961)
Improved Seed	-0.8589	0.9275	0.7253	-0.412
	(-2.678)*	(2.085)#	(2.442)*	(-1.342)
Schooling	0.6496	0.8064	0.6712	0.8169
	(2.174)#	(2.062)#	(2.282)#	(2.834)*
No. of crops planted	0.096	0.7698	0.7542	0.033
	-1.365	(2.671)*	(3.644)*	(2.478)*
Experience	0.43	0.8455	0.6559	0.573
	(2.433)*	(2.775)*	(3.572)*	(2.083)#
Production knowledge	0.37	0.977	0.7529	0.658
	(1.983)#	(3.285)*	(2.443)*	(2.561)*
N	216	52	114	154
R2	0.99	0.99	0.98	0.98

1/ Values in parenthesis are t-statistics. * and # indicate statistical significance at 1 and 5 percent respectively.

Table 8.9. Estimates of Causes of Variability in Crop Production[1] (Continued)

	Selale Region		Ada Region	
Variables	Test Farmers	Control Farmers	Test Farmers	Control Farmers
Intercept	-3.798	-0.67	-1.143	2.065
	(-2.697)*	(-0.866)	(-0.405)	-0.892
Extension	0.002	0.05	-0.664	0.017
education	-0.01	-0.252	(-0.449)	-0.192
Rainfall	-0.837	-0.888	-0.5213	-0.9456
Variability	(-2.141)#	(-3.253)*	(-2.163)#	(-3.92)*
Price Variability	. -0.4511	-0.8928	-0.634	-0.7707
	(-1.987)#	(-2.659)*	(-3.001)*	(-2.294)#
Soil type	-0.8951	-0.4887	-0.1923	-0.6414
	(-2.668)*	(-2.074)#	(-2.949)*	(-3.74)*
Steep plots	-0.0039	-0.2761	-0.8808	-0.1618
	(-0.006)	(-0.876)	(-0.466)	(-0.241)
Plots in valley or	0.957	0.668	-0.669	-0.99
flat lands	(2.687)*	(3.338)*	(-1.966)#	(-3.868)*
Land tenure	-0.356	-0.792	-0.737	-0.218
	(3.581)*	(-3.084)*	(-1.624)	(-2.986)*
Fertilizer	0.8318	0.3157	0.4924	-0.1243
	(2.544)*	(1.956)#	-1.89	(-0.189)
Pesticides	0.0981	0.193	0.742	0.1054
	-0.165	-0.62	(3.4)*	-0.145
Improved Seed	-0.6443	-0.3835	0.947	0.8073
	(-2.187)#	(-1.182)	(3.656)*	(2.45)*
Schooling	0.8645	0.4321	0.2261	0.4137
	-1.476	(2.278)#	(1.988)#	(2.059)#
No. of crops	0.75	0.951	0.6045	-0.395
planted	(3.319)*	(4.559)*	(1.966)#	(-0.596)
Experience	0.3	-0.124	0.88	0.393
	(3.361)*	(-2.602)*	(2.145)#	(2.917)*
Production	-0.21	0.326	0.43	0.275
knowledge	(-1.991)#	(2.889)*	(2.001)#	(2.007)#
N	88	127	25	26
R2	0.98	0.98	0.99	0.99

1/ Values in parenthesis are t-statistics. * and # indicate statistical significance at 1 and 5 percent respectively.

Table 8.10. Estimates of Causes of Variability in Milk Production

Variables	All Selale Farmers	All Ada Farmers	All Test Farmers	All Control Farmers
Intercept	-1.1596	-1.1552	-0.945	-3.559
	(-1.89)	(-3.01)*	(-1.952)#	(-4.229)*
Rainfall	-0.7111	-0.8391	-0.645	-0.526
variability	(-2.277)#	(-3.585)*	(-2.074)#	(-1.961)#
Adoption of	0.7982	0.9912	0.473	0.222
cross-bred	(1.967)#	(2.64)*	(2.624)*	-1.175
Veterinary	0.3623	0.5341	0.243	0.656
service	(2.474)*	-0.194	-1.765	(2.975)*
Extension	0.1531	0.9366	0.941	-0.219
education	-1.314	-1.193	(3.797)*	(-2.965)*
Region	0.0336	-0.5518	0.48	-0.661
	(2.00)#	(-1.60)	(2.476)*	(-3.58)*
Grow feed	-0.6047	-0.8297	-0.299	-0.93
	(-2.678)*	(-3.09)*	(-2.346)*	(-2.56)*
Use of	-0.7779	-0.9383	-0.927	-0.178
concentrate	(-4.785)*	(-2.743)*	(-2.476)*	(-4.352)*
Experience	0.7362	0.8981	0.31	0.485
	(2.425)*	-1.184	(2.586)*	(4.359)*
Land tenure	-0.858	-0.3714	-0.822	-0.556
	(-3.862)*	(-2.084)#	(-2.719)*	(-3.322)*
Use of Atela	0.3846	0.7129	0.679	0.47
	(3.374)*	(3.005)*	(2.238)*	-1.473
Schooling	0.457	0.871	0.761	0.64
	(1.974)#	(2.215)#	(1.988)#	(2.113)#
Production	0.871	-0.421	0.649	0.388
knowledge	(3.871)*	(-1.721)	(2.649)*	(1.968)#
N	216	52	114	154
R2	0.87	0.98	0.81	0.84

1/ Values in parenthesis are t-statistics.

* and # indicate statistical significance at 1 and 5 percent respectively.

Table 8.10. Estimates of Causes of Variability in Milk Production (Continued)

	Selale Region		Ada Region	
Variables	Test Farmers	Control Farmers	Test Farmers	Control Farmers
Intercept	-2.585	-4.846	-1.302	-0.642
	(-3.02)*	(-2.67)*	(-0.363)	(-4.153)*
Rainfall	-0.491	-0.418	-0.749	-0.723
variability	(-1.981)#	(-2.017)#	(-2.09)#	(-2.463)*
Adoption of	0.473	0.379	0.368	0.957
cross-bred	(2.234)#	(1.986)#	(2.796)*	(2.787)*
Veterinary	0.67	0.478	0.992	0.526
service	(3.04)*	(2.769)*	-0.767	(3.034)*
Extension	0.989	0.908	0.789	0.992
education	(2.209)#	(3.098)*	(2.552)*	(2.907)*
Region	0.64	0.702	-0.692	-0.491
	(2.533)*	(3.83)*	(-2.54)*	(-1.51)
Grow feed	0.187	0.118	-0.915	-0.098
	(3.184)*	-1.676	(-2.311)*	(-0.128)
Use of	0.044	0.108	0.451	0.729
concentrate	(2.033)#	(4.076)*	(3.125)*	(4.456)*
Experience	0.166	0.008	0.453	0.827
	(2.727)*	(4.256)*	-0.613	(2.528)*
Land tenure	-0.482	-0.419	-0.593	-0.569
	(-2.642)*	(-2.672)*	(-2.317)*	(-2.941)*
Use of Atela	0.001	0.051	0.584	0.764
	(2.094)#	-1.557	(2.521)*	-0.86
Schooling	0.251	0.737	0.519	-0.142
	(1.994)#	(2.107)#	(2.291)#	(-1.88)
Production	0.701	0.845	0.593	0.469
knowledge	(3.401)*	(2.845)*	(2.093)#	(1.969)#
N	88	127	25	26
R2	0.91	0.89	0.91	0.87

1/ Values in parenthesis are t-statistics.

* and # indicate statistical significance at 1 and 5 percent respectively.

exert a positive and significant effect on production of most crops. The impact of knowledge is not only larger than most variables but also indicates that it is locale-specific. That is, its influence is larger on the amount of crop produced by Ada compared to Selale farmers. The effect of extension education on most crops is not significant. This may suggest that either the contents or approaches of extension education are not benefiting production.

Comparison of production structure of test and control farmers within and between regions suggests similar patterns and magnitude of impacts of inputs on outputs as that obtained from comparison of regions. In general, the magnitude of the impact of inputs is larger in the Ada compared to the Selale region.[11] It implies that intervention strategies should result in larger increases in crop production in the Selale region to achieve successful adoption rates of crop technologies

Physical factors such as grazing area, concentrates, atela and number of cows exert significant effects on the amount of milk produced. The effect of these inputs are either smaller or negative in the Ada compared to the Selale region. Livestock production knowledge and schooling exert consistently significant effect on the amount of milk produced. The effects of livestock production knowledge and experience on the amount of milk produced, however, are higher in the Selale compared to the Ada region.

The statistical relationship between different combinations of technologies and inputs indicate that plot size, oxen, experience, production knowledge, and schooling positively and significantly influence production of cereals. Physical inputs and knowledge exert large, positive and statistically significant impacts on crop production when farmers use combinations of fertilizer and pesticides compared to other combinations. This implies that, given resources that are locally available, producers make strategic selection of technologies that not only guarantee subsistence requirements but also increase production with more certainty.

The second aspect of examining production decisions involves efficiency analysis. Measures of production efficiency are calculated using two methods. An aggregate measure of efficiency combining crop and livestock production is obtained from Data Envelopment Analysis (DEA). A disaggregated measure of efficiency is obtained from frontier production function.

The results of DEA indicate that Ada farmers are more efficient than Selale farmers. DEA results also reveal that test farmers of both study sites are more efficient than control farmers.[12] The results from frontier production function analysis indicate that Ada farmers exhibit relatively higher efficiency scores in cereal production than do Selale producers. Test farmers of both study sites show higher efficiency scores than control farmers. More Selale farmers are efficient in milk production relative to Ada producers.

Logit regressions are performed to identify factors that may explain differences in production efficiency. Most of the variables included in the regression model positively influence efficiency scores. The magnitude of the impacts of production knowledge and schooling are very high relative to other variables. Region contributes negatively to cereal production efficiency in the Selale and to milk production efficiency in the Ada region. It reflects comparative advantage, location and enterprise specific experiences. A combination of two innovations show a consistently large, positive and significant effect on all measures of production efficiency in both study sites. Previous discussions indicate that the combination of fertilizer and pesticides exert consistently high and positive impacts on the amount of production. It seems, therefore, that provision of these technologies to Ada farmers may produce higher efficiency. For the Selale region, however, appropriate selection has to be made between fertilizer, pesticides and cross-bred cows.

Risk-taking behavior of households and new technologies influence production efficiency. The relationship between risk, adoption and efficiency in the Selale and Ada regions of Ethiopia indicates that the degree of risk-aversion, probability of adoption, and production efficiency are higher among test farmers. It means that risk-averse behavior may not necessarily reduce the probability of adoption. It also implies that risk-averseness is related to "rational" choice of farmers, hence a reflection of knowledge possessed by households. Livestock diversification indices are lower among test farmers. It means that households who have adopted cross-bred cows reduce either classes of livestock or cattle to gain the maximum advantage from this technology. However, sensitivity indices are high among test farmers.

Production risk is examined using binary least square regression. The results indicate that extension contacts, rainfall and/or price variability and insecure tenure increase production risk.[13] Knowledge related variables such as experience, schooling and production knowledge consistently reduce variability in production. Farmers' own adjustment mechanisms, policies that increase access to technologies, reduce price variability, improve soil and water management techniques, and upgrade production knowledge of producers may help to reduce production risk.

8.9. Notes

1. Plots become soft, and are more easily exposed to soil and water erosion with increases in the frequency of plowing (more oxen input). Thus, higher oxen inputs may be associated with a decline in teff yield. Beans grown on soft seed beds tend to be weak. The plants do not resist wind drift, they may fall and decay.
2. See section 7.5.3., footnote 16, for the formula used to calculate plot characteristic.
3. In the Ada region, inputs used to produced chickpeas and the resulting outputs are added to field peas because relatively few number of farmers grow field peas in this region.
4. Labor can be hired in two ways. Laborers can enter into a contract to weed a given plot of land for a specific sum of money or be paid on a per day basis. If households chose to give contracts, they can spend 20 to 30 Ethiopian birr per hectare. Otherwise, the rate will be two dollars or a dollar plus lunch per day per person. A plot has to be sprayed two to three times. This will cost about 50 birr per hectare. Thus, it may be more advantageous for producers to hire labor than to use herbicides.
5. Schooling and extension education may not relate to local methods of production. For instance, in the absence of fertilizer farmers burn the soil which is not recommended by

"science education". Those who practice according to "scientific recommendation" may get lower crop yields than those who used traditional method (soil burning).

6. For details on the efficiency of peasants refer to Ellis (1988), Schultz (1964, 1978), Stevens, et al. (1988), and Barlett (1980).

7. This does not mean that increases in the number of technologies adopted would make farmers worse-off. Instead, given the current policies and socio-economic organization, farmers may fully exploit the potential of two innovations.

8. The fact that test farmers have received cross-bred cows doesn't mean that the probability of adoption is one (100 percent). They take risk when they purchased cross-bred cows. They did not invest their money being certain of the benefits or outcomes from the adoption of cross-bred cows. Furthermore, the logit model estimate not the actual probability but rather the odds ratio, the ratio adopting to not adopting a technology (see Chapter VII).

9. Improved seed increases crop yield variability in the Selale region. This is because farmers are exposed to this technology recently and that the improved cultivars have not performed better than the local cultivars. Plots located in a valley or flat lands increase yield variability if the soil structure is such that it retains a lot of moisture (e.g. vertisol). This is typical of the Ada region. That seems to be the reason for risk increasing effect of the locations of a plot in the Ada region compared to the Selale region.

10. Region reduces milk yield variability for the Selale area because of its suitability to livestock production compared to the Ada area. Growing feed, is partly, a reflection of access to grazing land. This factor is greatly influenced by the strength of friendship with executives of peasant associations (see chapter 5 and 6) and region potential. Thus, in a region where there is limited land for growing feed, this variable may increases variability in availability of feed and hence milk yield (e.g. Ada).

11. Infrastructure such as roads, schools, markets and access to inputs are easily accessible to Ada farmers compared to those in Selale. Region- and enterprise-specific experiences regarding crop production and the above factors may have contributed to this finding.

12. Test farmers had access to other production inputs, live closer to villages that are accessible to services such as insemination service and extension advice, gain better quality of information regarding production and marketing of crop and livestock outputs, among other things. Besides, they receive technologies which help to produce more and which may also cause re-organization in the pattern of resource allocation. Thus, they may be technically more efficient than control farmers.

13. See Chapter III, the section on householding and land tenure.

CHAPTER IX

STRUCTURAL RELATIONS BETWEEN SOCIO-ECONOMIC VARIABLES AND PRODUCTION EFFICIENCY

9.1. Introduction

Several variables directly or indirectly influence outcomes of household decisions. Most methods of household decision analysis are designed to investigate only the direct effect of factors on the outcome variable. In order to better understand the factors that contributed to consequences of decisions, however, it is important to investigate the impact of variables that directly and indirectly may influence an outcome variable. LISREL is a method that can be used for this purpose. In this study, LISREL analysis is performed to: i) test the validity of the conceptual model presented in Figure 1b (Chapter I) and ii) to examine the relationship among variables whose causal linkages are hypothesized a priori (Joreskog and Sorbom, 1989).

LISREL analysis, as the name indicates, is based on one strong assumption, that is, the causal relationship between exogenous (independent) and endogenous (dependent) variables are linear. The premise behind this assumption is that most real-world phenomena can be approximated with linear relationships and that exploring other forms of relationships (e.g., quadratic) may complicate empirical analysis and hide some important issues (Johnston, 1984; Koutsoyannis, 1984).

The LISREL model was formulated here for a large number of variables. However, it was not possible to attain statistical convergence. The estimates were not statistically acceptable, as indicated by large values of standard errors and beta coefficients, negative degrees of freedom and very high chi-square values (see Joreskog and Sorbom, 1989).

It has been pointed out in Chapters V to VIII that the influence of social networks and macro-integrating forces is reflected in differences in access to resources, economic and social status, and knowledge possessed. It appears reasonable to include these and other socio-economic variables in the LISREL model. The difficulty of testing the LISREL model with all variables that influence the processes and consequences of household decision-making necessitated choosing variables that satisfy specific criteria. Variables included in the LISREL model are those that: i) are essential to the production of both grain and livestock (e.g., land, feed area and worker:consumer ratio); ii) influence management style or human capital; and iii) reflect influences from access to resources and institutions (e.g., wealth). Three LISREL models are analyzed. Model I includes exogenous variables such as farm size, secular education, farming experience, extension education and worker:consumer ratios, and endogenous variables such as crop production knowledge, number of technologies adopted and crop production efficiency. The analysis is carried out by region, and within a region by test and control farmers. Model II is specified similarly to Model I except that wealth is added to the analysis. Model III presents results from LISREL analysis with variables similar to those in Model II but for Ada and Selale regions without categorizing farmers into test and control groups.

9.2. Results from Model I

The results of the analysis of causal relationships between exogenous and endogenous variables for test and control farmers of the Ada region are presented in Figures 9.1a and b. The results indicate that endogenous variables (e.g., production knowledge, number of technologies adopted and crop production efficiency) are positively influenced by most exogenous variables.[1] Production knowledge, farming experience and extension education exert relatively larger impacts on the number of technologies adopted and crop production efficiency of test compared to control farmers. The chi-square value of the model is smaller and this probability of obtaining a higher chi-square value is very low. It means that this is the best model to represent the conceptual framework depicted by Figure 1b (Chapter I).

LISREL analysis similar to that of Figures 9.1a and b for the Selale region is presented in Figures 9.2a and b. The results show that the number of technologies adopted, farm size, production knowledge, experience, extension education, and worker:consumer ratio exert larger and positive influences on crop production efficiency of test compared to control farmers. Worker:consumer ratio, extension education, production knowledge, and experience greatly influence the number of technologies adopted by test compared to control farmers of the Selale region. Comparison of Figures 9.1a and b, and 9.2a and b indicates that relatively more variables exert greater influences on adoption decisions and production efficiency of test compared to control farmers. Moreover, differences in the impact of crop production knowledge and number of technologies adopted on production efficiency are minimal in the Ada compared to Selale region.

Farm size is the most important single input for both crop and livestock production. The average farm size is fixed according to family size. Households are given user rights to land (Chapter II). The absence of ownership right has reduced the incentive of households to invest in land. In spite of this uncertainty, the contribution of land to crop production efficiency is high. The impacts of land and worker:consumer ratio are relatively higher among test compared to control farmers in both study sites.

One of the most important factors shaping the structure, function and decision-making processes of households is experiential knowledge. This knowledge is both a social and an individual product. It is a social product because certain ways of

Fig. 9.1a. Crop Production Efficiency & Socio-economic Variables (Ada Test Farmers)

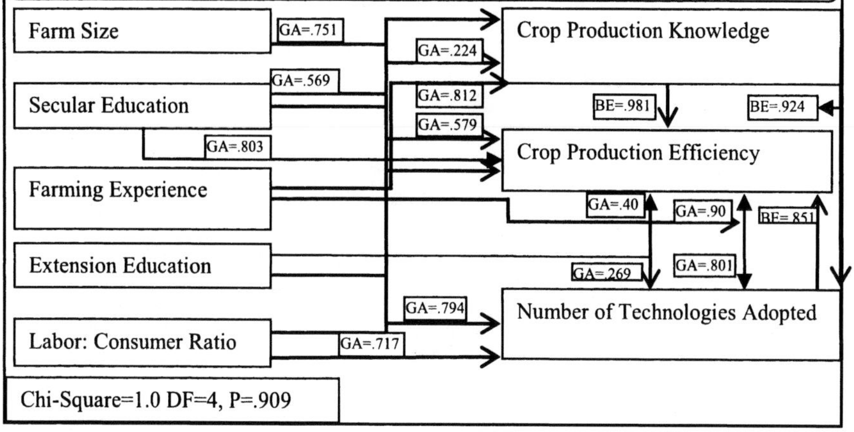

Fig. 9.1b. Crop Production Efficiency & Socio-economic Variables (Ada Test Farmers)

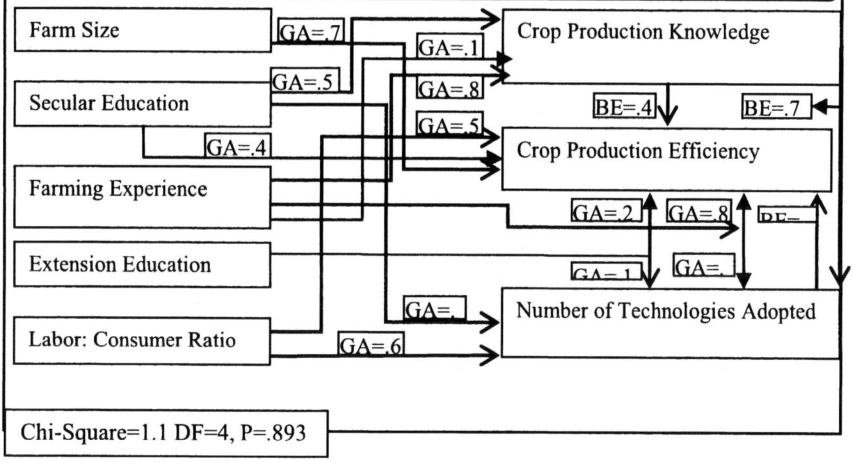

Fig. 9.2a. Crop Production Efficiency & Socio-economic Variables (Selale Test Far-

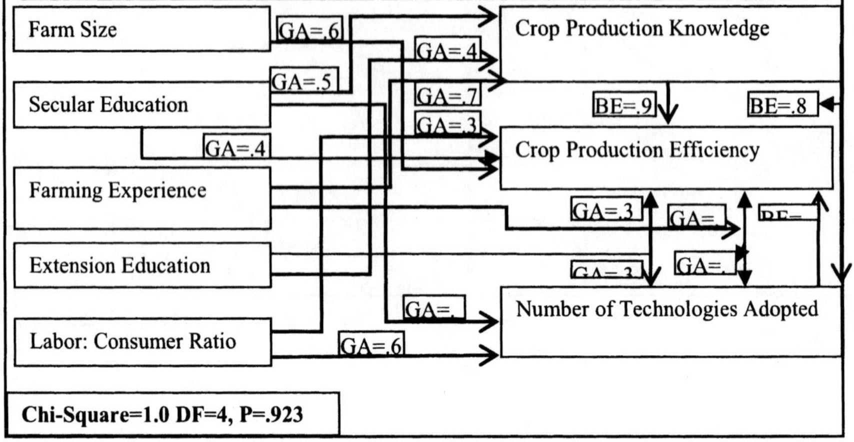

Fig. 9.2b. Crop Production Efficiency & Socio-economic Variables (Selale Control Farmers)

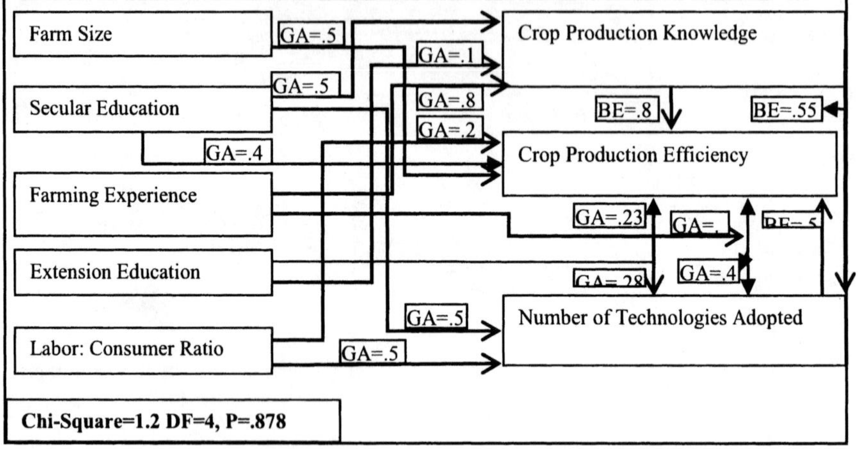

executing actions are standardized or shared. It is an individual product because households expand the body of their knowledge through information obtained from specialized or specific sources. The statistical results suggest that secular education and experience contribute significantly to crop production knowledge compared to extension education.

The contributions of secular education and experience to decisions regarding adoption of innovations and on crop production efficiency are consistently larger than other variables. Human capital, comprised of components such as vocational training, experience and skills from secular or sacred education, influences the capacity of economic agents to adjust to changes in the environment. Land-specific experience is an important factor affecting efficiency, where physical differences between parcels of plots are substantial. Location-specific experience influences the choice of farm adjustment mechanisms. If universal education is provided, those with land-specific and location-specific experience will have an extra advantage in adjusting to changes facing agriculture (Ekanyake, et al., 1989). As the results of this study suggest, crop production knowledge exerts significantly larger influences on crop production efficiency in the Ada compared to Selale region. In a similar vein, the effect of livestock production knowledge on milk production efficiency is higher among Selale producers than those in Ada.

The contributions of secular education and experience to decisions regarding adoption of innovations and on crop production efficiency are consistently larger than other variables. Human capital, comprised of components such as vocational training, experience and skills from secular or sacred education, influences the capacity of economic agents to adjust to changes in the environment. Land-specific experience is an important factor affecting efficiency, where physical differences between parcels of plots are substantial. Location-specific experience influences the choice of farm adjustment mechanisms. If universal education is provided, those with land-specific and location-specific experience will have an extra advantage in adjusting to changes facing agriculture (Ekanyake, et al., 1989). As the results of this study suggest, crop production knowledge exerts significantly larger influences on crop production efficiency in the Ada compared to Selale region. In a similar vein, the effect of livestock production knowledge on milk production efficiency is higher among Selale producers than those in Ada.

The effect of extension education is not as great as that of secular education or production knowledge. Becker (1990) argues that because weak adoption of yield-increasing technologies is explained by different opportunity costs for the labor-time spent by family members, special extension programs for family members with low off-farm employment opportunities are required to increase adoption of innovations. In the Ada and Selale regions, households have limited off-farm activities for self-employment. Thus, strengthening extension education to help producers understand innovations and to encourage investment in income-generating activities such as gardening and craft production may contribute to greater success in adoption of innovations and to increased production efficiency (see also Kebede, et al., 1990).

Various studies have demonstrated that, with increases in the number of consuming units within a household (low worker:consumer ratio), not only does production decline but also the probability of adoption of innovations may decrease (Feder, et al., 1985; Molnar, et al., 1983; Barlett, 1980). Conversely, with increases in the worker:consumer ratio, not only does the probability of adoption of innovations but also the efficiency with which they are used may increase. The results from Ada and Selale regions suggest that there may be a relatively moderate effect of the worker:consumer ratio on the number of technologies adopted and efficiency scores. If all the variables are grouped into physical (land and labor), knowledge (extension education, secular education, production knowledge and experience), and technologies, the combined effect of physical factors on production efficiency is less than skill or knowledge variables and technologies. The impact of variables on production efficiency, in ascending order, can be summarized as: physical factors, technologies and knowledge variables.

The conceptual frameworks for milk production efficiency for the test and control farmers of the Ada regions are presented in Figures 9.3a and 9.3b respectively. The results indicate that feed area and worker:consumer ratio exert larger influences on milk production efficiency. Secular education and the worker:consumer ratio greatly influence decisions to adopt cross-bred cows among test compared to control farmers of the Ada region.

The results of LISREL analysis for causal relationship shown in Figures 9.3a and b for the Selale region are presented in Figures 9.4a and b. The results indicate that adoption of cross-bred cows, feed area and experience greatly influence milk production efficiency of test compared to control farmers. Extension education, secular education and production knowledge exert larger influences on decisions regarding adoption of cross-bred cows.

Differences in the models of crop and livestock production efficiency can be seen with respect to pre-conditions or modernization conditions and the resource potentials of the two study sites. Technology transfer with the help of extension agents can produce positive results (Feder, et al. 1985). This study argues that in fact it is only when modernization conditions (for example milk collection centers, access to high demand centers, veterinary and A.I. services for milk production technology) are combined with adequate resource base (e.g., abundant source of feed) that extension services contribute to increases in milk production efficiency. For example, Ada is located near urban centers and marketing services. However, there is neither adequate grazing area nor cheaper ways of obtaining feed for milking cows. Selale, on the other hand, is located close to milk collection centers and has adequate sources of feed supply. Thus, the impact of extension education on adoption of cross-bred cows and milk production efficiency is small in the Ada compared to Selale region. Furthermore, feed area exerts a larger influence on the milk production efficiency of Selale compared to (that of) Ada farmers. On the other hand, the contribution of farm size on crop production efficiency is larger among Ada compared to Selale farmers.

Previous findings in the study of household decision-making argued that households try to avoid drudgery associated with activities which do not

Fig. 9.3a. Milk Production Efficiency & Socio-economic Variables (Ada. Test Farmers)

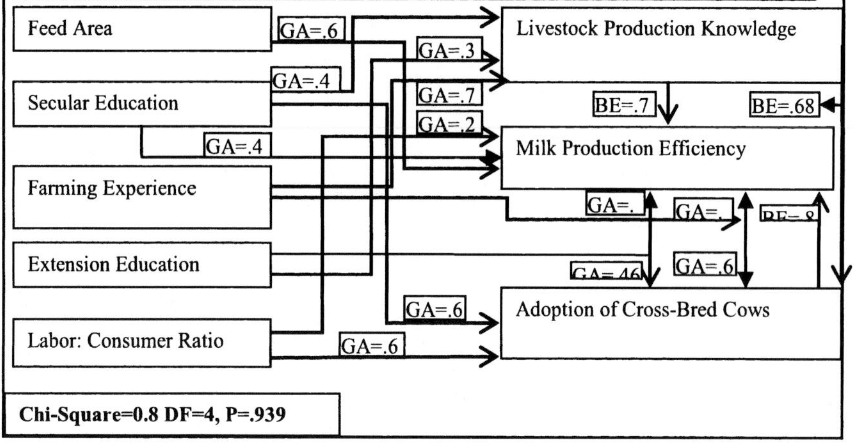

Fig. 9.3b. Milk Production Efficiency & Socio-economic Variables (Ada Control Farmers)

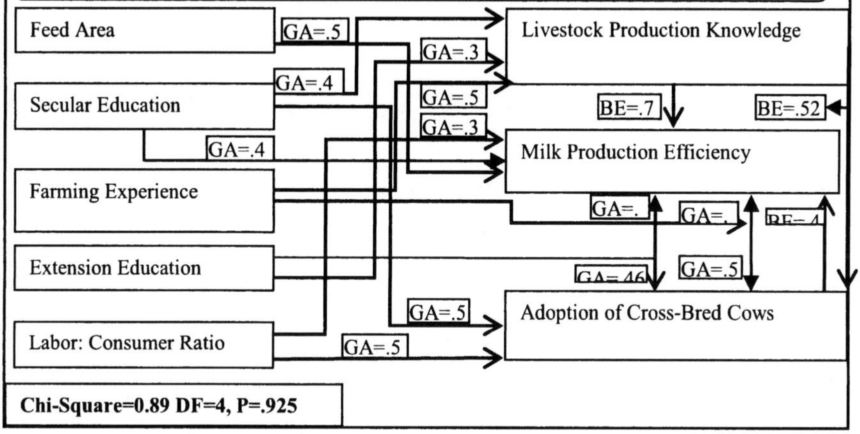

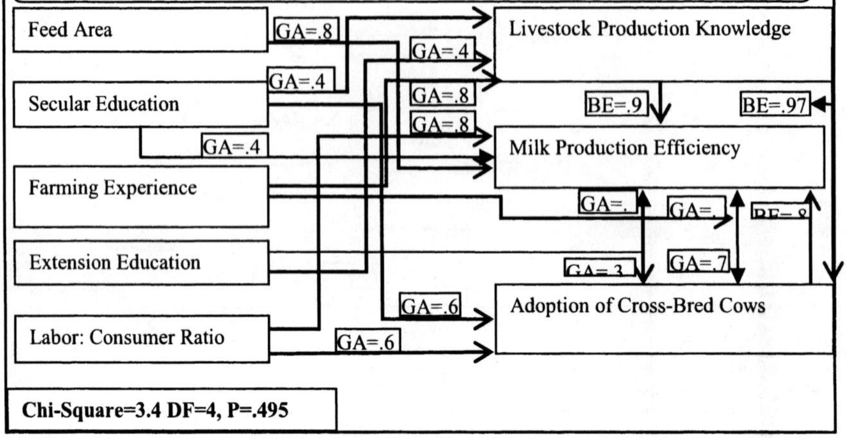

Fig. 9.4a. Milk Production Efficiency & Socio-economic Variables (Selale Test Farmers)

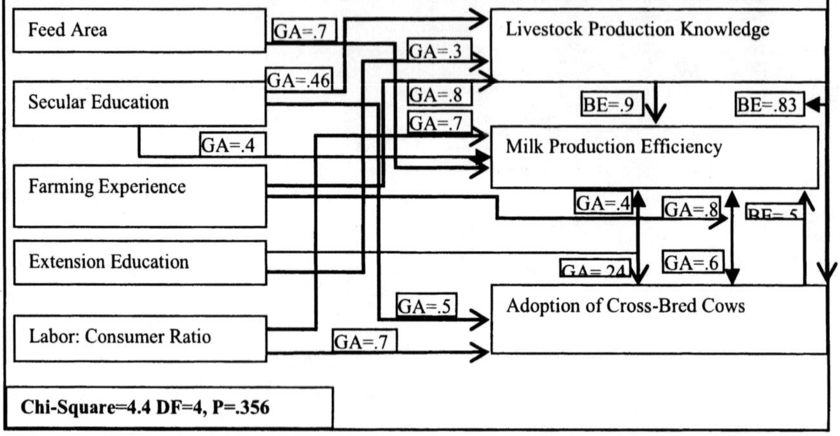

Fig. 9.4b. Milk Production Efficiency & Socio-economic Variables (Selale Control Farmers)

remunerate labor (Durrenberger, 1981; Barlett, 1980). For example, rearing cross-bred cows may not be a profitable venture for Ada producers. Cross-bred cows require a lot of labor for milking, feeding, watering, veterinary service and processing of milk into butter. The price of fresh milk is low. The amount of butter produced per liter of fresh milk from cross-bred cows is small compared to that obtained from local cows. Thus, labor used in feeding, caring and processing milk of cross-bred cows may not be remunerated adequately. This factor, ceteris paribus, may also be the reason for smaller contribution of labor to milk production efficiency. The results from Figures 9.3a and b show that the worker:consumer ratio contributes less than skill variables, such as experience and production knowledge, and physical factors, such as grazing area, to milk production efficiency.

9.3. Results from Model II

The second model involves the same set of exogenous and endogenous variables as the first model except that wealth is included. The results of LISREL analysis for test and control farmers of the Ada and Selale regions for crop production are presented in Appendices 11A and 11B respectively. The results show that secular education and farming experience exert larger influences on production knowledge. Extension education, worker:consumer ratio, wealth and production knowledge greatly influence adoption decisions of test compared to control farmers. The impacts of worker:consumer ratio, production knowledge and experience on crop production efficiency are higher among test compared to control farmers (Appendix 11A).

The contribution of worker:consumer ratio, experience, and extension education on adoption decisions is larger among test compared to control farmers. Worker:consumer ratio, extension education and knowledge greatly influence crop production efficiency of test compared to control farmers (Appendix 11B).

Causal relationships in the milk production efficiency model (type II) for test and control farmers of Ada and Selale regions are presented in Appendices 11a and b respectively. The results indicate that the impact of wealth, experience and extension education on adoption decisions is greater among test compared to control farmers of the Ada region. Production knowledge, experience and extension education greatly influence milk production efficiency of test compared to control farmers.

In the Selale region, worker:consumer ratio, wealth and production knowledge exert larger influences on the adoption decisions of test farmers. The magnitude of the effect of secular education, experience and extension education on production knowledge is larger among test compared to control farmers. Feed area, secular education, production knowledge, experience and adoption of cross-bred cows exert larger influences on the milk production efficiency of test than on control farmers.

The magnitude and direction of the contribution of knowledge variables (secular education, farming experience, production knowledge and extension education) for model II closely approximate results obtained from model I. With the exception of the impact of production knowledge, the effect of most exogenous variables is less than the values of the estimates obtained in the first model. The magnitude of the impact of wealth on decisions to adopt technologies is consistently larger than the effect of other variables. This finding confirms the hypothesis that households which are wealthy will adopt innovations more readily than those who are not. Wealthy farmers (have the means to) accumulate more information, and (consequently) with it increase their level of knowledge. Thus, they tend to be early adopters (Mason, et al., 1980; Becker, 1990).[2] Similar to model I, livestock-related variables (feed area, wealth, experience and livestock production knowledge) exert a larger influence on the milk production efficiency of Selale compared to Ada farmers. On the other hand, farm size, worker:consumer ratio, experience and crop production knowledge exert greater influence on the crop production efficiency of Ada compared to Selale farmers.

9.4. Results from Model III

LISREL analyses for crop and livestock enterprises by region are presented in Figures 9.5a and b. The results of the crop production model suggest that production knowledge, experience, wealth and secular education exert larger influences on adoption decisions. Experience and secular education strongly influence production knowledge. Crop production efficiency is largely determined by production knowledge, experience and secular education. The outcomes from the milk production model indicate that wealth, experience, and production knowledge greatly influence decisions to adopt cross-bred cows. The impacts of production knowledge, secular education and experience on milk production efficiency are greater than that of other variables.

The results of model III consolidate the findings from models I and II. The impacts of farm and feed size are lesser, while those of education, experience, production knowledge and wealth are greater in model III compared to those obtained in models I and II. The probability of obtaining a chi-square value larger than what is found (in) from this model is very low. The signs and magnitude of impacts of the estimates are acceptable. Thus, model III has performed quite well compared to models I and II.

LISREL models are useful to evaluate both the processes and the consequences of household decision-making. These models help to identify factors that need to be emphasized in the design of agricultural policies. For example in the study of farmers in the Ada and Selale regions, the two most important issues that should be addressed by policy makers are that; "how do we make use of the production knowledge of households"?, and "how do we design intervention strategies so that the know-how associated with technologies can be integrated into

Fig. 9.5a. Crop Production Efficiency & Socio-economic Variables
(Ada & Selale Regions)

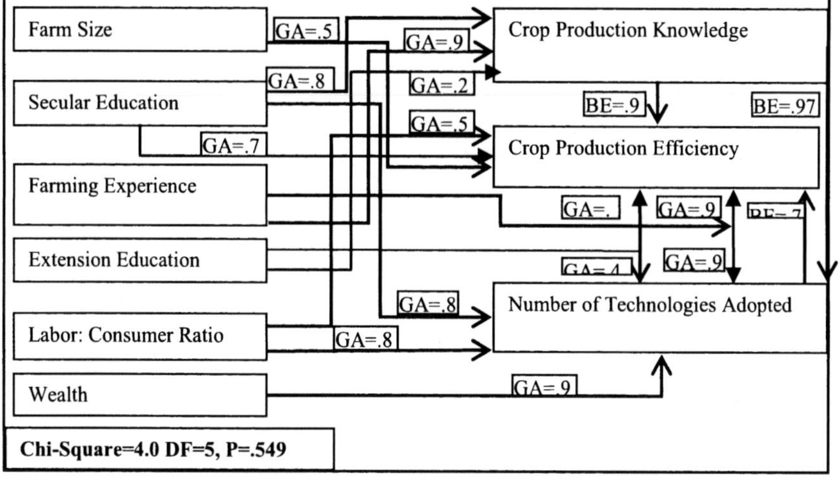

Fig. 9.5b. Milk Production Efficiency & Socio-economic Variables
(Ada & Selale Regions)

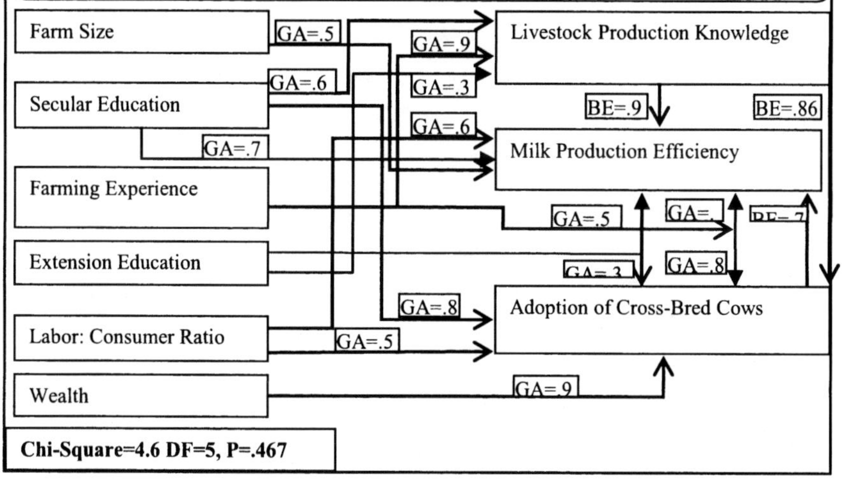

the broader sphere of experiential knowledge"? The contribution of intervention strategies that influence the skills and practices of producers would be greater than or at least equal to physical resources such as land and labor. Secular education, extension efforts and methods of utilizing indigenous knowledge may help to produce "sustainable" increases in food production among smallholders of the Ethiopian highlands.

9.5. Adoption of Mixes of Technologies and Production Efficiency

Households make strategic decisions in the adoption of selected mixes of new technologies. The manner in which they combine different innovations influences the efficiency of production. It has been shown that the impact of inputs on production efficiency is greater when producers adopt two innovations (fertilizer and pesticides). This is confirmed from a regression analysis (see Chapter VIII).

A comprehensive examination of the effect of inputs on production efficiency among households who have adopted various mixes of innovation can be obtained from LISREL analysis. LISREL analysis was conducted for crop and milk production efficiency, and for farmers who have not adopted new technologies, for those who have adopted one, two, three and four technologies. This analysis is performed by region. The results are presented in Appendix 11a &b. The findings of LISREL analysis indicated that the impacts of inputs on the efficiency with which crops or milk are produced is higher when producers adopt at least two technologies. That is, producers who have adopted two or more innovations exhibit higher production efficiency compared to those who adopted none or a single technology. This corresponds to the gradient approach to technological innovation.

9.6. Summary

The conceptual framework examined in this chapter defines the problem statement, and incorporates variables that influence the processes and consequences of household decision-making. It includes the most important physical resources (e.g., land and labor), variables that reflect the consequences of interaction between physical and non-physical resources such as institutions and knowledge (e.g. wealth and technology adoption), and measures of production efficiency.

This study hypothesizes that production efficiency is the derived outcome of the interaction between physical and non-physical resources. Differences in production efficiency may create differences on the amount of grain and livestock outputs produced by households, and their economic well-being. This measure of efficiency may vary depending on the combination of technologies adopted by producers.

The conceptual framework investigated by this study is found to be satisfactory in explaining the causal linkages between socio-economic variables in the production efficiency models which are based on the data collected from the Ada and Selale regions of Ethiopia. The results from the LISREL analysis suggest that variables that are related to cognitive ability or skills of households make the largest contribution to crop and milk production efficiency compared to physical factors such as land. Among variables that are hypothesized to influence production knowledge, the contribution of extension education is very small. The contribution of variables to production efficiency can be summarized in the following descending order: skill variables (experience, production knowledge and secular education), technologies adopted, physical factors, and extension education.

One of the problems faced by development projects in LDCs is to identify, given the scarcity financial and skill resources, a particular aspect of households that should be targeted to ensure increases in food production. The findings of this study indicate that the processes and consequences of production decision-making can be greatly influenced if intervention strategies design a method that enhance and/or utilize skill variables (especially indigenous production knowledge and schooling).

9.7. Notes

1. The beta coefficient (BE) indicating the effect of one endogenous variable on another, while the gamma coefficient (GA) indicates the effect of an exogenous variable on an endogenous variable.
2. Wealthy farmers tend to be closer to those with political influences, thus may have greater access to new technologies.

CHAPTER X

CONCLUSIONS AND RECOMMENDATIONS

10.1. Introduction

The processes and consequences of household decision-making have been investigated in two districts of the Central Highlands of Ethiopia. Selale and Ada districts are located respectively 130 km Northwest and 50 km East of Addis Ababa, the capital city of Ethiopia. Lack of peace in regions that are located north of Addis Ababa (capital city) in 1989-1991, differences in region specific history, resource potential and infrastructural facilities, and the need to compare regions that implemented different strategies to technological introduction, dictated the choice of these study sites (see Chapter II and III).[1]

The basic theme of the research is household decision-making. Households make various kinds of decisions in the spheres of production, consumption, marketing and reproduction, among other things. Several studies have demonstrated that the most important household decisions are those regarding production (Meillasoux, 1972, 1981; Schejtman, 1988; Spijkers-Zwart, 1979; Wilber, 1988; Meier, 1984).

Production decisions influence the reproduction of both producers and the farm. It is through the attainment of production objectives that most households gain status and increase their income. Furthermore, most intervention strategies in the peasant economy target agricultural production. Therefore, production decisions have been selected as the domain through which to investigate household decision-making in the Ada and Selale regions of Ethiopia.

Production decisions may involve only traditional inputs. However, these decisions may not reveal the nature of risks involved and strategies that households employ to reduce the severity of unfavorable events under situations where new technologies are introduced. Moreover, several studies have argued that with the current trend of population growth rate, increases in food production can be attained only with increases in the productivity of resources employed (Hayami and Ruttan, 1985; Dommen, 1988; Stevens, et al., 1988). Therefore, analysis of production decisions can provide better evidence to policy makers if the analysis involves households who use both traditional and new technologies.

In studying households in the Central highlands of Ethiopia, attention should be given to the structure of their economy. Farmers in the Central Ethiopian highlands practice mixed farming. Therefore, household studies should integrate crop and livestock enterprises in their analysis. The technologies investigated must be related to both crop and livestock enterprises. Consequently, three crop and one livestock production technologies are identified for detailed investigation. The technologies include fertilizer, pesticides, improved crop varieties and cross-bred cows.

The decline in food production among smallholders in the Ethiopian highlands is partly attributed to declines in the size and fertility of arable lands. The large number of livestock population in the Ethiopian highlands has contributed to overgrazing and soil erosion (Constable, 1985). Introduction of exotic genes into local breeds is expected to increase productivity of local breeds. Thus, it may serve as an incentive to smallholders to reduce the numerous but less productive local breeds. If this anticipated outcome is achieved, changes will be observed in the composition of the crop-livestock farming system. Income generated from one enterprise is hypothesized to generate money that can be invested in another enterprise. Consequently, crop and livestock production-increasing technologies should be incorporated into the study of household decision-making.

The objectives of the study of household decision-making in the Selale and Ada regions include examination of household organization, the structural relationships between household goals and strategies, differences in information and knowledge among farmers, and the institutional bases of household decision-making. These objectives relate to the processes that take place prior to decision-making. The second part of this study involves an evaluation of the consequences of household decision-making for adoption of innovations, risk-taking behavior, production efficiency and production risk. To attain these research aims, several kinds of qualitative and quantitative analyses have been employed. The conclusions emerging from this study are summarized below.

10.2. Conclusions

Relevant theories of development and strategies followed by the governments of Ethiopia in the pre- and post-1974 period are discussed in Chapter II. All strategies of development implemented in Ethiopia favored large scale farming, deprived smallholders of ownership rights to basic resources, such as land, and targeted agricultural technology as their major component. Intervention strategies which have not involved peasants in the design and implementation of technologies also have not succeeded. This failure has been reinforced by uncertainties of land tenure, farming practices that did not help to conserve and increase productivity of resources, unbounded intervention of government into social and economic organization of peasants and variability of rainfall.

Over the past several decades, changes in governments have only brought changes in the methods of surplus extraction. In the feudal regime (pre-1974), the land tenure system forced farmers to give one- to three-quarters of their crop output to landlords. Under the socialist regime (between 1974 to 1991), peasants gained only user rights and were forced to sell up to twenty percent of what they produced at low market prices. Farmers paid taxes and contributions that reduced their income by as much as twenty percent (Kebede, 1988).

Government interventions through controlled market prices, technologies which are not feasible, and radical land reform without adequate infrastructure and access to services, reinforced by variability in rainfall, have contributed to declining food production in Ethiopia since the early 50's. Expansion of farm size and increasing productivity of resources are the only means of attaining food self-sufficiency (Griffin, et al., 1985; Belete, 1989; Belete, et al., 1991). The arable lands of the highlands cannot support the existing large number of livestock and human population.[2] Respectively, the density of human and livestock population per square kilometer is 74 persons and 55 tropical livestock unit.[3] Therefore, possibilities to increase agricultural production in the highlands have to make use of improved methods of production.

New agricultural technologies have been introduced in Ethiopian agriculture since the early 1950's. However, household studies in Ethiopian agriculture have only examined factors that influenced adoption of those technologies (Waktola, 1980; Ayana, 1985). They have not examined the processes and consequences of the adoption of technologies. Their findings, therefore, cannot provide adequate evidence for the design of appropriate intervention strategies. Prior to examining household decision-making, it was necessary to evaluate the socio-economic nexus within which technologies are introduced. This issue is investigated in Chapter III.

Chapter III outlined controversies surrounding the definition of a household. It argued that these controversies arise because of disagreements regarding the structure and function of households. To facilitate analysis and interpretation of decision-making in the Ethiopian setting, a household is defined as: "People who live under the same roof, sharing in consumption, production, marketing and other decisions; and contribute to the stability of the household" (see Chapter I).[4]Furthermore, this chapter has pointed out that prior to generalizing the definition of a household, it is essential to investigate similarities in the structure and functions of households across regions or societies. If similarities cannot be established, a pragmatic approach of a location-specific definition should be adopted which takes into account of historical developments.

The socio-cultural description of households and the regions showed that i) Selale farmers specialize in livestock husbandry while those in Ada in crop production, ii) patterns of decision-making among households of Selale are more diffused than in Ada, and iii) the effect of government intervention through land reform and agricultural marketing policy is greater in the region which has relatively less infrastructure and is located farther from the capital city (e.g., Selale).

The socialist regime of Ethiopia changed the traditionally dispersed village life of peasants to that of "pseudo-city" residential patterns. This policy resulted in the strengthening cooperation among households who are not blood-related. Village-level Local relationships have become more impersonal and self-centered. Traditional value systems of households in the study sites became weaker. Farmers spent a lot of time travelling from these pseudo-cities to their fields, resulting in a loss of effective working hours (see Belete, 1989). Farmers were contacted more frequently by extension agents, government officials and development agents.

Agricultural technologies have been introduced into a peasant economy where: i) the system of values, forms of cooperation and village life are disrupted, ii) production is declining, and iii) the state is seen as an invader. Households in this kind of economy may reject technologies or intervention strategies. Prior to examining the consequences of household decision-making, it is necessary to investigate the goals and strategies, and access to and ownership of physical (e.g., land) and non-physical resources (e.g., information, knowledge).

To examine household decision-making in the Ada and Selale regions of Ethiopia, several qualitative and quantitative methods were reviewed in Chapter IV. Appropriate models were selected to study households in the study regions. Chapter IV discussed the controversies regarding methods of decision analysis employed in studying peasant economies. These methods include qualitative analysis of information and knowledge, inter- and intra-household decision-making, the institutional basis of decision-making, and patterns and determinants of adoption decisions. Furthermore, methods have been selected to examine the risk-taking behavior, the impact of input on the amount of outputs produced and the efficiency with which the outputs are produced, and production risk among households of the Ada and Selale regions. A linear structural relation (LISREL) model was chosen to investigate a model that incorporates parameters involved in the processes and consequences of household decision-making (see Chapters I and IX).

Household goals and strategies, information, production knowledge, and intra- and inter-household decision-making were investigated in Chapter V. The results indicated that strategies that were proved to be successful tend to have a stronger relationship to household goals. Strategies that are directly related to livestock husbandry are ranked higher in the hierarchy of goals of Selale than Ada farmers. On the other hand, strategies that help boost crop production contributed more significantly to the attainment of goals of Ada than of Selale farmers. Producers rank strategies hierarchically, and goals ranked high in the hierarchy are valued high on subsequent goals (e.g., securing subsistence on livestock husbandry).

Chapter V has shown that the primary goal of households in both study sites is securing subsistence food requirements or reducing the chances of failure in crop and livestock production. Securing subsistence requirements is ranked high by at least eighty percent of producers, followed by livestock husbandry. This result implies that households follow the strategy of safety-first. That is, households implement strategic decision-making criteria and employ various kinds of adjustment mechanisms to avoid the risk of falling below subsistence requirements. These mechanisms or strategies include planting more than one crop, raising more than one class of livestock, pursuing various cropping patterns, and so on. These strategies are categorized as risk-minimizing and are ranked high on the goals of securing subsistence and livestock husbandry compared to opportunistic and long-range strategies.

The implementation of strategies to attain household goals requires gathering, processing and integrating information with social experience. Households gather information from formal (e.g., extension agents) and informal (e.g., colleagues) sources. The extent to which information is subjected to conscious processing depends on the framework of thinking. This framework is shaped by indigenous knowledge or social experience and schooling.

Formal sources of information are less accessible to households compared to informal sources. Information from extension agents tends to favor peasant associations that are close to cities and service cooperatives. Peasant associations whose officials have strong connections to extension agents and are elected at sub-district, district or regional levels, gain greater access to information compared to peasant associations whose officials are not elected at sub-district or district levels. Similarly, producers who have developed stronger friendship with politicians and extension agents received more information than those who have not. This is particularly true of farmers which are relatively wealthy. Moreover, test farmers gain greater access to sources of information compared to farmers who have not develop stronger friendship with influential people (e.g., politicians).

Variation in access to information is anticipated to be reflected by differences in production knowledge. Production knowledge is examined using answers given to problem solving tests and causal explanations of selected production problems. The results from problem solving tests indicate that test farmers and politicians received higher scores on those tests than control farmers. Households in the Selale and Ada areas obtained higher scores in livestock and crop production problems respectively. This means that production knowledge is locale-specific. Furthermore, production knowledge varies by age and sex. Intra-household division of labor is influenced by differences in production and marketing knowledge of household members. As opposed to production knowledge which increases with the age of farmers, higher scores on marketing knowledge are achieved by younger farmers. Households living in areas closer to big cities and market centres such as in Ada gain higher scores on marketing knowledge than those who do not live in the vicinity of cities and commercial centers (e.g., Selale farmers).

The results of causal attribution or explanation of production problems indicated that young and non-schooled households provided few causes and solutions to problems compared to older non-schooled farmers. Older literate households identified several causes and solutions to production problems compared to younger literate and illiterate farmers. These results indicate that education enables households to relate production problems to experience and outside information. Thus, households can undertake a comprehensive examination of production problems.

Households are assigned duties according to their age, sex, and knowledge. Stratification of labor by age and sex corresponds to the timing of farming operation and the growth stages of crops and livestock. In general, decisions in the Selale area are made jointly by the husband and wife moreso than in the Ada region.[5] The contribution of children to crop production decisions is low in the Selale region but high in the Ada region. The opposite holds for decisions regarding livestock production.

Marketing of crop products are made jointly or only by the husband. However, milk production decisions are the domain of the wife. Intra-household decision-making assumes different patterns when decisions involve marketing outputs obtained from risky and expensive technologies (e.g., cross-bred cows). Decisions that require exchanges of labor are shared by relatives and friends (e.g., plowing and harvesting). Decisions that are critical to the survival of households (e.g., use of income and sales) entertain little input from people living outside the house. The choice of goals and strategies, access to information, intra- and inter-household decision-making are influenced by institutions that operate within and outside the village.

Chapter six examined the institutional basis of household decision-making. It is argued that peasants of the Selale and Ada regions operate in an environment that includes inter-household, micro-macro and macro-integrating institutions. These institutions require contributions or sharing of labor, crop output, tools, livestock and livestock by-products and production risk. Consequently, processes of sharing or making community contributions influence household goals, resource allocation strategies and patterns of decision-making. Institutions influence households differently depending on family size, number of relatives, participation in political or religious activities and markets. The most important resources affected by institutions are labor, land, information, production knowledge and livestock. These influences induce changes in allocation strategies. The effect of institutions on these resources are reflected in crop and livestock output as well as in the efficiency of production.

Institutions are analyzed at three levels. These include the micro-institutions (e.g., the coffee ceremony), the micro-macro interface (e.g., social networks) and macro-integrating forces (e.g., markets). The coffee ceremony is an institution where all forms of cooperation begin. At a village level, social relationship between farmers are not as strong as those between neighbors. Village-level institutions tend to be more specific in their function and structure.

The micro-macro interface institutions that influence households' economic and social organization and their decisions include: i) production, ii) insurance or credit and ii) distribution. The effect of these institutions on household decision-making can be examined with respect to the availability of labor. The results for Selale and Ada regions indicate that households spend about half of the year participating in social networks. Households in the Selale region spend several days in distributive institutions, those of Ada in insurance or credit institutions. The distributive nature of institutions carry with it the power of gift and reciprocity. Thus, the strength of relationships among Selale farmers are stronger than those of the Ada region. This finding may imply that the primary risk- minimizing strategies of households in the Ada region include economic management while Selale farmers use social institutions.

The functions of inter-household and micro-macro interface institutions are affected by local markets, marketing boards, extension agents, political organizations and development projects. Institutions of such nature are, for the purpose of this study, called "macro-integrating" forces. One example of macro-integrating forces is agricultural policy. The responses of Selale and Ada farmers to the elimination of forced procurement and low market prices for grains, and the possibilities of expanding farm size indicate that they will increase the area allocated to crops which are widely used as staples (barley) or command high monetary value (teff). With liberalization of the marketing system, households may be willing to adopt innovations that help to increase production of these crops.

One method of examining the impact of institutions on household characteristics is to identify a variable whose magnitude is greatly affected by institutions. In this study, wealth and worker:consumer ratio were selected as a proxy to reflect the effects or consequences of the interaction between institutions and physical resources owned or used by households. Worker:consumer ratios and wealth influence the structure, function and stability of households. These parameters vary by family-cycle. The results of this research indicate that increases in family size are associated with increases in wealth and the worker:consumer ratio. The increase in the magnitude of these variables is higher among the test than among control farmers. Issues examine din chapters II to VI show that social, historical, regional, farmers-specific aspects shaped by institutions. The interaction of institutions with resources, goals, strategies, knowledge is reflected by decisions regarding resources allocation strategies. These strategies may involve adoption of technologies, willingness to take risks, adjustment mechanisms to reduce the effect of production risk, and selection of combination of production inputs that enable production of food for family requirements.

Chapter seven investigated the impact of social, economic and technical factors on decisions to adopt new technologies. Knowledge-related variables, such as schooling, experience and extension contacts, exert positive and significant effects on adoption of most technologies. Economic variables, such as wealth, off-farm income and farm/feed size, increase the probability of adoption of technologies. High input prices, however, reduce the probability of adoption. Social variables, such as the number of relatives and the high worker:consumer ratio, and technical variables, such as productivity of technologies, exert positive impacts on adoption of most of the innovations in both study sites. The worker:consumer ratio does not exert greater impact on crop production technologies than on cross-bred cows. Contrary to the hypothesized differential impact of the worker:consumer ratio on technologies that require a large labor input, the impact of this variable relates more to the objective of securing adequate food grain than to the requirements of new technologies. Cross-bred cows require greater labor input per head compared to local cows. The empirical result shows that the impact of worker:consumer ratio is greater on crop production than on milk production.

The influence of production knowledge and experience on adoption decisions is not only positive but greater than most variables. The results also confirms the hypothesis that the impact of these variables is greater on technologies that are proven to have a more certain outcome (e.g., fertilizer and pesticides) than on technologies which are either expensive or risky (e.g., cross-bred cows and improved seed).

Adoption of cross-bred cows in the Ada region is negatively influenced by any variables compared to the Selale region. Adoption of improved crop varieties are negatively influenced by several variables compared to fertilizer and pesticides. It is, therefore, plausible to infer that in regions where modernizing conditions or preconditions are unavailable and the possibilities are limited, technologies may not be successfully adopted.[6] For example, Ada is more conducive for crop production while Selale for livestock production.

Households' goals and strategies in the Ada and Selale regions can be represented hierarchically (Chapter V). Similarly, decisions regarding production, marketing and consumption activities tend to assume sequential patterns. Previous studies of adoption of technologies have found that farmers' decision patterns are sequential (Kebede, et al., 1990; Molnar, et al., 1983; Ahmed and Ruttan, 1988). The study in the Ada and Selale regions, however, indicates that this sequentiality of adoption holds for technologies which have either proved to be successful or do not involve huge financial investments. Typical examples are fertilizer and pesticides as opposed to improved crop varieties and cross-bred cows.

With respect to prospects for the diffusion of innovations, the results of this research indicate that with increases in the number of adopting farmers, the probability of adoption of cross-bred cows, pesticides and fertilizer increases. The primary goal of households in the study sites is to secure adequate food for their families. This goal can be attained through adoption of crop production technologies. If households secure their food requirements, they may be willing to take the risk of adopting other innovations. The findings of this study indicate that the probability of adoption of cross-bred cows increases sharply among farmers who have adopted most of the available crop production technologies.

Studies about the effects of the green revolution in India have found that unequal access to new crop varieties and other complementary inputs have resulted in increased income disparity (Adelman, et al., 1973; Hayami and Ruttan, 1985; Ahmed and Ruttan, 1988). The research in the Selale and Ada regions of Ethiopia indicates that with increases in a household's wealth, the number of technologies adopted increases. Furthermore, differences in the amount of wealth possessed by households who adopted three or more technologies is less than among those households who adopted less than three technologies. This conclusion, however, holds true among households which are comparable with respect to major parameters except in the ownership of cross-bred cows.[7]

Awareness of new innovations helps to increase the probability of adoption. The results of this study indicate that with increases in the number of technologies adopted, the index of awareness not only increases but also shows less variability. A greater worker:consumer ratios helps to increase the chances of securing family food requirements. With increases in this ratio, therefore, the number of technologies adopted increases. The difference in worker:consumer ratios among farmers who have adopted three or more innovations tend to be less than those who have adopted one or two technologies.

As opposed to the usual assertion that relatives serve as an insurance for households which have adopted innovations, increases in the number of relatives are associated with a decline in the number of technologies adopted (e.g., in Ada region).[8] However, in the Selale region increases in the number of technologies adopted are associated with increases in the number of relatives. This is probably due to the stronger relationship between and cooperation among relatives in the

Selale region.

One of the most important characteristics of households in economies operating under conditions of uncertainty is their risk-averse behavior. It is difficult to obtain accurate empirical values that reflect this inherent characteristics of peasants. Approximations range from those using hypothetical gambles to those that are based on actual decisions. This behavior strongly influences all kinds of household decisions.

Risk-averse behavior of households in the Selale and Ada regions is examined using information from actual production decisions. Risk sensitivity indices that measure the amount of money required to invest per unit of land or cow without risking family food requirements are determined. Moreover, crop and livestock diversification indices are constructed. It should be noted that information gathered from actual decisions may indicate "rationality" as well as risk-averse behavior of households.[9] Thus, interpretation of results from actual decision to characterize the behavior of peasants should be made with great care.

The results of sensitivity and diversification indices indicate that risk-taking behavior varies with the concentration of farmers in specific enterprises and regional potentials for either crop or livestock production. With respect to crop sensitivity indices, most Ada farmers are in the low risk-averse group while Selale farmers are moderate to high risk averters. The opposite holds true for livestock production indices. Producers are willing to take risk in enterprises where they have location and enterprise-specific experiences. These experiences are reinforced by availability of resources for expansion of a particular type of farming or enterprise (Ekanyake, et al., 1989). Thus, Ada and Selale farmers can take greater crop and livestock production risks respectively.

With respect to crop diversification indices, most producers in the Ada region are relatively more risk-averse than Selale farmers. The majority of study farmers are either moderately or extremely risk-averse when the index of classification is livestock diversification. However, when each sub-class of cattle are considered as a separate class (e.g., cow, ox, heifer, steer and calves), Selale farmers emerge with a higher index than Ada farmers.

Livestock diversification indices are lower among test farmers. This implies that households which have adopted cross-bred cows reduce either classes of livestock or cattle to gain the maximum advantage from this technology. However, sensitivity indices are high among test farmers. This means that households require substantial amounts of money to avoid falling below subsistence requirements when they invest large sums of money in a technology. With increases in the number of technologies adopted, both kinds of risks tend to increase.

The process of characterizing risk-taking behavior does not indicate factors that shape this behavior of households. The second aspect of examining risk-taking behavior involves identifying factors that are strongly correlated with it. The results of relating risk-taking behavior to socio-economic factors indicate that most variables exert a negative (risk-reducing) effect on crop and livestock sensitivity or diversification indices in both regions. The impact of crop and livestock production knowledge on measures of risk-taking behavior, however, are larger than other variables among Ada and Selale farmers respectively.

Chapter VII also investigates the impact of risk-taking behavior on decisions to adopt new technologies. Risk-averse behavior of households reduces the probability or chances of adopting technologies in both Selale and Ada regions. Households may be willing to take more risks if they receive insurance from social networks and governments or are rich. Farmers of the Ethiopian highlands are exposed to crop and livestock risks caused by vagaries in the amount and distribution of rainfall. Thus, social networks may not adequately insure households from unfavorable events. Consequently, governmental or non-governmental assistance should accompany innovations to mitigate the effect of risk and increase the probability of adoption of technologies. Such interventions would help to increase food production.

Chapter eight examined production decisions and risk. Analysis of production decisions involves evaluating the impact of inputs on the level of outputs and the efficiency with which crops and milk are produced. Prior to examining the importance of inputs and production efficiency, the structure of production of test and control farmers by study sites were compared. The results indicate that adoption of technologies has resulted in greater increases in production among test farmers compared to control farmers. Productivity in milk production is higher among Selale farmers than among Ada farmers. However, Ada farmers are more productive in crop production compared to Selale farmers.

The impact of inputs on the magnitude of outputs is examined in three categories. These include inter-region, intra-region and inter-farmer (all test versus all control) comparisons. Examination of the impact of inputs on the level of outputs of Selale and Ada farmers suggest that physical factors such as land, labor, oxen and seeding rate exert positive and significant impacts on the amount of crop production realized. Management-related variables such as experience, schooling and crop production knowledge exert positive and significant effects on the production of most crops. The impact of crop production knowledge is not only greater than most variables but also indicates that it is location-specific, that is, its influence is larger on the amount of grain produced by Ada compared to Selale farmers. The effect of extension education on most crops is not significant. This may suggest that either the content or approach of extension education are not geared towards increasing agricultural production.

The pattern and magnitude of the impacts of inputs on the amount of grain produced is similar across farmers and regions. However, the magnitude of the impact of inputs on crop production are larger in the Ada compared to the Selale region. Crop production technologies in the Selale region can be successfully adopted if increases in output are substantially larger compared to output obtained from using traditional inputs.

Physical factors such as grazing area, concentrates, atela (by-product of beer brewing) and number of cows exert significant effects on the amount of milk produced. However, the effects of these inputs are either smaller or negative in the Ada compared to the Selale region. The major reasons for this finding are differences in the resource base, their enterprise-specific experience and the availability of preconditions (infrastructure) (see chapter III). Similar to crop production functions, livestock production knowledge and schooling exert consistently significant effects on milk production. However, the impact of livestock production knowledge and experience on the amount of milk produced is greater among Selale farmers when compared with Ada farmers.

This study hypothesizes that producers who have adopted different combinations of technologies are more efficient compared to those who have adopted one or none of the technologies. The results from the Selale and Ada regions suggest that physical inputs and knowledge exert large and significant positive impacts on production when farmers use combinations of fertilizer and pesticides. This implies that, given resources that are locally available, producers make strategic selection of technologies that not only guarantee subsistence requirements but also increase production with more certainty.

The differential impact of inputs and combinations of technologies on the level of output is hypothesized to produce differences in the efficiency with which outputs are produced. Production efficiencies are calculated using two methods. An aggregate measure of efficiency combining crop and livestock production is obtained from Data Envelopment Analysis (DEA). A disaggregated measure of efficiency is obtained from a frontier production function.

The results of DEA indicate that Ada farmers are more efficient than Selale farmers. Examination of information, experience, knowledge and resource bases of Selale and Ada regions indicate that Ada farmers may be more efficient in crop production, while those of Selale lead in livestock production efficiency. The aggregate efficiency scores, however, indicate that the region that concentrates more in crop production is efficient compared to the region specializing in livestock production. This implies that Ada farmers produce crop and livestock outputs with a minimal outlay of inputs. Compared to Selale, the relative scarcity of arable and grazing land in Ada may force farmers to use these resources more efficiently. DEA results also reveal that test farmers of both study sites are more efficient than control farmers. Several studies in crop-livestock production systems have argued that intervention strategies in one enterprise help boost production in a complementary enterprise. Increases in crop production as a result of fertilizer and pesticides, for example, increases household income which can be invested in livestock production. The converse also holds true. This may be one of the reasons for higher aggregate efficiency scores of test compared to control farmers.

The results from frontier production function analysis indicate that Ada farmers exhibit relatively higher efficiency scores in cereal production compared to Selale producers. Test farmers of both study sites show higher efficiency scores than control farmers. Most Selale farmers are efficient in milk production compared to Ada producers. These findings confirm the hypothesis that the success of intervention strategies is determined not only by recognition of appropriate combinations of technological packages, but also by the comparative advantages of regions with respect to production of specific crops or livestock.

Characterization of farmers with respect to differences in production efficiency scores does not indicate the kinds of strategies that should be designed to enable increases in agricultural production by most households. Logit regression analysis was performed to identify factors that may explain differences in production efficiency among the study farmers. Most of the variables included in the regression model positively influence efficiency scores. The magnitude of the impacts of knowledge- related variables (i.e., production knowledge and schooling) is very high relative to other variables. The region variable contributes negatively to cereal production efficiency in the Selale and to milk production efficiency in the Ada region. This finding reflects the comparative advantage, location- and enterprise-specific experience. In other words, it favors livestock and crop production in the Selale and Ada regions respectively.

The results of the regression analysis also indicate that different combinations of innovations affect production efficiency of households differently. A combination of two innovations show a consistently large, positive and significant effect on all measures of production efficiency in both study sites. It seems that with current policies and production constraints, households can attain maximum efficiency from adopting two innovations. From the results of Chapter VIII, provision of fertilizer and pesticides to Ada farmers may help to increase crop production efficiency.

Group discussions in the Selale region indicate that fertilizer, pesticides and cross-bred cows are ranked high on their list of priorities. Conscious evaluation has to be made to recommend the appropriate combination of technologies in the Selale region. Several farmers burn the soil, over-turn green plants and leave stubbles to decay on the farm as methods of maintaining soil fertility. Thus, fertilizer is ranked third in their list of preferred production technologies. Nevertheless, the majority of farmers indicate that they prefer to purchase pesticides and cross-bred cows.

The relationship between risk, adoption and efficiency in the Selale and Ada regions of Ethiopia indicates that the degree of risk-aversion and production efficiency are higher among test farmers. This implies that risk embodied in new technologies or risk-taking behavior of farmers does not necessarily reduce the probability of adoption.[10]

Chapter VIII also has investigated the production risk of households in the Selale and Ada regions using binary least squares regression. The results indicate that rainfall and price variability, soil and plot characteristics, and uncertain land tenure are positively and significantly associated with yield deviations. On the other hand, black soils, location of a plot, use of fertilizer, pesticides, improved seed, schooling, planting more than one crop, experience and production knowledge are negatively associated with deviations in crop outputs. Most variables in the milk production risk model exert risk-reducing effects. Rainfall variability, region and availability of concentrates increase variability of milk production. In summary, knowledge-related variables (schooling and production knowledge) consistently exert large production risk-reducing effects while variabilities in rainfall and/or price and insecure tenure exert risk-increasing effects. Policies targeted at variables that reduce production risk, especially knowledge-related, prices and rainfall, can contribute to stabilization of production.

The last chapter presented the conceptual model which incorporates variables that influence the processes and consequences of household decision-making. Four types of models are tested. The first two models are the same except that one of the models includes wealth in the analysis. The third model is examined by region rather than by test and control group as in the first two models. The last model divides farmers by the number of technologies they adopted. The statistical results indicate that the models are well suited to represent the conceptual framework for decision-making. Accordingly, the findings of these models consolidate results obtained in previous chapters. The results suggest that knowledge- or skill-related variables contribute significantly to production efficiency compared to physical factors such as land or labor. When the contributions of education, experience and extension education to production knowledge are larger, the impact of these factors on production efficiency tends to be greater. Extension education contributes the least to production efficiency compared to physical factors (i.e., land and labor) and skill factors (i.e., education, experience, and production knowledge). The contribution of factors to production efficiency can be ranked in the following descending order: skill variables (e.g., experience, secular education and production knowledge), consequences of access to resources or institutions (e.g., wealth), technologies adopted, physical factors (e.g., land and labor) and extension education. Analysis of linear structural relations for farmers who have not adopted any technology, those who have adopted one, two, three and four technologies indicate that the impact of inputs on production efficiency was greater among farmers who have adopted at least two technologies.

10.3. Recommendations

This is the first household study that incorporates the processes and consequences of decision-making in Eastern Africa. The methods employed in this research are applicable to the study of households who practice crop and livestock production in other countries. The findings of this study are expected to be useful for planners in East Africa and in other LDCs where households who practice mixed farming.

The research reported in this dissertation is the first attempt to provide evidence about important parameters of decision-making that should be recognized by policy makers in order that smallholders can attain sustainable increases in food production. The land tenure system in Ethiopia was modified and grain markets were liberalized in 1990. The research findings are expected to reflect important criteria used by peasants in decision-making, which may not have been previously observed, prior to these changes. Based on field research in the Ada and Selale regions, and prior experience with technological interventions in the Ethiopian highlands, the following recommendations are submitted for consideration by policy makers:

1. Feasible intervention strategies should be based not only on the consequences of household decision-making, but most importantly on the processes that led to these consequences. Holistic study of the processes and consequences of decision-making requires the incorporation of qualitative and quantitative methods of analysis. Qualitative analysis should involve examination of the cognitive processes, socio-historical accounts, agricultural production, and the social behavior and organization of peasants. If intervention strategies are to have their greatest impacts on household food production prospects, the households' knowledge of the economy and farming systems, and institutions should carry as much weight as the physical resources in the design, generation and diffusion of technologies. Development strategies that are based on either processes or consequences are doomed to fail (see for example Ahmed and Ruttan, 1988; Lele, 1976)
2. Information, extension and secular education help to increase the cognitive ability of producers. The contribution of information to cognition is enhanced by a householder's social experience. Secular education and differential access to information cause differences in production knowledge. Experience and production knowledge greatly influence the selection of goals and strategies, decisions to adopt technologies, production efficiency and risk when compared with the impact of secular and extension education. This implies that the content and/or approach of secular and extension education are not entirely compatible with the experience of households. Redesigning the contents of formal sources of information (e.g., extension education) and the agricultural education programs of high schools, colleges and universities to focus on problems of the peasant economy may increase the contribution of education to household incomes, production and standard of living.
3. The traditional approach of providing technologies as a package is not feasible. Assessment of the goals and strategies, resource base and experience of peasant economies should precede the design of policies regarding technological intervention. With the current agricultural policies and the resource base of the Ethiopian highlands, farmers can benefit more from increased production if they adopt at least two technologies. Combinations of technologies vary depending on whether the region specializes in crop or livestock production. For example in areas specializing in crop production (e.g., Ada), fertilizer and pesticides/herbicides seem the appropriate combination. In regions that specialize in livestock production (e.g., Selale) and have adequate labor to use fertility-maintaining methods (e.g., soil burning and manuring), pesticides/herbicides and cross-bred cows are recommended.
4. Policy makers should examine the data from which research reports are generated before designing interventions strategies. For instance, the results of this study show that farmers who received cross-bred cows are

not better-off in all parameters, as has been reported by Finnida/ILCA (Finnida, 1989; Gryseels, et al., 1985, 1986). Farmers who purchased cross-bred cows and were better-off were compared with randomly selected farmers, most of whom were poor. Differences between the farmers compared were attributed to the Finnida project, without accounting for differences before the project provided cross-bred cows. Reports of this nature send wrong signals to policy makers about the effect of these kinds of projects.

5. The availability of preconditions or modernization conditions is critical to whether or not innovations will be adopted. To insure sustained use of technologies, policy makers should assess not only the requirements for adoption but also factors or pre-conditions necessary for the utilization of outputs of technologies.
6. Farmers of Selale and Ada regions rely on their knowledge and their resource bases to minimize the effect of undesirable events. Nevertheless, when they are exposed to uncontrollable events (e.g., low rainfall) strategies should be designed to make rehabilitation faster. Early warning systems should target methods to increase skills of producers (e.g., via an appropriate educational curriculum) and technologies such as soil and water conservation techniques to reduce the impact of production risks.
7. This study has demonstrated that non-physical resources such as knowledge-related variables are important to the survival of households and maintenance of the environment. In fact, losses of indigenous knowledge and institutions have social and psychological, as well as physical and economic costs. Avoidance of this loss should aim at preservation and rejuvenation of indigenous knowledge and institutions, thus helping to ensure sustainable development among the world's peasant communities.

10.4. Notes

1. The current transitional government of Ethiopia is headed by the TPLF, under the coalition called the Ethiopian Democratic Revolutionary Front (EPDRF), that ousted the socialist government in May 1991 (see Chapter II).
2. Ethiopia has the largest livestock herd in Africa and stands eighth in the World with respect to livestock population (Constable, 1985).
3. One tropical livestock unit is equivalent to an animal weighing 250 kilogram.
4. Households in the Ethiopian highland can be identified by the compound or locale (see Chapters I to III).
5. However, the pattern of decision making tends to be enterprise-specific; in both regions, crop production decisions are made by the husband, livestock production by the wife.
6. Furthermore, in farming systems where traditional technologies perform better than new technologies (e.g. local versus new crop cultivars), the probability of adoption will be low

and few farmers will be willing to adopt.

7. If access to innovations is limited to those who can afford to purchase them, then the income differential between the rich and the poor increases. On the other hand, if land markets are liberalized, the gap between the poor and the rich may increase (see Tecle, 1975; Belete, 1989).

8. This finding also may imply that in areas where extended families are few, farmers tend to adopt several technologies because there are few relatives with whom resources are shared.

9. For example, farmers plant several crops to reduce the risk of not securing adequate subsistence requirement or to maximize the use of family-owned resources (e.g. Norman, 1978). Planting several crops also indicates greater value of the diversification index (e.g. greater risk-averse behavior).

10. This finding seems to contradict the results obtained from the statistical relationship between risk-taking behavior and the probability of adoption in Chapter VII.* However, the results in Chapter VII refer to the magnitude of net impacts rather than graphical exposition of the general relationship between risk and probability of adoption. Furthermore, farmers who are rational in their patterns of decision-making may also be risk-averse. Therefore, it is possible to expect this kind of finding.

APPENDICES

Appendix 1. A Comprehensive Conceptual Framework for the Study Sites

Figure 1. Socio-Economic Factors and Production Efficiency: A Holistic Approach

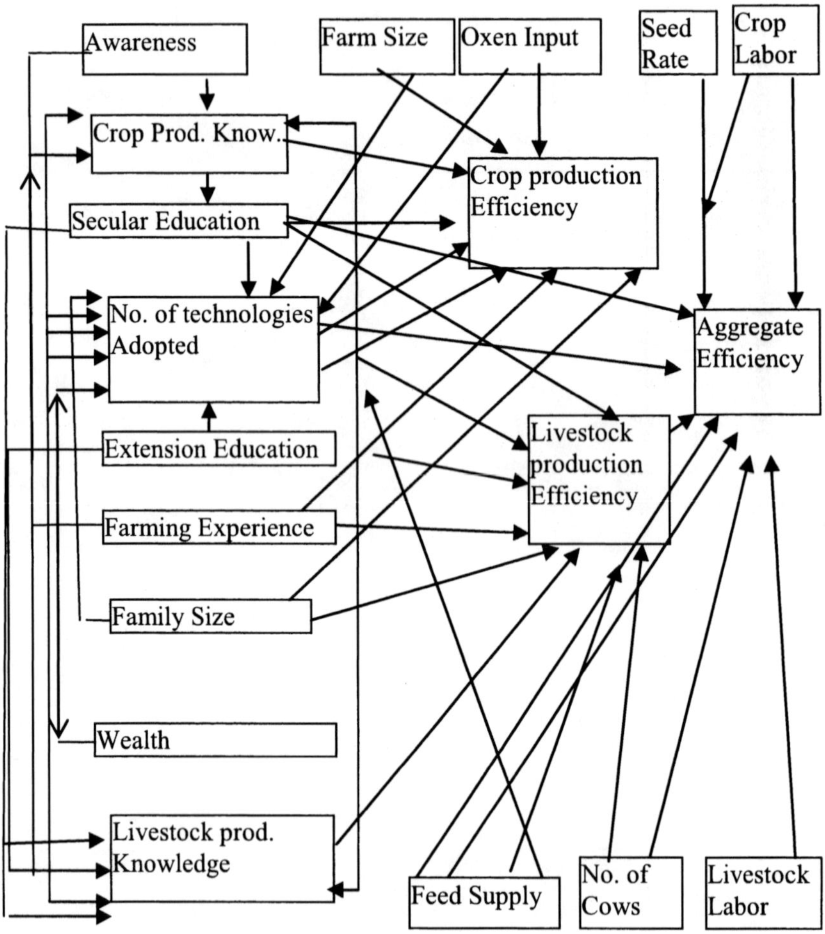

Appendix 2. Patterns of Rainfall In Ada and Selale Regions

Fig. 2a. Mean Annual Rainfall (mm) for Ada Regioon, 1951-1989

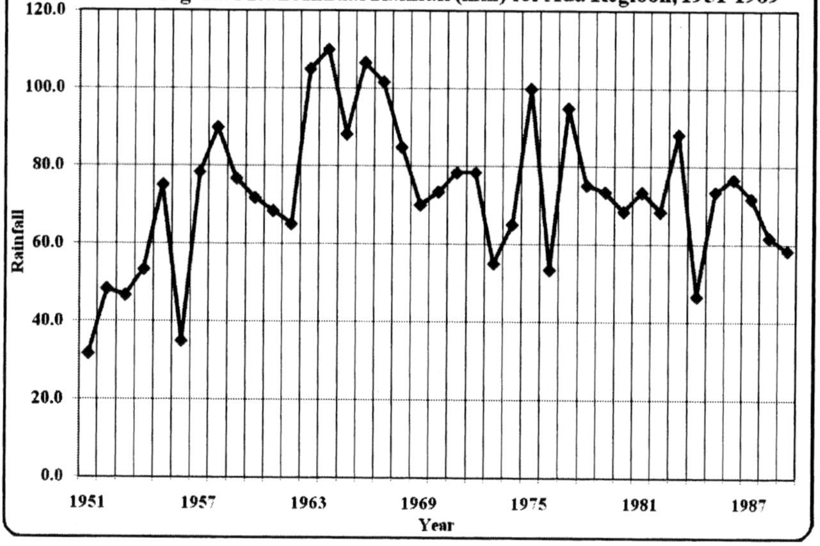

Fig. 2b. Mean Monthkly Rainfall (mm) for Selected Cities in Selale and Ada Regions

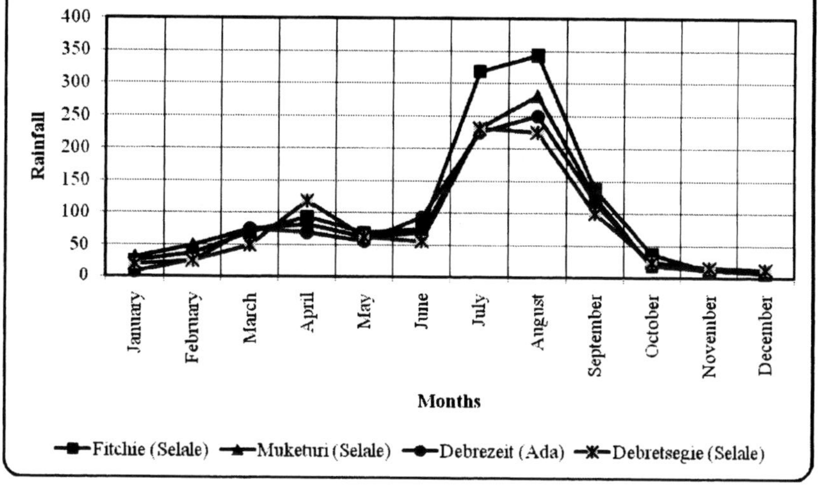

Appendix 3. Map of Ethiopia before 1992

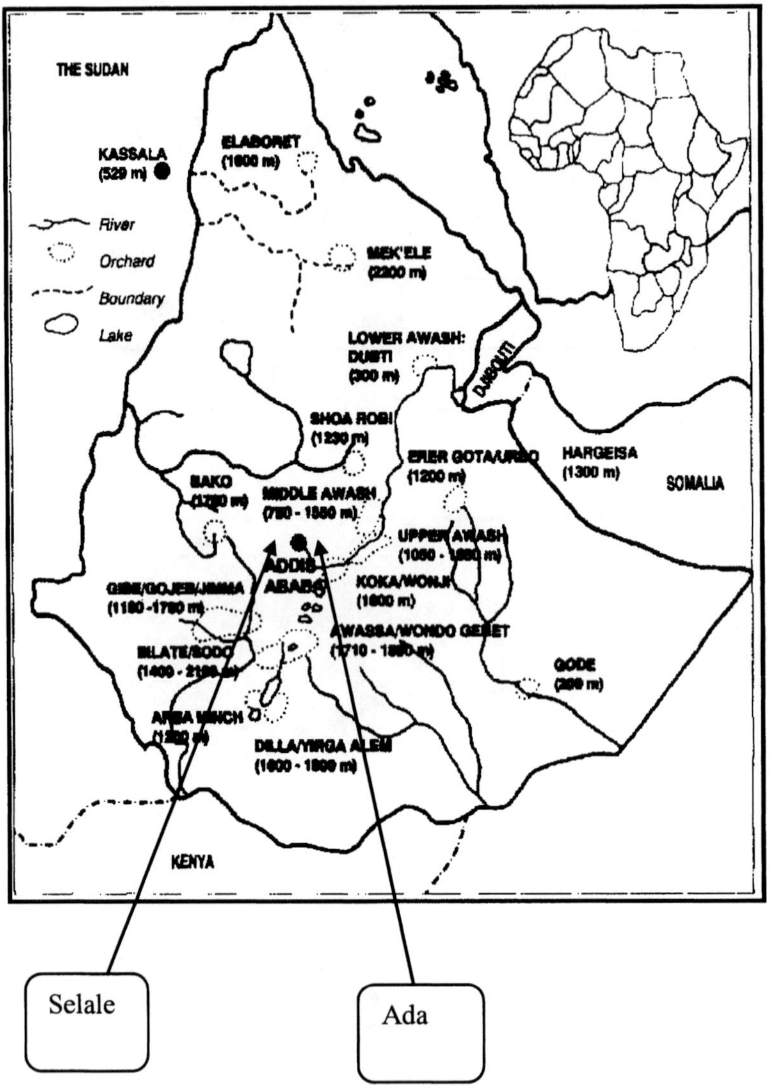

Appendix 4a. Questionnaire to Recruit Research Assistants

In Addition to tests on ability to understand and interpret the questionnaire between Amharic, Oromigna and English, the following questions were asked to select research assistants.

1. Why are you interested to work with us?
2. How do you see farmers vis-a-vis your knowledge of agriculture?
3. How do you respond if the farmer
i) Did not show up on the day of appointment
ii)if he get bored with your questionnaire
4. If even after I and the people at the ministry briefed with the study farmers, they may still suspect that we are working for the government in a way which may not benefit them. For instance collecting their income and expenses for tax purposes. You may record the responses of farmers. They may suspect that it has to something more than just studying their farming environment. What would you do to make sure that what you are doing is just for academic purposes and that the results of the study would be used to design methods to improve his livelihood?
5. How do you present yourself to the farmer with respect to
i) the kind of clothing you put on
ii) the manner in which you asked him to give you information
iii) observing whether or not he is tired or not whenever you are interviewing him/her? why?
6. In order to complete the whole questionnaire, you might have to talk the same farmer at least 4 or five times. How do you explain to the farmer if he show signs of not being enthusiastic?
7. Do you appreciate being a farmer? why?

Appendix 4b. Participatory-Observatory Method of Research

Participatory and observatory studies were performed by the principal investigator (Y. Kebede). Few farmers were selected for this kind of field research. Moreover, these methods were mainly intended to examine strategies, goals and knowledge of households.

Participatory:

I spent five days every two months, beginning from the plowing season to the harvesting season. Four households in Wayu, Gulele biro and Anokere peasant associations were visited. During these visits, I looked after the cattle with the herd boy/girl in the grazing fields and did exactly what he/she did. I would ask the herder, where to direct the cattle?, When I got a response, I would ask why I should direct the cattle to that particular section of the field and not to other sections. I would leave my voice activated recorder on at all times so that she/he would not know that I was recording what they were saying. Sometimes listened music with them. I also plowed with the father or head of the household. Coming from a farming family, I knew how to plow and it helped me gain their confidence. However, I sometimes made mistakes intentionally to observe their response and record their reaction. In general, I participated in the day-to-day activities of male members of the selected farmers to have first hand information on their way of life.

Observatory:

Observatory study involved sudden and on-the-spot checking of what households were doing. Most study farmers are located in Selale. Furthermore, travelling 175 kilometer (distance from Selale to Ada) was difficult. Thus, forty-five farmers were selected from Selale and only five from Ada Fifty people participated in the observatory study. After completing my daily interview with one of the study farmers, I would drop at any farmer's house or farm whom I was not interviewing on that particular day, and observe activities that members of the household were performing. If necessary, I would pose questions in a manner that looked like that I was curious or was joking. I commented how important and challenging their work is, to gain their confidence and develop a good rapport. In case I wanted to ask or discuss some issues with farmers, I always had my recorder on. I wrote my observation as soon as I completed my daily activity.

Appendix 5a. Model of Adoption Decision

Let's denote the ith household preference of adopting a new technology (ies), for a vector of economic, social, cultural, and physical or technical factors (X_I) by $U_{NI}(X_I)$ and the preference of adopting the traditional technology (or continuing current practices) by $U_{TI}(X_I)$. Then, the preference for adopting the new and old technology(ies) can be defined as a linear relationship of the form:

$$U_{NI}(X_I) = A_{NXI} + E_{NI} \quad\text{.....................}\ (1)$$

$$U_{TI}(X_I) = A_{TXI} + E_{TI} \quad\text{.....................}\ (2)$$

Where X_I is a vector of independent variables representing the economic, social, cultural, technical and physical characteristics of the ith household, E_{TI} and E_{NI} are random disturbances with zero mean associated with the adoption of traditional and new technology(ies) respectively. If the index of adoption for the ith household is denoted by Y_I, it will take a value between zero and one if the farmer [is willing to] adopt one, two or more technology(ies) and zero otherwise.

The probability that the ith household randomly selected from a population, with given characteristics and environment represented by X, will adopt the new technology (ies) can be expressed as a function of X_I as follows:

$$P_I(Y_{I=1}) = P_I(U_{NI} > U_{TI}) \quad\text{..............................}\ (3)$$

$$= P_I(A_{NXI} + E_{NI} > A_{TXI} + E_{TI})$$

$$= P_I[X_{I(AN-AT)} > E_{TI} - E_{NI}]$$

$$= P_I(ßX_I > R_I) = F(ßX_I) \quad\text{.........................}\ (4)$$

where P_I is the probability function, R_i is (E_{TI} - E_{NI}) is a random disturbance term, ß is (A_N-A_T) a vector of unknown parameters which can be interpreted as the net influence of the vector of independent variables on adoption of the new technology(ies) and $F(ßX_I)$ is cumulative distribution function F evaluated at $ßX_I$ (Rahm and Huffman, 1984) .

According to the logit model, the probability of adopting a new technology(ies) by the ith farmer, given economic, social, cultural, technical and physical characteristics (X_I), $P_I(N\,|\,X_I)$, is specified as:

$$P_I(N\,|\,X_I) = \exp(ßX_I)/\{1+\exp(ßX_I)\}, \text{ where } -4 < ßX_I < 4 \text{ ,.. } (5)$$

The probability of adoption of the traditional technology, $P(T\,|\,X_I)$, is therefore,

$$P_I(T\,|\,X_I) = 1 - P(N\,|\,X_I)$$

$$= 1 - [\exp(ßX_I)/\{1+\exp(ßX_I)\}]$$

$$= 1/(1+\exp(ßX_I) \quad\text{...................}\ (6)$$

The relative probabilities of adopting versus not adopting a new technology are given by:

$$P_i[(N|X_i)/(T|X_i)] = \exp(X_iß)*[\{1+\exp(X_iß)\}/\ \{1+\exp(X_iß)\}]$$

$$= \exp\ (X_iß) \quad\text{.......................}\ (7)$$

taking the logarithms of both sides,

$$\ln\ [P_i(N|X_i)/P_i(T|X_i)] = X_iß \quad\text{.......................}\ (8)$$

The empirical estimation of the dependent variable in equation (8) is the logarithm of odds of choice not the actual probability (see Pindyck & Rubinfeld, 1981).

Appendix 5b. The Average Production Function and Local Units of Measurement

The model is specified as follows;

$$Y = f(X, Z)$$

$$Y = \prod_{i=1}^{K} \beta_0 X_i^{\beta_i} \exp\left(\alpha_o + \prod_{j=1}^{N} \alpha_j Z_j + U\right) \quad \text{..... (1)}$$

where Y represents output, X_i the i'h continuous input, and Z_j the j'h discrete input (0-1 variables). The multiplicative function in (1) can be written in log-linear form as follows:

$$\text{Log } Y = \delta_o + \Sigma \beta_i \text{ Log } (X_i) + \Sigma \alpha_j Z_j + U \quad \text{.......... (2)}$$

where δ_o is log (β_0) + α_o.

Local Units of Measurements of Area and their Metric Equivalent

Land with	One unit of "Fer"	"Timad/kert" equals to = ? M^2	"Gemede"
Red soil	1400	2327.5	1156
Stony (4% slope)	2870	2506.9	2870
Slope 15-25%		1283.6	
10-20% stony	1441	1441	2882
Black soil	1660	1660	1660
black/stony	3130	3130	4696
Average	2100	2058.2	2653

Metric Equivalent Units of Selected Local Units of Measurement:

Local Unit	Equivalent Metric Unit (kg) for barley/wheat	Teff	cheese	butter
Kuna	3.5	5		
Enkib	35	50		
Dawula	70	100		
Silicha	50			
Sahin			1	
Sini				.1
Kubaya				.6
Tasa				1.2
Kg equivalent of a Donkey load of hay				64
Kg equivalent of a Donkey load of straw				52

Appendix 5c. Data Envelopment Analysis

The linear fractional program, analysis for each farm, was carried for 270 producers (see Sexton, 1986; Smith, 1990; Seiford, et al., 1990). The objective function of the fractional linear program, the efficiency ratio, can be written as:

$$\text{Maximize } Z_k = \frac{\sum_{r=1}^{l} U_{rk}Y_{rk}}{\sum_{i=1}^{m} V_{ik}X_{ik}} \quad \ldots\ldots\ldots\ldots\ldots\ldots (1)$$

Subject to

$$\frac{\sum_{r=1}^{l} U_{rk}Y_{rj}}{\sum_{i=1}^{m} V_{ik}X_{ij}} \; \# \; 1; \; j = 1,2,3, \ldots, 270$$

$$U_{rk} \geq 0; \; r=1$$

$$V_{ik} \geq 0, \quad 0; \; i = 1,2,3,4,\ldots, 10$$

This linear fractional program utilizes the ratio of total weighted output over total weighted units for farm j and searches for the combination of weights which maximizes output. This program requires firm j to choose weights subject to the constraints that no other farm could produce an output-input ratio (efficiency) greater than one if it used the same weights. It is also assumed that the weights cannot be negative or should be strictly positive.

The fractional linear program in equation (1) can be rewritten as an ordinary linear program and simplex algorithm is used to solve it. To understand how the simplex method works reference can be made to Heady (1963) and Horton, et al. (1989). The transformed linear program (DEA) is formulated (primal) as :

$$\text{(DEA)} \quad \text{Maximize } Z_k = \sum_{r=1}^{s} U_{rk}Y_{rk} \quad \ldots\ldots\ldots\ldots\ldots (2)$$

Subject to

$$\{p_{kj}\} \; \sum_{r=1}^{s} U_{rk}Y_{rj} - \sum_{i=1}^{m} V_{ik}X_{ij} \; \# \; 0 \; ; \; j = 1,2,3,\ldots, 270$$

$$\{q_k\} \; \sum_{i=1}^{m} V_{ik}X_{ik} = 1$$

$$U_{rk} \geq 0; \; r=1$$

$$V_{ik} \geq 0; \; i = 1,2,3,4,\ldots, 10$$

Maximization problems can be translated to minimization. The minimization counterpart of the maximization problem is called dual. The values in the brackets represent dual variables. By duality theory, solutions to the primal are the same as those satisfying the dual constructs. The dual variables can be used to generate multipliers which will help individual farms attain superior levels of efficiency. For instance p_{kj} represent the comparison group of the jth farm of the sample households. The performance of the jth farm is compared with the output level of farm k. This condition is satisfied when,

$$\sum_{j=1}^{n} Y_{rj}p_{kj} \geq Y_{rk} \quad r=1$$

p_{kj} is the efficiency of firm j. Thus, it must lie between 0 and 1. Also, q_k is a measure of efficiency corresponding to the optimal value of Z_k. The most efficient farm in using the inputs is found by solving the following in terms of q_k.

$$\sum_{j=1}^{n} p_{kj}X_{ij} \leq q_k X_{jk} \; ; i=1,2,3,4, ..., 10$$

The value of q_k will be between 0 and 1. Generally, if firm j is efficient, then the corresponding dual variable, $q_k = 1$.

Hence, the minimization problem, i.e. the dual the maximization problem can be written as follows,

Dual Minimize $h_k = q_k$ .... (3)

Subject to

$$\{Urk\} \quad \sum_{j=1}^{n} p_{kj}Y_{rj} \geq Y_{rk} \; ; r=1$$

$$-\sum_{j=1}^{n} p_{kj}X_{ij} + q_k X_{jk} \geq 0; \; i=1,2,3,4, ..., 10$$

$$q_k, p_{kj} \geq 0; \; j=1,2,3,..., 270$$

The minimization program searches for the most efficient combination of farms which can produce the same level of output as farm k given the potential output combinations. The reciprocal of the fractional linear program gives the inefficiency ratio. To understand how this is accomplished, see Charnes, et al (1978) and Sengupta (1987, 1989).

Empirical Formulation of the Models of Data Envelopment Analysis

a) Definition of Variables

Y3= VALUE OF CROP PRODUCED

Y61= VALUE OF MILK PRODUCED

Y7= VALUE OF LIVESTOCK

Y8= OFF-FARM INCOME
V1= CROP AREA
V2= FERTILIZER AND PESTICIDES EXPENDITURE
V3= OXEN INPUT FOR CROP PRODUCTION
V4= AMOUNT OF SEED USED FOR CROP PRODUCTION
V5= LABOUR USE FOR CROP PRODUCTION
V6= NUMBER OF COWS
V7= ROUGHAGES FED TO COWS (KG OR LOAD OF DONKEYS)
V8= LABOUR INPUT FOR MANAGEMENT OF COWS
V9= AMOUNT OF ATELA USED
V10= FEED AND VETERINARY EXPENDITURE
V16= AMOUNT OF CONCENTRATES FED TO COWS
V17= FARMING EXPERIENCE
V18= AREA OF STUBBLE FEEDING
V21= GRAZING AREA

b) Prices Used to Value Grain (Ethiopian birr)

Region	TEFF	BARLEY	BEAN	WHEAT
ADA	95	60	70	83
SEL	88	58	65	77

Region	PEAS	OAT	AVERAGE
ADA	58	73	
SEL	55	55	66

c) The Model

```
MAX     1716Y3  +  3293Y61  + 918.125Y7 +  408Y8
SUBJECT TO
  2)   20000V1  + 41.875V2  +    68V3   +   162V4
  +       123V5  +     4V6  +  101.5V7   +    23V8
  +       600V9  +    552V10 +  26.06V16  +   29V17
  +    20000V18 +    5000V21  = 1
  3)     1914Y3  +    3068Y61 +  1450.5Y7   +  1000Y8
  -    22400V1  - 331.45V2  -     56V3   -  225V4
  -      131V5  -      4V6  -  34.75V7   -   20V8
  -      360V9  -    488V10 -  10.05V16  -    5V17
  -    20000V18 -   2100V21 < = 0
  .
  .
272)    1716Y3 +    3293Y61  + 918.13Y7    + 408Y8
  -    20000V1  -  41.875V2  -     68V3   -  162V4
  -      123V5  -     4V6  -  101.5V7   -  23V8
  -      600V9  -    552V10  -  26.06V16 -    29V17
  -    20000V18 -    5000V21  < = 0
END
```

d) Sample Solution

OBJECTIVE FUNCTION VALUE

1) 1.00000000

VARIABLE	VALUE	REDUCED COST
Y3	.000090	.000000
Y61	.000224	.000000
Y8	.000387	.000000
V1	-.000021	.000000
V2	.000715	.000000
V5	.002364	.000000
V17	.036364	.000000
V18	.000049	.000000
V21	.000073	.000000

ROW	SLACK OR SURPLUS	DUAL PRICES
2)	.000000	1.000000
272)	.000000	1.000000

NO. ITERATIONS= 21

Appendix 5d. Stochastic Frontier Function

The stochastic frontier production function, assuming the Cobb-Douglas form, can be written as:[1]

$$\ln Y_i = \ln\alpha_0 + \Sigma \beta_i \ln X_{ij} + \varepsilon_j \quad \ldots\ldots\ldots\ldots\ldots (1)$$

where

$$\varepsilon_j = V_j - U_j \quad \ldots\ldots\ldots\ldots\ldots\ldots\ldots\ldots (2)$$

is a composed error term with V_i – iid $N(0,\sigma v^2)$ and U_j – iid*$N(0,\sigma u^2)$*. V is the two sided random error, U is the one-sided efficiency component distributed half-normally, Y is output, α is the intercept, ß is an unknown parameter, X's are inputs (Aigner, et al., 1977; Meeusen, et al., 1977). V and U are assumed to be uncorrelated and ε is assumed to be independent of X. The symmetric component, V, permits random variations in output due to factors outside the control of the firm such as weather.

Let's assume that the variances of u and v are σ_v^2 and σ_u^2. That is,

$\sigma^2 = \sigma_v^2 + \sigma_u^2$,

and

$\tau = \sigma_u^2/\sigma^2,\ 0 \leq \tau \leq 1,$ .... (3)

is the total variation in output from the frontier which is attributable to technical

1. Kopp and Smith (1980) argued that the choice of functional specification has a very small impact on measured efficiency. The Cobb-Douglas production function is a compromise between a complex production process and a complex estimation technique.

(in)efficiency.

For a sample of N observations, the log likelihood function can be written as :

$$LnL = N\ln(2/\pi)^{1/2} + N\ln\sigma^{-1} + \sum_{j=1}^{N} \ln(1-F^{*}(\varepsilon_j \lambda \sigma^{-1}) - [1/(2\sigma^2) \sum_{j=1}^{N} \varepsilon j^2) \quad \text{.... (4)}$$

where $\varepsilon_j = \ln y_j - \ln\beta_0 + \sum_{i=1}^{m} \beta_i \ln x_{ij}$, $\sigma = (\sigma^2 u + \sigma^2_v)^{1/2}$, and $\lambda = \sigma_u/\sigma_v$.

Jondrow, et al., (1982) have demonstrated that individual firm measures of technical efficiency can be calculated from the error terms ε_j . Specifically, they are the expected value of U_j conditional on ε_j, that is:

$$E[u_j/\varepsilon_j) = \frac{\sigma_u \sigma_v}{\sigma} \left[\frac{f(\varepsilon_j \lambda/\sigma)}{F(\varepsilon_j \lambda/\sigma)}\right] \frac{\varepsilon j \lambda}{\sigma} \quad j=1,2,...,n \quad (5)$$

Where f(.) and F(.) are the standard normal density and standard distribution functions evaluated at ($\varepsilon_j \lambda/\sigma$) and $\sigma^2 = \sigma_u^2 \sigma_v^2/\sigma^2$. Using equation (2) and (5), values of uj and vj can be obtained after replacing ε_j, σ and λ by their estimates. The measure of technical efficiency (TE_j) is calculated as:

$$\ln \hat{TE_j} = \exp(-E[U_j \mid \varepsilon_j]) \qquad i=1,2,...,n$$

so that $0 \leq TE_j \leq 1$.

OR

The appropriate technical efficiency index of each individual firm is equal to:

$$\ln TE_j = U_j = Ln\hat{y}_j - Lny_j \text{ or}$$

$$TEj = \hat{Y}j/Yj$$

Appendix 5e. Linear Structural Relations (LISREL)

The full LISREL model is defined by the following three equations:

Structural Equation Model: $\mu = \&\mu + \Gamma\varepsilon + u$

Measurement Model for Y : $Y = \Omega_y \mu + e$

Measurement Model for X : $X = \Omega_x \varepsilon + \delta$

where μ and ε are vectors of latent dependent and independent variables, & (mxm) and Γ (mxn) are coefficient matrices and u is a random vector of residuals. The elements of & represent the direct effects of μ-variables on other μ-variables and the elements of Γ represent direct effects of ε variables on μ-variables. It is assumed that u is uncorrelated with ε and that I-& is non-singular.

Vectors μ and ε are not observed, but instead vectors y and x are observed, such that:

$$Y : Y = \Omega_y \mu + e$$
$$X : X = \Omega_x \varepsilon + \delta$$

where e and δ are vectors of uncorrelated error terms (errors of measurement or measure-specific components). These equations represent the multivariate regressions of y on μ and of x on ε, respectively. Y and X are referred as observed variables and μ and ε as latent variables (for detailed statistical description, see Jeroskog, et al., 1989).

Actual Empirical Formulation of the LISREL Model

```
MATRIX DATA
  / VARIABLES= X1 TO X16
  /CONTENTS=CORR
    (Correlation matrix here)
END DATA
LISREL
  / DA NI=16 NO=128 MA=KM
  / LA
    /'"FAMILY"EDUCA"EXPER"EXTEN"AWAR2"RELA"WEALTH"'
  /'"CROPKN"AREACRO"OXCRO"SEEDCRO"LABOCR"'
  /'"CROPDIV"NOTECH"CEREFF"WCratio"'
/ SE
  /CROPKN NOTECH CEREFF EDUCA EXPER EXTEN WEALTH
    AREACRO OXCRO Worker:Cons./
/ MO NY=3 NX=7 GA=FI BE=SD PS=DI
/ FR GA 1 1 GA 1 2 GA 1 3 GA 2 1 GA 2 2 GA 2 3 GA 2 4
  GA 2 7
/ FR GA 3 1 GA 3 2 GA 3 3 GA 3 5 GA 3 6 GA 3 7
/ FR BE 2 1 BE 3 2 BE 3 1
/ OU SE TV RS AD=80
            NUMBER OF INPUT VARIABLES 16
            NUMBER OF Y - VARIABLES   3
            NUMBER OF X - VARIABLES   7
            NUMBER OF ETA - VARIABLES 3
            NUMBER OF KSI - VARIABLES 7
            NUMBER OF OBSERVATIONS  128
```

Appendix 5f. Description of Households who participated in the Observatory-participatory Study

Farmer #1 (Mr. Tesfaye Kebede)

Mr. Tesfaye Kebede has five children. His sons are Abebe (age 16) and Temesgen (age 10). His daughters are Alemitu (age 14), Zinash (age 12), and Lemlem (age 7). Names in Ethiopian culture identify i) character of an individual born, ii) his looks, iii) how desperate parents were to have a baby, iv) aspirations, thankfulness, v) names of grandparents whose influence had a profound effect on the current status of the family, and so on. In this case, Abebe, Temesgen, Alemitu, Zinash and Lemlem means flower, thank God, my world, and green respectively. His wife Turiye means "you are good" and husband Kebede (means heavy).

Farmer #2 (Mr. Abebe Dilnesaw)

Mr. Dilnesaw is a migrant with his parents from the Northern tip of the province of Shoa from where many kings of Ethiopia came. He is perfectly bilingual (Amahrigna and Oromigna). He has six children. Ato Bekele, Gashaw, Weizero Genet, Ato Belay, Weizero Saba and Yenenesh. The meaning of their names respectively is to grow, my protector or shield, paradise or garden of Aden, better than everybody, after the Queen of Sheba and you are mine. The first three are married and live in the neighbourhood. Mr. Dilnesaw lives with his three children age 21, 19 and 17. Belay is engaged and married by now, so is Saba. Yenenesh was in her final year of high school at the time of the interview.

Similar to Mr. Kebede's case, I initiated discussions with the head, wife and children separately and when they are at home. I spent quite a good number of days with them (including sleeping in their house). Mr. Kebede has five years of sacred education and he was a celebrated deacon, as is his son Belay. Furthermore, Mr. Dilnesaw has four years of secular education which during that time was considered to be quite a good achievement. His ambitions even at the age of 62 are extraordinary. The decision criteria he employs are quite fascinating. The discussions in a form of questions and answers are given below.

Appendix 6.Results of Logit Regression, LISREL and Frontier production Function Model By Test and Control Farmers

	Results from Logit						
	Probability of Adoption of					Frontier Production Function	
Test=1/ Control=0	Cross-bred cows	Fertilizer	Pesticides	Improved Seed	LISREL DEA	Cereal	Milk
Selale Region							
0	0.898	0.385	0.815	0.498	1	0.943	0.727
0	0.837	0.586	0.748	0.266	1	0.971	0.714
0	0.977	0.682	0.637	0.348	1	0.891	0.574
0	0.781	0.488	0.662	0.288	0.997	0.973	0.585
0	0.93	0.386	0.804	0.474	0.467	0.888	0.512
0	0.518	0.485	0.635	0.326	0.797	0.838	0.778
0	0.878	0.623	0.66	0.364	1	0.948	0.403
0	0.9	0.609	0.699	0.42	1	0.936	0.734
0	0.89	0.484	0.728	0.419	1	0.937	0.582
0	0.549	0.738	0.709	0.416	0.655	0.968	0.002
0	0.727	0.541	0.642	0.317	0.719	0.615	0.768
0	0.673	0.817	0.664	0.274	1	0.844	0.617
0	0.522	0.426	0.73	0.411	0.711	0.85	0.718
0	0.813	0.842	0.668	0.369	0.785	0.736	0.829
0	0.97	0.822	0.855	0.318	0.676	0.843	0.734
0	0.622	0.76	0.687	0.313	0.785	0.866	0.929
0	0.906	0.842	0.849	0.271	0.807	0.952	0.84
0	0.873	0.803	0.524	0.375	1	0.947	0.548
0	0.253	0.89	0.608	0.36	1	0.904	0.903
0	0.687	0.421	0.744	0.382	1	0.776	0.601
0	0.539	0.802	0.706	0.385	1	0.82	0.875
0	0.573	0.74	0.718	0.339	0.999	0.898	0.679
0	0.867	0.605	0.804	0.316	0.897	0.885	0.005
0	0.917	0.652	0.833	0.297	1	0.816	0.837
0	0.921	0.72	0.736	0.379	1	0.806	0.838
0	0.896	0.64	0.733	0.474	1	0.921	0.129
0	0.806	0.544	0.664	0.345	1	0.759	0.585
0	0.77	0.713	0.74	0.302	1	0.98	0.563
0	0.721	0.593	0.598	0.284	0.747	0.898	0.22
0	0.878	0.843	0.568	0.422	0.989	0.889	0.836
0	0.755	0.394	0.662	0.166	1	0.812	0.671
0	0.883	0.566	0.771	0.387	0.877	0.801	0.683

Appendix 6.(continued)

	Results from Logit						
	Probability of Adoption of					Frontier production Function	
Test=1/ Control=0	Cross-bred cows	Fertilizer	Pesticides	Improved Seed	LISREL DEA	Cereal	Milk
0	0.959	0.641	0.658	0.364	0.869	0.808	0.352
0	0.863	0.734	0.648	0.399	0.988	0.863	0.61
0	0.94	0.45	0.67	0.224	1	0.826	0.484
0	0.968	0.522	0.723	0.315	1	0.938	0.756
0	0.93	0.635	0.637	0.329	1	0.842	0.79
0	0.844	0.691	0.696	0.423	0.787	0.831	0.764
0	0.541	0.642	0.787	0.326	0.977	0.87	0.699
0	0.493	0.536	0.784	0.222	0.409	0.754	0.817
0	0.907	0.484	0.848	0.366	0.901	0.908	0.723
0	0.83	0.67	0.85	0.339	0.506	0.853	0.824
0	0.796	0.565	0.781	0.293	0.999	0.854	0.223
0	0.564	0.851	0.584	0.353	1	0.976	0.847
0	0.878	0.673	0.798	0.312	1	0.852	0.851
0	0.936	0.612	0.902	0.492	1	0.949	0.74
0	0.043	0.562	0.504	0.357	1	0.874	0.106
0	0.592	0.719	0.682	0.399	0.781	0.895	0.862
0	0.96	0.497	0.677	0.242	1	0.7	0.145
0	0.743	0.728	0.663	0.263	1	0.964	0.548
0	0.961	0.701	0.72	0.398	1	0.846	0.744
0	0.915	0.86	0.609	0.283	1	0.903	0.623
0	0.894	0.389	0.734	0.223	0.685	0.71	0.361
0	0.923	0.878	0.713	0.332	0.521	0.976	0.648
0	0.754	0.624	0.659	0.316	0.891	0.87	0.771
0	0.96	0.8	0.661	0.42	1	0.888	0.133
0	0.892	0.714	0.731	0.488	0.999	0.688	0.744
0	0.886	0.693	0.676	0.409	1	0.774	0.836
0	0.947	0.693	0.788	0.414	1	0.9	0.764
0	0.768	0.808	0.486	0.38	0.656	0.676	0.714
0	0.149	0.43	0.615	0.425	1	0.858	0.854
0	0.893	0.506	0.671	0.335	0.878	0.817	0.683
0	0.913	0.407	0.769	0.32	1	0.855	0.666
0	0.694	0.491	0.728	0.421	1	0.871	0.713
0	0.574	0.889	0.822	0.576	1	0.931	0.572
0	0.778	0.839	0.6	0.443	1	0.647	0.786

Appendix 6.(continued)

	Results from Logit						
	Probability of Adoption of					Frontier production Function	
Test=1/ Control=0	Cross-bred cows	Fertilizer	Pesticides	Improved Seed	LISREL DEA	Cereal	Milk
0	0.804	0.578	0.72	0.395	0.669	0.878	0.898
0	0.501	0.706	0.583	0.48	0.529	0.881	0.815
0	0.868	0.432	0.629	0.377	0.455	0.75	0.728
0	0.084	0.83	0.608	0.327	1	0.974	0.774
0	0.456	0.531	0.573	0.232	0.952	0.802	0.637
0	0.886	0.887	0.634	0.312	0.97	0.934	0.632
0	0.653	0.853	0.719	0.295	1	0.882	0.737
0	0.948	0.504	0.662	0.28	1	0.949	0.478
0	0.546	0.666	0.541	0.385	0.882	0.821	0.817
0	0.988	0.762	0.7	0.544	1	0.826	0.826
0	0.741	0.536	0.517	0.289	1	0.904	0.821
0	0.871	0.896	0.641	0.413	0.841	0.883	0.898
0	0.993	0.711	0.64	0.286	0.709	0.565	0.482
0	0.949	0.533	0.776	0.397	0.681	0.895	0.56
0	0.591	0.81	0.635	0.371	0.999	0.905	0.294
0	0.892	0.763	0.54	0.444	0.96	0.919	0.779
0	0.416	0.769	0.595	0.412	0.677	0.777	0.8
0	0.735	0.836	0.747	0.424	0.706	0.894	0.678
0	0.8	0.659	0.597	0.366	1	0.884	0.683
0	0.632	0.629	0.696	0.458	0.981	0.774	0.793
0	0.917	0.648	0.604	0.404	1	0.899	0.198
0	0.895	0.491	0.793	0.494	0.884	0.961	0.581
0	0.531	0.431	0.619	0.195	1	0.889	0.464
0	0.836	0.611	0.737	0.463	0.891	0.922	0.723
0	0.935	0.606	0.817	0.456	0.78	0.825	0.809
0	0.801	0.904	0.547	0.472	0.885	0.83	0.537
0	0.722	0.7	0.551	0.365	1	0.849	0.323
0	0.745	0.666	0.771	0.437	0.877	0.971	0.68
0	0.799	0.409	0.64	0.185	0.89	0.904	0.822
0	0.868	0.705	0.704	0.374	0.73	0.96	0.781
0	0.531	0.656	0.657	0.431	0.903	0.961	0.577
0	0.958	0.714	0.817	0.472	1	0.908	0.735

Appendix 6.(continued)

	Results from Logit						
	Probability of Adoption of					Frontier production Function	
Test=1/ Control=0	Cross-bred cows	Fertilizer	Pesticides	Improved Seed	LISREL DEA	Cereal	Milk
0	0.304	0.525	0.769	0.153	0.952	0.867	0.909
0	0.96	0.445	0.702	0.468	0.823	0.919	0.736
0	0.883	0.699	0.642	0.469	1	0.721	0.306
0	0.969	0.55	0.791	0.54	1	0.85	0.754
0	0.353	0.255	0.702	0.16	1	0.85	0.702
0	0.453	0.761	0.539	0.367	0.875	0.909	0.684
0	0.929	0.624	0.779	0.357	0.887	0.905	0.329
0	0.859	0.596	0.672	0.316	0.843	0.857	0.39
0	0.779	0.673	0.689	0.327	0.579	0.932	0.929
0	0.606	0.606	0.613	0.451	0.809	0.836	0.71
0	0.549	0.678	0.667	0.273	1	0.785	0.577
0	0.108	0.604	0.576	0.493	0.786	0.723	0.824
0	0.865	0.709	0.783	0.288	0.78	0.921	0.034
0	0.902	0.84	0.559	0.472	0.809	0.791	0.83
0	0.666	0.626	0.603	0.441	0.546	0.945	0.857
0	0.337	0.652	0.642	0.323	0.901	0.916	0.846
0	0.83	0.414	0.555	0.278	1	0.914	0.269
0	0.411	0.786	0.714	0.392	0.563	0.925	0.227
0	0.611	0.884	0.604	0.38	0.676	0.689	0.359
0	0.313	0.821	0.76	0.443	0.787	0.892	0.611
0	0.904	0.728	0.752	0.47	0.887	0.628	0.723
0	0.935	0.916	0.601	0.489	0.999	0.952	0.832
0	0.331	0.798	0.8	0.176	1	0.904	0.606
0	0.406	0.821	0.612	0.43	0.598	0.85	0.579
0	0.944	0.634	0.678	0.296	0.385	0.963	0.206
0	0.741	0.82	0.671	0.447	0.767	0.8	0.511
0	0.467	0.892	0.83	0.428	1	0.889	0.797
0	0.937	0.743	0.91	0.468	0.887	0.934	0.723
1	0.068	0.716	0.542	0.383	1	0.927	1
1	0.673	0.829	0.739	0.372	0.624	0.922	0.204
1	0.027	0.616	0.581	0.387	1	0.941	0.913
1	0.379	0.82	0.499	0.378	1	0.872	0.874
1	0.07	0.627	0.86	0.349	0.878	0.95	0.68
1	0.822	0.288	0.587	0.332	1	0.916	0.928

Appendix 6.(continued)

	Results from Logit						
	Probability of Adoption of					Frontier production Function	
Test=1/ Control=0	Cross-bred cows	Fertilizer	Pesticides	Improved Seed	LISREL DEA	Cereal	Milk
1	0.023	0.575	0.733	0.408	1	0.995	0.828
1	0.405	0.655	0.71	0.277	0.901	0.893	0.938
1	0.177	0.52	0.726	0.498	0.891	0.875	0.922
1	0.899	0.89	0.712	0.378	1	0.928	0.655
1	0.072	0.658	0.521	0.369	1	0.91	1
1	0.433	0.855	0.76	0.415	0.991	0.983	0.782
1	0.842	0.768	0.781	0.422	1	0.877	0.494
1	0.65	0.691	0.724	0.329	0.891	0.963	0.294
1	0.088	0.426	0.773	0.175	1	0.879	0.968
1	0.55	0.883	0.773	0.266	0.702	0.918	0.933
1	0.756	0.778	0.564	0.425	1	0.936	0.788
1	0.048	0.277	0.752	0.302	1	0.957	0.959
1	0.583	0.736	0.621	0.357	0.872	0.896	0.862
1	0.777	0.599	0.555	0.284	1	0.965	0.911
1	0.841	0.916	0.694	0.47	0.999	0.964	0.977
1	0.121	0.483	0.791	0.484	0.547	0.836	0.918
1	0.035	0.685	0.772	0.475	0.689	0.887	0.707
1	0.517	0.833	0.704	0.238	1	0.988	0.423
1	0.03	0.508	0.742	0.519	1	0.709	0.895
1	0.046	0.56	0.637	0.417	1	0.928	0.958
1	0.842	0.525	0.771	0.502	0.677	0.518	0.993
1	0.647	0.338	0.793	0.226	1	0.924	0.862
1	0.541	0.85	0.809	0.328	0.704	0.722	0.877
1	0.043	0.379	0.725	0.483	1	0.962	0.895
1	0.252	0.93	0.83	0.31	1	0.278	0.887
1	0.037	0.559	0.668	0.537	1	0.9	0.622
1	0.166	0.689	0.828	0.128	1	0.762	0.926
1	0.029	0.47	0.732	0.386	1	0.877	0.786
1	0.006	0.453	0.58	0.483	0.787	0.886	0.891
1	0.132	0.809	0.601	0.385	1	0.909	0.779
1	0.116	0.573	0.747	0.556	0.883	0.866	1
1	0.061	0.564	0.474	0.203	1	0.649	1
1	0.47	0.431	0.679	0.252	1	0.775	0.506

Appendix 6.(continued)

	Results from Logit						
	Probability of Adoption of					Frontier production Function	
Test=1/ Control=0	Cross-bred cows	Fertilizer	Pesticides	Improved Seed	LISREL DEA	Cereal	Milk
1	0.112	0.233	0.521	0.257	1	0.868	0.773
1	0.038	0.53	0.534	0.226	0.713	0.622	0.992
1	0.711	0.429	0.467	0.259	1	0.662	0.212
1	0.393	0.702	0.665	0.227	0.657	0.834	0.657
1	0.673	0.518	0.492	0.263	0.968	0.701	0.915
1	0.036	0.88	0.526	0.442	1	0.82	0.961
1	0.151	0.499	0.657	0.272	1	0.897	0.971
1	0.722	0.799	0.774	0.37	0.565	0.864	0.754
1	0.071	0.747	0.711	0.386	0.888	0.862	0.819
1	0.163	0.87	0.705	0.433	1	1	0.494
1	0.016	0.727	0.553	0.369	0.871	0.684	0.992
1	0.899	0.838	0.663	0.557	1	0.909	0.894
1	0.908	0.504	0.795	0.415	1	0.882	0.886
1	0.455	0.699	0.71	0.447	1	0.736	0.953
1	0.007	0.805	0.462	0.444	0.803	0.852	0.726
1	0.729	0.665	0.666	0.395	0.889	0.931	0.379
1	0.056	0.49	0.702	0.501	0.677	0.823	0.898
1	0.543	0.772	0.741	0.464	0.779	0.948	0.637
1	0.043	0.458	0.74	0.463	0.788	0.888	0.879
1	0.531	0.831	0.648	0.383	0.999	0.798	0.574
1	0.669	0.753	0.633	0.302	0.703	0.938	0.885
1	0.108	0.622	0.77	0.173	1	0.872	0.793
1	0.646	0.734	0.627	0.257	0.676	0.879	0.765
1	0.043	0.48	0.821	0.484	0.881	0.891	0.848
1	0.776	0.824	0.77	0.263	1	0.934	0.726
1	0.18	0.506	0.788	0.299	1	0.998	0.911
1	0.299	0.537	0.652	0.295	1	0.575	0.968
1	0.071	0.623	0.553	0.378	1	0.875	0.865
1	0.791	0.603	0.816	0.404	0.657	0.856	0.402
1	0.727	0.82	0.658	0.374	0.988	0.812	0.92
1	0.179	0.718	0.638	0.285	1	0.974	0.891
1	0.609	0.619	0.466	0.414	0.745	0.95	0.917
1	0.174	0.64	0.586	0.222	1	0.832	0.734
1	0.484	0.661	0.763	0.4	0.755	0.94	0.847

Appendix 6.(continued)

	Results from Logit						
	Probability of Adoption of					Frontier production Function	
Test=1/ Control=0	Cross-bred cows	Fertilizer	Pesticides	Improved Seed	LISREL DEA	Cereal	Milk
1	0.176	0.811	0.66	0.213	0.408	0.812	0.711
1	0.452	0.87	0.719	0.323	1	0.882	0.788
1	0.883	0.745	0.705	0.342	0.968	0.896	0.348
1	0.014	0.574	0.732	0.485	0.878	0.861	0.645
1	0.199	0.533	0.691	0.3	0.703	0.934	0.878
1	0.102	0.609	0.581	0.245	1	0.704	0.489
1	0.029	0.537	0.75	0.229	0.408	0.847	0.962
1	0.406	0.829	0.714	0.417	0.697	0.788	0.909
1	0.475	0.619	0.607	0.301	1	0.838	0.867
1	0.031	0.435	0.666	0.563	0.877	0.896	0.842
1	0.163	0.631	0.673	0.218	1	0.973	0.968
1	0.201	0.795	0.708	0.384	1	0.967	0.43
1	0.669	0.622	0.651	0.264	1	0.875	0.892
1	0.534	0.815	0.733	0.451	0.768	0.86	0.463
1	0.19	0.737	0.856	0.277	1	0.842	0.98
1	0.057	0.74	0.725	0.235	1	0.894	0.789
Ada Region							
0	0.859	0.643	0.57	0.471	0.644	0.981	0.757
0	0.975	0.584	0.694	0.467	0.547	0.923	0.752
0	0.943	0.309	0.67	0.452	1	0.917	0.788
0	0.906	0.412	0.656	0.348	0.746	0.779	0.803
0	0.493	0.667	0.586	0.448	0.812	0.978	0.8
0	0.982	0.401	0.612	0.405	0.921	0.893	0.937
0	0.532	0.209	0.577	0.292	0.878	0.947	0.957
0	0.962	0.673	0.766	0.284	1	0.854	0.82
0	0.844	0.287	0.565	0.258	0.532	0.923	0.765
0	0.883	0.563	0.643	0.345	1	0.907	0.984
0	0.789	0.427	0.459	0.378	1	0.897	0.788
0	0.952	0.528	0.624	0.386	1	0.975	0.598
0	0.706	0.404	0.485	0.354	0.823	0.883	0.959
0	0.861	0.562	0.576	0.372	1	0.896	0.706
0	0.363	0.588	0.601	0.489	0.845	0.887	0.779
0	0.582	0.343	0.537	0.324	0.484	0.844	0.798

Appendix 6.(continued)

	Results from Logit						
	Probability of Adoption of					Frontier production Function	
Test=1/ Control=0	Cross-bred cows	Fertilizer	Pesticides	Improved Seed	LISREL DEA	Cereal	Milk
0	0.947	0.35	0.461	0.473	1	0.852	0.984
0	0.969	0.348	0.555	0.421	1	0.824	0.737
0	0.743	0.372	0.468	0.374	1	0.942	0.576
0	0.684	0.443	0.534	0.281	1	0.975	0.69
0	0.86	0.443	0.63	0.461	1	0.827	0.805
0	0.643	0.607	0.452	0.3	0.725	0.873	0.791
0	0.91	0.712	0.481	0.654	0.39	0.685	0.574
0	0.773	0.419	0.668	0.399	1	0.919	0.685
0	0.882	0.44	0.536	0.43	0.761	0.74	0.787
0	0.717	0.363	0.608	0.345	1	0.855	0.839
0	0.295	0.752	0.42	0.4	0.512	0.463	0.871
0	0.922	0.341	0.629	0.406	0.876	0.767	0.762
1	0.288	0.359	0.53	0.401	0.778	0.961	0.969
1	0.037	0.556	0.498	0.462	1	0.863	0.959
1	0.125	0.219	0.567	0.312	0.774	0.815	0.892
1	0.134	0.505	0.49	0.424	0.786	0.753	0.914
1	0.126	0.603	0.59	0.393	0.896	0.922	0.947
1	0.263	0.312	0.57	0.418	1	0.948	0.933
1	0.018	0.508	0.574	0.314	1	0.898	0.914
1	0.668	0.439	0.563	0.445	1	0.993	0.886
1	0.215	0.508	0.609	0.325	0.984	0.833	0.876
1	0.095	0.179	0.559	0.411	1	0.976	0.907
1	0.186	0.376	0.622	0.453	1	0.811	0.851
1	0.277	0.257	0.687	0.385	0.798	0.326	0.923
1	0.591	0.422	0.554	0.272	0.999	0.877	0.553
1	0.037	0.343	0.489	0.423	1	0.957	0.895
1	0.21	0.449	0.686	0.461	1	0.882	0.883
1	0.054	0.398	0.506	0.569	1	0.846	0.871
1	0.314	0.217	0.656	0.306	0.539	0.882	0.713
1	0.73	0.442	0.389	0.425	0.765	0.651	0.933
1	0.132	0.434	0.556	0.433	0.522	0.787	0.83
1	0.266	0.556	0.775	0.422	1	0.959	0.934
1	0.713	0.406	0.644	0.533	0.999	0.763	0.728
1	0.081	0.332	0.402	0.293	1	0.795	0.991

Appendix 6.(continued)

	Results from Logit						
	Probability of Adoption of					Frontier production Function	
Test=1/ Control=0	Cross-bred cows	Fertilizer	Pesticides	Improved Seed	LISREL DEA	Cereal	Milk
1	0.111	0.195	0.703	0.397	0.569	0.827	0.662
1	0.532	0.433	0.363	0.329	1	0.981	0.863
1	0.608	0.385	0.606	0.367	1	0.845	0.754
1	0.748	0.415	0.754	0.374	0.967	0.965	0.73

Appendix 7. Results of Farm Specific Efficiency Scores from Frontier Production Function

Test=1/Control=0	Teff	Wheat	Barley	Bean	Peas	Oat
Selale Region						
0	0.945	1	0.991	0.197	0.982	0.996
0	0.91	0.77	0.99	0.862	0.992	0.978
0	0.62	0.983	0.947	0.272	0.993	0.148
0	0.887	1	0.983	0.944	0.979	0.993
0	0.673	1	0.998	0.961	0.995	0.997
0	0.794	0.977	0.854	0.791	0.99	0.993
0	0.716	0.87	0.638	0.767	0.983	0.994
0	0.964	0.934	0.969	0.785	0.929	0.993
0	0.764	0.91	0.814	0.293	0.992	0.171
0	0.616	0.704	0.983	0.77	0.916	0.839
0	0.8	1	0.983	0.21	0.999	0.993
0	0.867	0.944	0.358	0.864	0.99	0.996
0	0.755	1	0.977	0.235	0.994	0.989
0	0.862	0.919	0.741	0.838	0.989	0.995
0	0.729	1	0.953	0.896	0.978	0.991
0	0.642	0.014	0.733	0.812	0.995	0.992
0	0.886	0.66	0.97	0.925	0.989	0.995
0	0.699	1	0.987	0.939	0.986	0.995
0	0.932	1	0.966	0.666	0.994	0.996
0	0.812	0.884	0.889	0.824	0.99	0.993
0	0.72	0.974	0.982	0.808	0.996	0.994
0	0.782	1	0.829	0.845	0.989	0.992
0	0.848	0.798	0.896	0.862	0.988	0.992
0	0.854	0.612	0.928	0.789	0.994	0.743
0	0.804	0.988	0.614	0.829	0.987	0.351
0	0.88	1	0.962	0.842	0.996	0.995
0	0.713	0.994	0.936	0.057	0.999	0.991
0	0.363	1	0.952	0.969	0.989	0.724
0	0.917	1	0.985	0.832	0.992	1
0	0.816	1	0.506	0.76	0.985	0.786
0	0.871	0.346	0.485	0.814	0.988	0.992
0	0.838	0.963	0.984	0.746	0.989	0.993
0	0.848	0.321	0.028	0.532	0.996	0.751
0	0.816	1	0.967	0.815	0.993	0.994
0	0.907	1	0.987	0.906	0.992	0.994
0	0.804	1	0.972	0.8	0.994	0.966
0	0.888	0.986	0.986	0.713	0.994	0.637
0	0.781	1	0.956	0.794	0.996	0.994
0	0.899	0.962	0.722	0.49	0.957	0.992

Appendix 7 (Continued)

Test=1/ Control=0	Teff	Wheat	Barley	Bean	Peas	Oat
0	0.372	0.983	0.955	0.747	0.983	0.992
0	0.931	0.846	0.989	0.758	0.923	0.586
0	0.42	1	0.948	0.808	0.976	0.997
0	0.886	0.778	0.985	0.833	0.986	0.76
0	0.939	0.747	0.611	0.876	0.973	0.992
0	0.891	1	0.965	0.839	0.992	0.992
0	0.833	1	0.998	0.721	0.989	0.999
0	0.876	1	0.608	0.757	0.961	0.994
0	0.935	1	0.985	0.809	0.927	0.215
0	0.96	0.41	0.988	0.817	0.994	0.074
0	0.872	1	0.96	0.816	0.996	0.994
0	0.965	0.623	0.987	0.927	0.94	0.992
0	0.888	0.862	0.684	0.341	0.972	0.99
0	0.885	1	0.971	0.927	0.993	0.994
0	0.734	1	0.983	0.926	0.995	0.994
0	0.996	1	0.921	0.858	0.988	0.991
0	0.717	0.973	0.99	0.722	0.969	0.989
0	0.978	0.27	0.985	0.818	0.984	0.995
0	0.734	0.891	0.604	0.804	0.99	0.991
0	0.722	1	0.986	0.745	0.988	0.993
0	0.989	1	0.989	0.958	0.943	0.994
0	0.613	1	0.931	0.749	0.991	0.994
0	0.697	0.296	0.182	0.673	0.994	0.994
0	0.824	1	0.992	0.818	0.983	0.014
0	0.866	0.811	0.947	0.893	0.986	0.993
0	0.895	0.81	0.921	0.767	0.934	0.992
0	0.886	1	0.909	0.313	0.973	0.992
0	0.886	0.964	0.935	0.54	0.992	0.993
0	0.661	1	0.96	0.842	0.981	0.786
0	0.915	1	0.954	0.395	0.995	0.968
0	0.744	1	0.923	0.789	0.985	0.991
0	0.914	1	0.992	0.893	0.986	1
0	0.918	1	0.954	0.868	0.996	0.994
0	0.911	0.986	0.983	0.899	0.968	0.993
0	0.803	1	0.983	0.891	0.987	0.993
0	0.954	0.99	0.902	0.885	0.981	0.995
0	0.816	0.839	0.884	0.766	0.993	0.992
0	0.916	0.776	0.991	0.058	0.974	0.995
0	0.977	1	0.979	0.144	0.998	0.995
0	0.878	1	0.993	0.437	0.982	0.995
0	0.689	1	0.993	0.48	0.994	0.993

Appendix 7 (Continued)

Test=1/Control=0	Teff	Wheat	Barley	Bean	Peas	Oat
0	0.737	0.953	0.829	0.864	0.985	0.991
0	0.788	0.832	0.969	0.783	0.991	0.798
0	0.877	1	0.99	0.877	0.987	0.992
0	0.661	1	0.973	0.793	0.968	0.994
0	0.888	1	0.902	0.94	0.979	0.994
0	0.694	0.527	0.983	0.707	0.999	0.997
0	0.836	1	0.948	0.719	0.993	0.995
0	0.738	0.406	0.765	0.844	0.998	1
0	0.829	1	0.93	0.592	0.971	0.827
0	0.862	0.842	0.93	0.578	0.996	1
0	0.879	0.979	0.931	0.647	0.988	0.494
0	0.705	1	0.986	0.86	0.955	0.992
0	0.817	1	0.995	0.186	0.996	0.994
0	0.827	1	0.984	0.209	0.997	0.843
0	0.892	1	0.993	0.502	0.988	0.985
0	0.923	0.371	0.912	0.79	0.993	0.952
0	0.958	0.717	0.985	0.885	0.949	0.994
0	0.964	1	0.935	0.691	0.985	0.993
0	0.898	1	0.985	0.375	0.978	0.993
0	0.956	0.987	0.622	0.868	0.986	0.92
0	0.974	1	0.993	0.396	0.982	0.995
0	0.835	1	0.705	0.693	0.993	0.994
0	0.935	0.889	0.988	0.32	0.958	0.994
0	0.65	0.594	0.982	0.375	0.996	0.991
0	0.899	1	0.894	0.512	0.992	0.992
0	0.786	0.832	0.949	0.77	0.996	0.994
0	0.844	1	0.976	0.022	0.994	0.371
0	0.66	0.645	0.913	0.738	0.992	1
0	0.801	0.992	0.986	0.451	0.996	0.993
0	0.731	1	0.984	0.684	0.995	0.993
0	0.786	0.996	0.823	0.22	0.997	0.992
0	0.889	0.838	0.676	0.912	0.981	0.748
0	0.868	0.707	0.769	0.344	0.994	0.996
0	0.704	1	0.99	0.847	0.888	0.994
0	0.886	0.773	0.918	0.77	0.992	0.993
0	0.901	1	0.97	0.848	0.994	0.537
0	0.79	0.351	0.998	0.822	0.993	0.998
0	0.626	1	0.758	0.53	0.998	0.994
0	0.894	0.41	0.524	0.529	0.994	0.727
0	0.852	1	0.973	0.817	0.995	0.995
0	0.889	0.883	0.98	0.857	0.96	0.993

Appendix 7 (Continued)

Test=1/ Control=0	Teff	Wheat	Barley	Bean	Peas	Oat
0	0.744	1	0.914	0.85	0.996	0.998
0	0.776	1	0.994	0.558	0.996	0.994
0	1	0.418	0.937	0.891	0.987	0.994
0	0.779	1	0.995	0.818	0.985	0.995
0	0.949	1	0.833	0.659	0.985	0.915
0	0.872	1	0.991	0.464	0.974	0.994
1	0.878	0.963	0.997	0.947	0.958	0.584
1	0.921	0.399	0.962	0.559	0.991	0.993
1	0.932	0.468	0.995	0.595	0.996	0.694
1	0.865	0.949	0.971	0.853	0.989	0.359
1	0.89	1	0.979	0.814	0.984	0.992
1	0.921	0.581	0.978	0.425	0.985	0.992
1	0.965	0.991	0.811	0.62	0.973	0.976
1	0.923	0.877	0.988	0.814	0.986	0.993
1	0.811	1	0.984	0.646	0.994	0.995
1	0.938	0.861	0.989	0.767	0.967	0.993
1	0.884	1	0.988	0.702	0.993	0.993
1	0.879	0.945	0.954	0.896	0.953	0.394
1	0.905	0.756	0.793	0.343	0.988	0.992
1	0.872	1	0.988	0.868	0.897	0.992
1	0.931	0.953	0.97	0.941	0.972	0.594
1	0.942	1	0.996	0.83	0.982	0.45
1	0.808	1	0.673	0.941	0.982	0.995
1	0.791	1	0.936	0.975	0.985	0.737
1	0.893	1	0.976	0.478	0.995	0.995
1	0.802	0.799	0.985	0.861	0.975	0.99
1	0.976	0.875	0.594	0.925	0.956	0.995
1	0.808	0.987	0.95	0.902	0.994	0.994
1	0.788	1	0.599	0.834	0.994	0.928
1	0.738	0.42	0.936	1	0.971	0.994
1	0.89	1	0.976	0.849	0.991	0.992
1	0.728	1	0.936	0.84	0.985	0.992
1	0.732	1	0.99	0.835	0.987	0.997
1	0.991	1	0.698	0.816	0.96	0.511
1	0.932	1	0.97	0.558	0.992	0.992
1	0.972	0.866	0.976	0.923	0.984	0.316
1	0.885	1	0.843	0.517	0.995	0.994
1	0.795	1	0.953	0.75	0.994	0.906
1	0.904	1	0.984	0.593	0.971	0.993
1	0.859	1	0.836	0.321	0.996	0.862
1	0.758	0.922	0.982	0.712	0.99	0.95

Appendix 7 (Continued)

Test=1/Control=0	Teff	Wheat	Barley	Bean	Peas	Oat
1	0.644	1	0.922	0.454	0.989	0.994
1	0.838	1	0.947	0.898	0.996	0.993
1	0.804	1	0.996	0.897	0.988	0.5
1	0.885	0.981	0.35	0.766	0.994	0.992
1	0.869	0.295	0.961	0.662	0.992	0.99
1	0.722	0.787	0.998	0.837	0.998	0.462
1	0.804	0.877	0.992	0.096	0.996	0.991
1	0.749	0.671	0.929	0.807	0.992	0.96
1	0.879	1	0.99	0.851	0.994	0.996
1	0.782	1	0.997	0.859	0.994	0.608
1	0.882	0.984	0.782	0.903	0.979	0.281
1	0.793	1	0.969	0.619	0.99	0.995
1	0.791	1	0.729	0.879	0.989	0.786
1	0.886	0.668	0.964	0.787	0.979	1
1	0.803	0.96	0.994	0.836	0.997	0.995
1	0.929	0.995	0.671	0.445	0.996	0.996
1	0.935	1	0.984	0.864	0.991	0.995
1	0.808	0.998	0.994	0.731	0.973	0.2
1	0.821	1	0.661	0.89	0.991	0.992
1	0.891	1	0.965	0.589	0.918	0.995
1	0.716	0.925	0.994	0.821	0.991	0.995
1	0.914	1	0.985	0.964	0.984	0.614
1	0.781	0.771	0.759	0.76	0.984	0.992
1	0.837	1	0.927	0.825	0.999	0.995
1	0.92	0.886	0.816	0.757	0.975	0.991
1	0.941	1	0.985	0.834	0.978	0.692
1	0.935	0.542	0.346	0.766	0.993	0.992
1	0.844	1	0.907	0.904	0.995	0.992
1	0.877	1	0.757	0.857	0.96	0.992
1	0.975	0.574	0.964	0.853	0.982	0.944
1	0.609	1	0.991	0.218	0.999	0.992
1	0.774	0.967	0.736	0.492	0.99	0.994
1	0.573	1	0.984	0.869	0.999	1
1	0.897	1	0.987	0.419	0.99	0.996
1	0.898	1	0.958	0.746	0.972	0.998
1	0.8	0.988	0.612	0.851	0.929	0.99
1	0.741	0.995	0.758	0.158	0.997	0.138
1	0.823	1	0.901	0.837	0.977	0.994
1	0.887	1	0.74	0.764	0.984	0.899
1	0.943	0.964	0.969	0.869	0.994	0.993
1	0.84	0.487	0.598	0.895	0.991	0.995

Appendix 7 (Continued)

Test=1/ Control=0	Teff	Wheat	Barley	Bean	Peas	Oat
1	0.841	1	0.951	0.889	0.993	0.994
1	0.77	0.917	0.951	0.845	0.991	0.935
1	0.567	0.889	0.194	0.803	0.992	0.165
1	0.868	0.989	0.998	0.289	0.997	0.991
1	0.715	0.737	0.906	0.434	0.981	0.993
1	0.875	1	0.958	0.914	0.998	0.994
1	0.881	0.684	0.981	0.952	0.991	0.996
1	0.81	1	0.985	0.869	0.983	0.994
1	0.918	1	0.973	0.515	0.983	0.999
1	0.906	0.926	0.905	0.673	0.994	0.995
1	0.779	1	0.992	0.784	0.999	0.995
1	0.815	1	0.682	0.918	0.989	0.584
1	0.929	1	0.862	0.869	0.99	0.311
Ada Region						
0	0.973	1	0.974	0.805	0.889	
0	0.318	1	0.939	0.725	0.96	
0	0.967	0.853	0.973	0.766	0.979	
0	0.923	1	0.937	0.807	0.944	
0	0.963	1	0.992	0.861	0.956	
0	0.918	0.5	0.956	0.789	0.963	
0	0.673	1	0.995	0.842	0.921	
0	0.909	0.448	0.987	0.468	0.992	
0	0.466	0.996	0.958	0.807	0.906	
0	0.857	1	0.955	0.747	0.993	
0	0.91	0.815	0.905	0.861	0.96	
0	0.8	0.983	0.869	0.905	0.929	
0	0.82	0.776	0.987	0.855	0.951	
0	0.882	0.975	0.84	0.705	0.975	
0	0.847	1	0.804	0.831	0.905	
0	0.886	0.565	0.933	0.767	0.967	
0	0.783	0.421	0.99	0.824	0.989	
0	0.787	0.968	0.975	0.811	0.986	
0	0.835	1	0.971	0.923	0.929	
0	0.901	1	0.883	0.945	0.987	
0	0.946	0.917	0.968	0.799	0.969	
0	0.962	0.697	0.677	0.639	0.949	
0	0.838	0.899	0.987	0.834	0.964	
0	0.935	0.815	0.701	0.692	0.916	
0	0.715	0.845	0.965	0.808	0.965	
0	0.955	1	0.992	0.841	0.974	
0	0.477	0.945	0.975	0.797	0.976	

Appendix 7 (Continued)

Test=1/ Control=0	Teff	Wheat	Barley	Bean	Peas	Oat
0	0.806	0.995	0.915	0.841	0.965	
1	0.792	0.988	0.797	0.912	0.964	
1	0.873	1	0.947	0.634	0.988	
1	0.711	0.512	0.951	0.883	0.964	
1	0.746	1	0.96	0.743	0.974	
1	0.94	0.657	0.964	0.787	0.981	
1	0.866	0.996	0.981	0.696	0.97	
1	0.931	1	0.705	0.593	1	
1	0.998	0.227	0.149	0.926	0.9	
1	0.901	0.951	0.972	0.633	0.993	
1	0.988	0.539	0.893	0.919	0.932	
1	0.767	1	0.937	0.775	0.98	
1	0.758	1	0.988	0.839	0.997	
1	0.92	0.953	0.963	0.705	0.977	
1	0.901	1	0.944	0.896	0.959	
1	0.83	0.931	0.911	0.921	0.953	
1	0.816	1	0.989	0.875	0.991	
1	0.856	1	0.91	0.78	0.977	
1	0.655	0.977	0.556	0.859	0.993	
1	0.273	1	0.832	0.913	0.983	
1	0.955	0.994	0.987	0.57	0.905	
1	0.713	0.982	0.774	0.849	0.976	
1	0.752	0.309	0.988	0.83	0.982	
1	0.579	0.995	0.983	0.658	0.98	
1	0.978	0.462	0.969	0.942	0.953	
1	0.959	1	0.446	0.847	0.967	
1	0.859	1	1	0.907	0.923	

Appendix 8a. Intra-Regional Analysis of Production Decisions

Table A. Estimates of Average production Function for Test and Control Farmers[1/]

	Teff				Barley			
	Selale		Ada		Selale		Ada	
Variables	Test	Control	Test	Control	Test	Control	Test	Control
Intercept	1.13	0.412	1.47	3.017	0.857	0.503	2.504	0.69
	(2.512)*	-1.001	-0.923	-1.508	(2.06)#	-1.14	-1.79	-0.713
Plot	0.127	0.082	0.159	0.286	0.06	0.016	0.015	0.147
	(3.133)*	(2.4)*	-1.42	(1.991)#	-1.53	-0.48	-0.201	-1.674
Oxen	0.143	0.161	0.377	0.273	0.38	0.35	0.344	0.258
	-1.287	(2.111)#	(2.175)#	(1.983)#	(3.96)*	(4.2)*	-1.563	(2.109)#
Seed	0.4	0.407	0.845	0.568	0.38	0.286	0.54	0.201
	(2.261)*	(3.109)*	(3.029)*	(2.117)#	(2.4)*	(2.1)#	(2.141)#	-1.37
Labor	0.926	0.893	1.266	0.633	0.31	0.74	1.104	0.844
	(2.077)#	(2.383)*	(3.688)*	(2.863)*	(1.96)#	(2.2)#	(2.023)#	(2.309)*
Fertilizer	0.03	0.117	0.496	0.412	0.014	0.04	0.454	0.787
	-1.144	-0.448	(2.131)#	(2.692)*	-0.37	-0.74	(2.827)*	(2.388)*
Pesticides	0.236	0.206	0.248	0.056	0.425	0.13	0.085	0.009
	(1.986)#	(1.984)#	(1.945)#	-1.339	(2.25)*	(1.9)#	-0.588	-0.035
Plot characteristics	-0.214	-0.391	-0.100	0.134	0.275	0.12	0.265	0.032
	(-1.98)#	(-2.184)#	(-1.973)#	(1.986)#	(1.98)#	-0.7	(2.727)*	-0.189
Black	0.111	0.238	0.601	0.306	-0.129	-0.12	0.029	0.138
	-1.069	(2.647)*	(3.659)*	-1.568	(-1.517)	(-1.4)	-0.124	-0.723
Experience	0.536	0.769	0.922	0.885	0.42	0.49	0.752	0.971
	(2.062)#	(3.273)*	(1.986)#	(2.311)*	(1.9)#	(1.9)#	(1.983)#	(2.79)*
Extension	0.196	0.011	0.484	0.331	0.136	-0.01	0.002	-0.312
education	(2.14)#	-0.343	(2.234)*	(2.35)*	-0.91	(-0.26)	-0.017	(-0.677)
Production	0.459	0.367	0.92	0.775	0.83	0.547	0.739	0.582
knowledge	(2.49)*	(2.676)*	(2.617)*	(1.985)#	(1.99)#	(2.1)#	(1.949)#	(1.956)#
Schooling	0.129	0.429	0.091	0.148	0.52	0.42	0.103	0.244
	(1.99)#	(2.021)#	(1.957)#	(1.947)#	(2.02)#	(2.3)*	-1.279	(1.981)#
N	88	127	25	26	88	127	25	26
R2	0.767	0.71	0.899	0.719	0.674	0.595	0.809	0.802

Table A. Estimates of Average production Function for Test and Control Farmers[1/] (Continued)

	Bean				Wheat			
	Selale		Ada		Selale		Ada	
Variables	Test	Control	Test	Control	Test	Control	Test	Control
Intercept	1.185	0.813	0.045	1.32	0.351	0.698	3.583	1.208
	(2.218)#	(3.127)*	-0.031	(3.08)*	-0.472	-1.061	(3.401)*	-0.709
Plot	0.313	0.303	0.295	0.48	0.127	0.17	0.041	0.308
	(2.94)*	(1.989)#	(3.843)*	(2.92)*	(3.487)*	(3.607)*	-0.808	(3.522)*
Oxen	0.134	0.09	0.258	0.284	0.031	0.279	0.368	0.938
	(2.695)*	(1.971)#	(1.98)#	(1.97)#	-0.175	(1.958)#	(2.159)#	-1.594
Seed	0.09	0.213	0.192	0.17	0.186	-0.051	0.17	-0.510
	-1.773	(2.883)*	(1.99)#	(1.98)#	(1.97)#	(-0.492)	-1.098	(-1.262)
Labor	1.0039	1.126	1.06	1.171	0.115	-0.025	0.357	-0.810
	(2.14)#	(2.549)*	(2.22)#	(2.76)*	-0.792	(-0.21)	(2.17)#	(-1.632)
Fertilizer					0.024	0.008	0.441	0.311
					-0.482	-0.197	(2.155)#	(2.13)#
Pesticides					0.05	0.154	0.757	0.144
					-0.49	-0.982	(2.651)*	(2.21)#
Plot	0.423	0.278	0.482	0.417	-0.268	0.115	-0.118	-0.398
	(3.322)*	(2.412)*	-1.49	(3.43)*	(-1.501)	-0.669	(-0.483)	(-1.354)
Black	0.059	0.025	0.274	0.068	0.057	0.008	0.34	0.127
	-0.88	-0.422	(2.42)#	-1.004	-0.458	-0.079	(2.461)*	-0.578
Experience	0.575	0.175	0.62	0.897	0.759	(0.451	0.741	0.926
	(2.48)*	(1.976)#	(2.44)*	(2.11)#	(2.345)*	(1.969)#	(1.945)#	(2.517)*
Extension	0.035	-0.007	0.499	0.568	0.147	0.072	0.371	-0.053
	-1.158	(-0.401)	(1.96)#	(1.95)#	-1.891	(2.044)#	-1.872	(-0.539)
Production	0.42	0.503	0.825	0.908	0.573	0.747	0.975	0.825
	(2.243)*	(1.947)#	(1.95)#	(1.97)#	(3.325)*	(1.945)#	(4.711)*	(2.949)*
Schooling	0.316	0.201	0.327	0.212	0.113	0.079	0.388	0.271
	(2.001)#	(1.959)#	(2.01)#	(1.95)#	-1.322	(1.972)#	(2.341)*	(2.176)#
N	88	127	25	26	88	127	25	26
R2	0.808	0.821	0.8	0.947	0.564	0.653	0.944	0.927

Table A. Estimates of Average production Function for Test and Control Farmers[1/] (continued)

	Peas				Cereal			
	Selale		Ada		Selale		Ada	
Variables	Test	Control	Test	Control	Test	Control	Test	Control
Intercept	0.157	0.889	1.925	0.03	-1.463	0.441	0.88	1.528
	-0.38	(3.02)*	(2.195)#	-0.057	(-3.881)*	(2.114)#	-1.53	(2.11)#
Plot	0.161	0.13	0.055	0.364	0.661	0.655	0.971	0.366
	(4.34)*	(3.88)*	-0.852	(2.656)*	(3.407)*	(2.596)*	(1.98)#	(2.02)#
Oxen	-0.008	0.172	0.254	0.183	0.049	0.228	0.288	0.358
	(-0.09)	(2.02)#	-1.207	-1.508	-0.321	(2.63)*	(2.33)*	(1.98)#
Seed	-0.041	-0.043	0.258	0.629	0.239	0.117	0.56	0.379
	(-0.62)	(-0.654)	(2.812)*	(3.319)*	(2.239)*	-1.688	(2.48)*	(1.96)#
Labor	-0.026	0.034	0.341	0.293	0.01	0.228	0.734	0.207
	(-0.29)	-0.511	(1.985)#	(2.235)*	-0.084	(2.278)*	(2.14)#	-0.674
Fertilizer					0.041	0.001	0.021	0.033
					(2.908)*	-0.072	-0.63	-0.857
Pesticides					0.058	0.016	0.067	0.453
					-1.324	(2.038)#	-0.65	-1.493
Plot	0.059	0.131	-0.493	0.041	-0.026	0.098	0.365	0.537
	-0.45	-1.09	(-2.199)#	-0.332	(-0.237)	-1.444	-1.12	(2.02)#
Black	-0.011	0.014	0.067	0.277	-0.045	-0.050	0.271	-0.631
	(-0.13)	-0.207	-0.387	(4.054)*	(-0.233)	(-0.362)	-0.43	(-1.643)
Experience	0.591	0.368	0.56	0.42	0.64	0.448	0.849	0.789
	(1.96)#	(1.95)#	(2.012)#	(1.966)#	(3.40)*	(1.984)#	(2.59)*	(2.87)*
Extension	0.461	0.141	0.112	0.062	0.105	-0.021	0.269	0.242
	(2.66)*	-1.636	(1.95)#	-1.823	(2.35)*	(-0.799)	(2.37)*	(1.97)#
Production	0.636	0.245	0.467	0.407	0.896	0.861	0.949	0.896
	(2.86)*	(1.96)#	(1.988)#	(1.954)#	(2.896)*	(2.767)*	(2.99)*	(3.39)*
Schooling	0.466	0.469	0.244	0.457	0.501	0.281	0.39	0.26
	(2.19)#	(1.97)#	(1.955)#	(2.004)#	(2.676)*	(1.947)#	(2.14)#	(1.95)#
N	88	127	25	26	88	127	25	26
R2	0.844	0.835	0.833	0.973	0.587	0.574	0.602	0.736

Appendix 8b. Analysis of Production Decisions by Test-Control Group

Table A. Estimates of Average production Function for Test and Control Farmers [1/]

	Teff		Barley		Bean		Wheat	
Variables	Test	Control	Test	Control	Test	Control	Test	Control
Intercept	0.915	0.391	1.164	0.629	1.021	0.921	0.672	0.694
	(2.252)*	-1.037	(3.269)*	-1.669	(3.51)*	(3.92)*	-1.041	-1.242
Plot	0.109	0.062	0.059	0.034	0.288	0.313	0.122	0.17
	(3.663)*	(2.102)#	(2.152)#	-1.146	(2.92)*	-1.66	(2.045)#	(2.506)*
Oxen	0.193	0.169	0.328	0.303	0.116	-0.076	0.099	0.325
	(2.319)*	(2.921)*	(3.357)*	(2.525)*	(2.24)*	(-1.77)	-0.692	(2.524)*
Seed	0.448	0.428	0.41	0.282	0.115	0.208	0.095	0.696
	(2.994)*	(3.141)*	(2.605)*	(1.99)#	(2.34)*	(1.96)#	-0.967	(2.742)*
Labor	0.97	0.085	0.959	0.743	0.971	1.133	0.695	0.268
	(2.485)*	-1.24	(2.142)#	(2.12)#	(1.96)#	(3.96)*	-1.594	(2.246)*
Fertilizer	0.014	0.039	0.038	0.038			0.002	0.004
	-0.72	-1.887	-1.317	-0.852			-0.068	-0.116
Pesticides	0.009	0.051	0.062	0.095			0.05	0.211
	-0.095	-0.384	-0.841	-0.937			-0.55	-1.477
Plot	-0.241	-0.364	0.298	0.136	0.567	0.319	-0.210	0.016
	(-1.636)	(-2.425)*	(2.367)*	-0.987	(4.95)*	(3.15)*	(-1.455)	-0.113
Black	0.191	0.239	0.135	-0.088	0.01	0.028	0.119	-0.002
	(2.267)*	(2.997)*	-1.863	(-1.233)	-0.15	-0.526	-1.181	(-0.025)
Experience	0.769	0.956	0.442	0.355	0.547	0.415	0.42	0.232
	(2.069)#	(3.353)#	-1.845	-0.786	(2.12)#	(2.29)*	(2.21)#	(2.617)*
Extension	0.253	0.019	0.128	0.213	0.023	0.007	0.054	-0.061
	(1.978)#	-0.624	-1.839	-1.477	-0.79	-0.342	-1.238	(-1.943)
Production	0.521	0.454	0.585	0.318	0.62	0.664	0.52	0.889
	(2.013)#	(2.686)*	(3.574)*	(2.251)#	(2.26)*	(2.57)*	(1.955)#	(1.949)#
Schooling	0.481	0.209	0.224	0.089	0.513	0.433	0.376	0.478
	(2.135)#	(2.006)#	(1.975)#	-1.322	(2.03)#	(2.49)#	(2.023)#	(1.996)#
N	114	154	114	154	114	154	114	154
R2	0.761	0.694	0.704	0.605	0.759	0.803	0.583	0.704

1/ Values in parenthesis are t-statistics.

* and # indicate statistical significance at 1 and 5 percent respectively.

Table A. Estimates of Average production Function for Test and Control Farmers (continued)[1]

	Peas		Oats		Cereal	
Variables	Test	Control	Test	Control	Test	Control
Intercept	0.456	0.958	0.891	-0.095	1.921	0.388
	-1.151	(3.624)*	(2.41)*	(-0.396)	(2.249)*	-1.426
Plot	0.148	0.155	0.12	0.044	0.661	0.637
	(4.285)*	(3.229)*	(2.24)*	-0.903	(4.854)*	(4.866)*
Oxen	0.001	0.147	0.662	0.6	0.196	0.287
	-0.015	-1.896	(3.63)*	(3.22)*	-1.488	(3.55)*
Seed	0.722	0.895	0.218	0.164	0.156	0.086
	(1.969)#	(2.491)*	-1.71	-1.85	-1.498	-1.254
Labor	0.03	0.015	0.346	0.018	0.32	0.103
	-0.379	-0.24	(3.62)*	-0.219	(3.288)*	(2.136)#
Fertilizer					0.011	0.004
					-1.014	-0.405
Pesticides					0.065	0.001
					-1.734	-0.031
Plot	0.043	0.09	0.158	0.094	-0.115	-0.187
	-0.365	-0.884	-1.48	-1.001	(-2.07)#	(-1.501)
Black	0.007	0.06	-0.046	-0.080	0.097	0.109
	-0.094	-1.044	(-0.68)	(-1.599)	-1.005	-1.662
Experience	0.405	0.185	0.579	0.47	0.755	0.542
	(2.059)#	-1.603	(2.41)*	(2.58)*	(3.573)*	(2.649)*
Extension	0.061	-0.028	0.02	0.004	0.114	-0.019
	-1.762	(-1.24)	-0.65	-0.233	(2.65)*	(-0.716)
Production	0.438	0.759	0.681	0.391	0.268	0.686
	(1.992)#	(2.848)*	(2.60)*	(1.95)#	(1.97)#	(2.853)*
Schooling	0.548	0.688	0.655	0.437	0.358	0.476
	(1.973)#	(2.046)#	(2.36)*	(1.95)#	(1.987)#	(2.091)#
N	114	154	88	127	114	154
R2	0.722	0.84	0.936	0.951	0.535	0.556

1/ Values in parenthesis are t-statistics.

* and # indicate statistical significance at 1 and 5 percent respectively.

Appendix9a. LISREL Results for Model Type II (Ada, Crop)

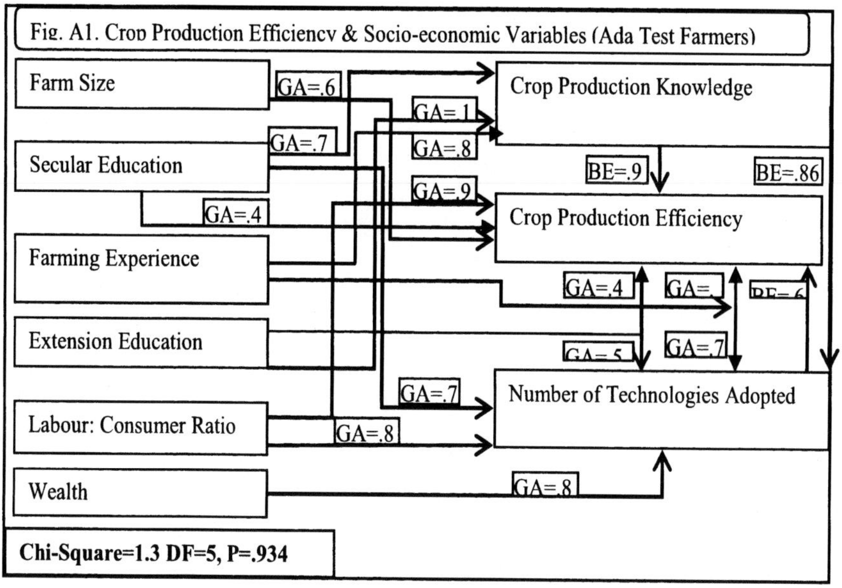

Fig. A1. Crop Production Efficiency & Socio-economic Variables (Ada Test Farmers)

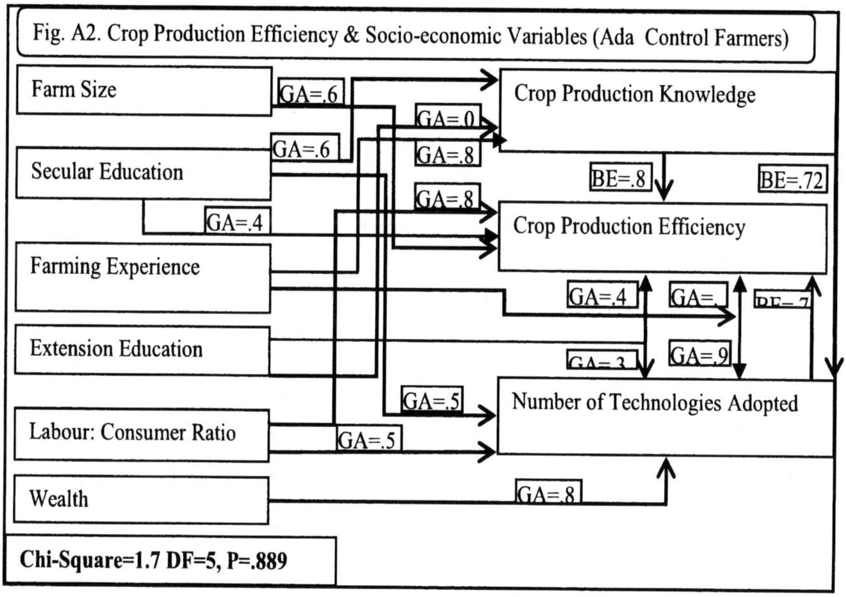

Fig. A2. Crop Production Efficiency & Socio-economic Variables (Ada Control Farmers)

Appendix 9b. LISREL Results for Model Type II (Selale, Crop)

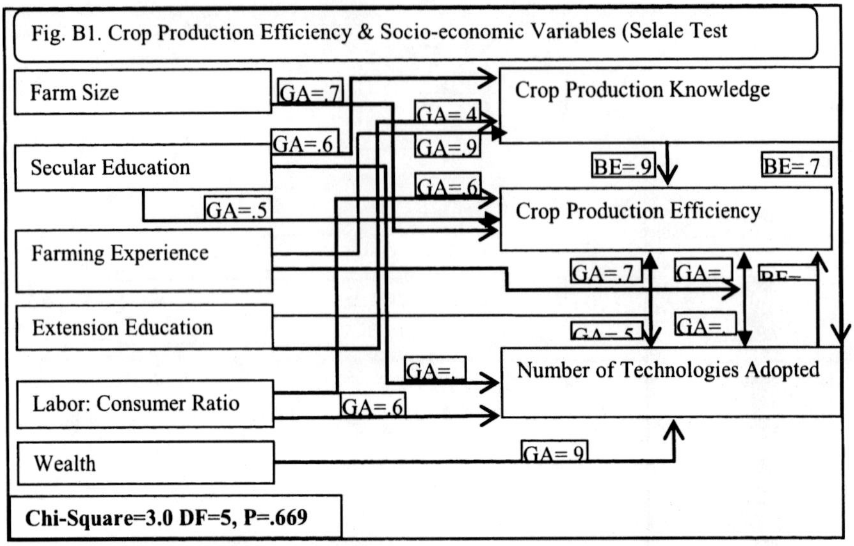

Fig. B1. Crop Production Efficiency & Socio-economic Variables (Selale Test

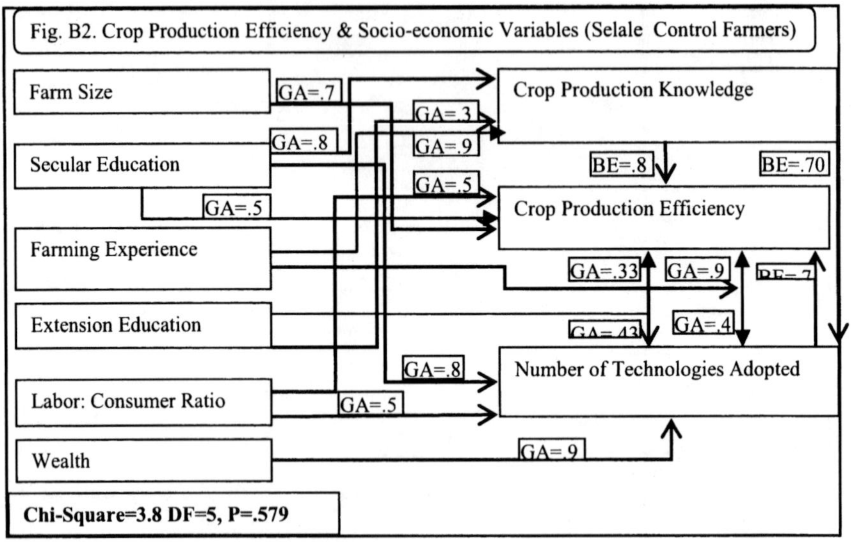

Fig. B2. Crop Production Efficiency & Socio-economic Variables (Selale Control Farmers)

Appendix 9c. LISREL Results for Model Type II (Ada, Milk)

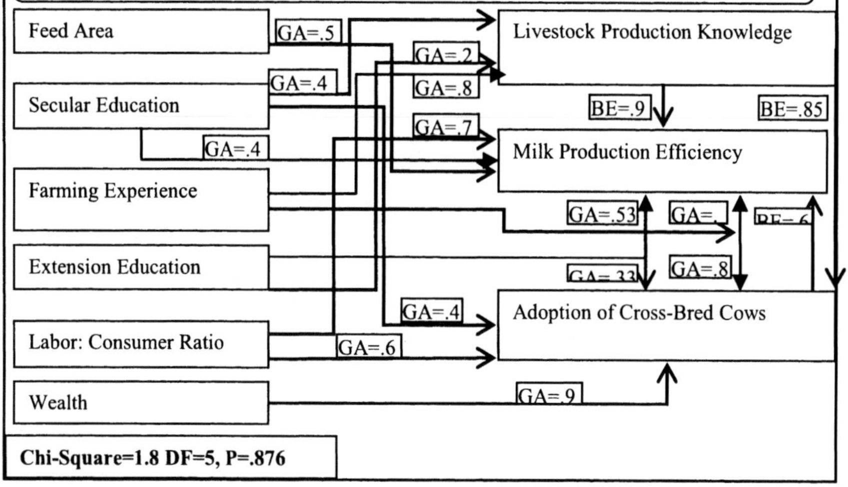

Fig. C1. Milk Production Efficiency & Socio-economic Variables (Ada Test Farmers)

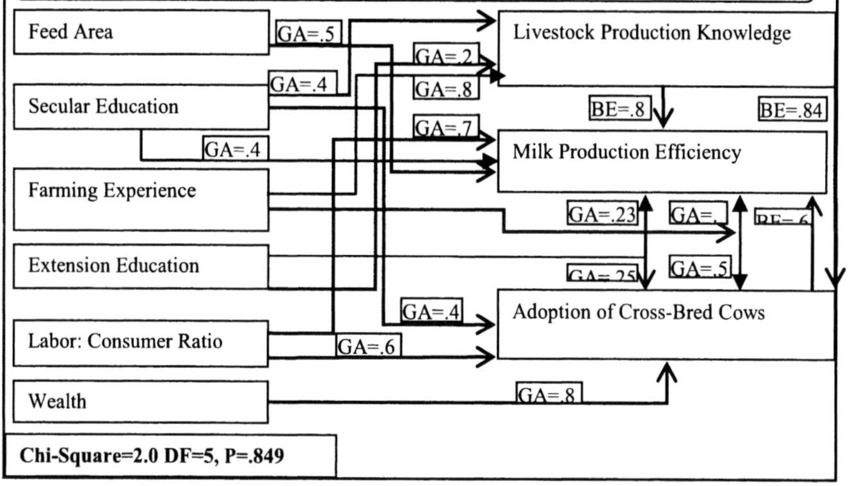

Fig. C2. Milk Production Efficiency & Socio-economic Variables (Ada Control Farmers)

Appendix 9d. LISREL Results for Model Type II (Selale, Milk)

In Addition to tests on ability to understand and interpret the questionnaire between

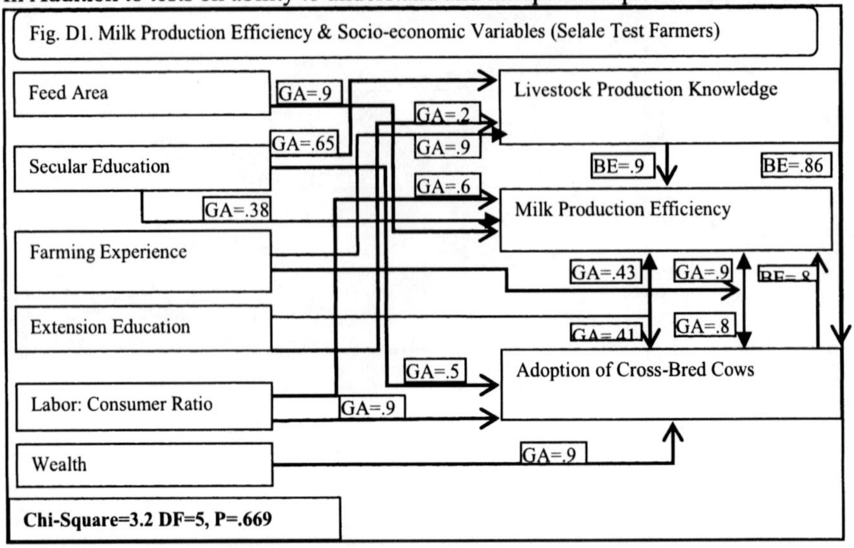

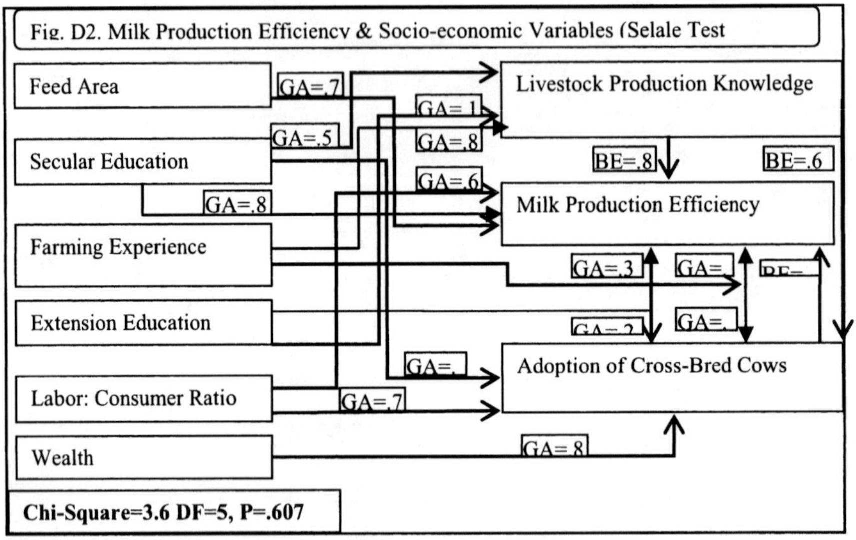

Appendix 9e. LISREL Analysis of Selective Choice of Technologies and Crop Production Efficiency (Ada)

Fig. E1. Crop Production Efficiency & Socio-economic Variables (Non-Adopters, Ada)

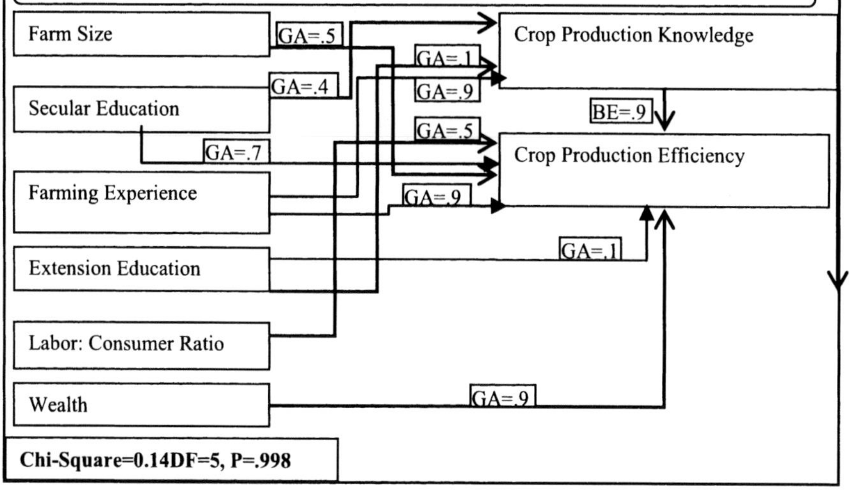

Fig. E2. Crop Production Efficiency & Socio-economic Variables (No. of Technologies adopted=1, Ada)

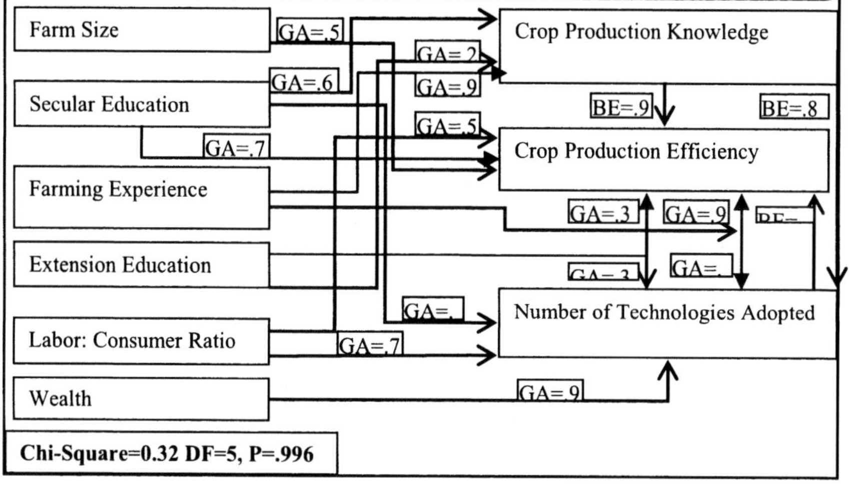

Fig. E3. Crop Production Efficiency & Socio-economic Variables
(No. of Technologies Adopted=2, Ada)

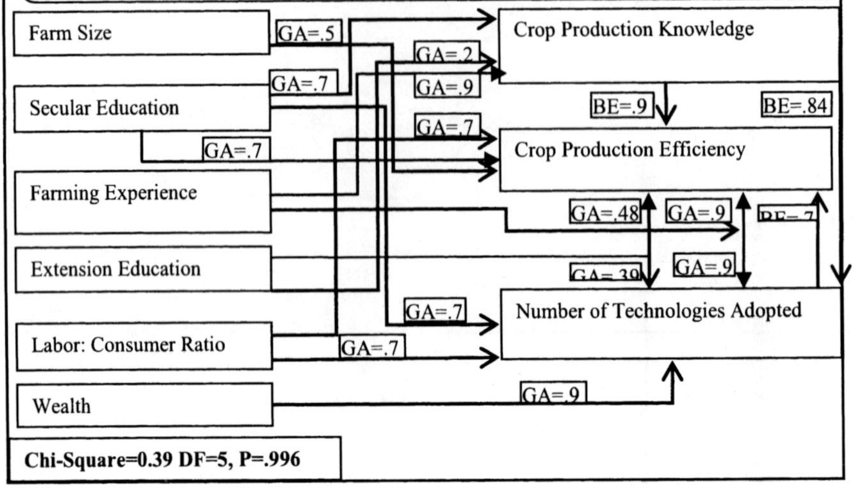

Fig. E4. Crop Production Efficiency & Socio-economic Variables
(No. of Technologies Adopted=3, Ada)

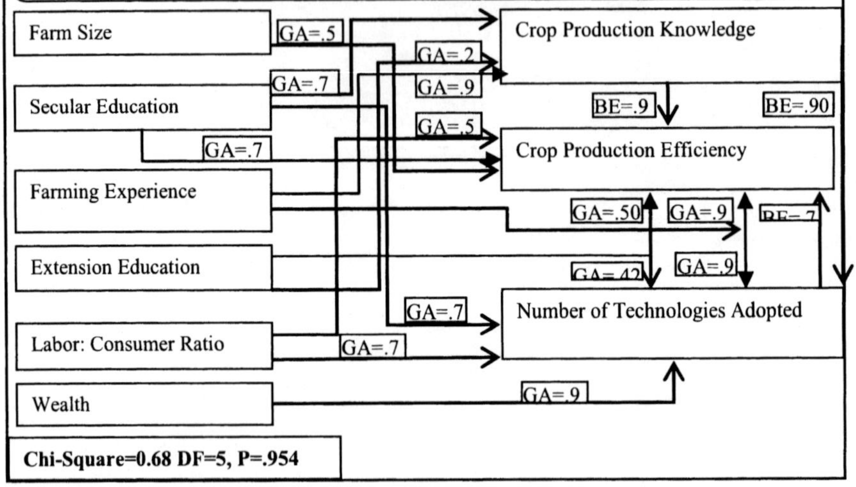

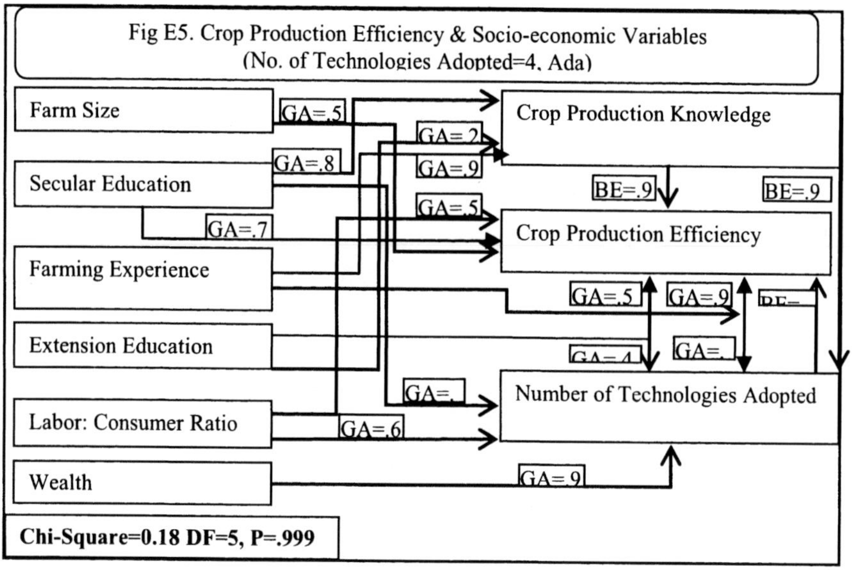

Appendix 9f. LISREL Analysis of Selective Choice of Technologies and Milk Production Efficiency (Ada)

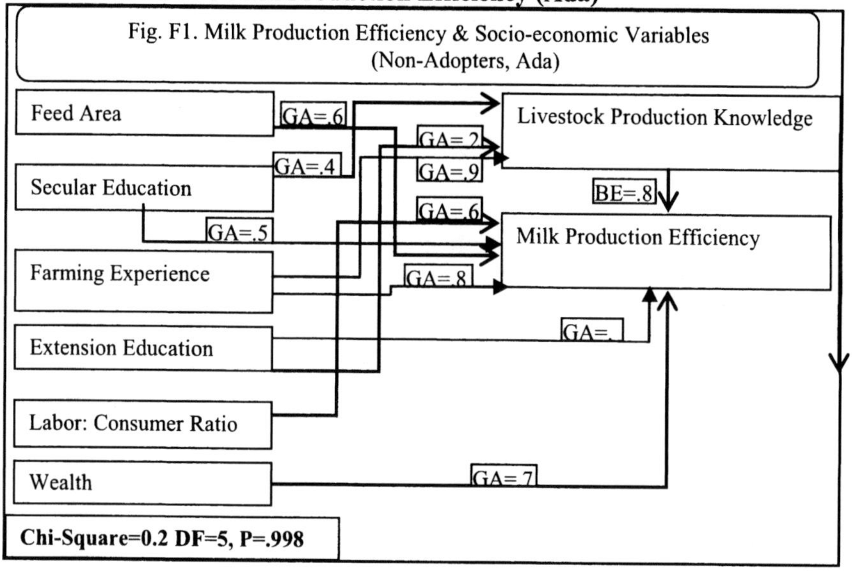

Fig. F2. Milk Production Efficiency & Socio-economic Variables (No. of Technologies Adopted=1, Ada)

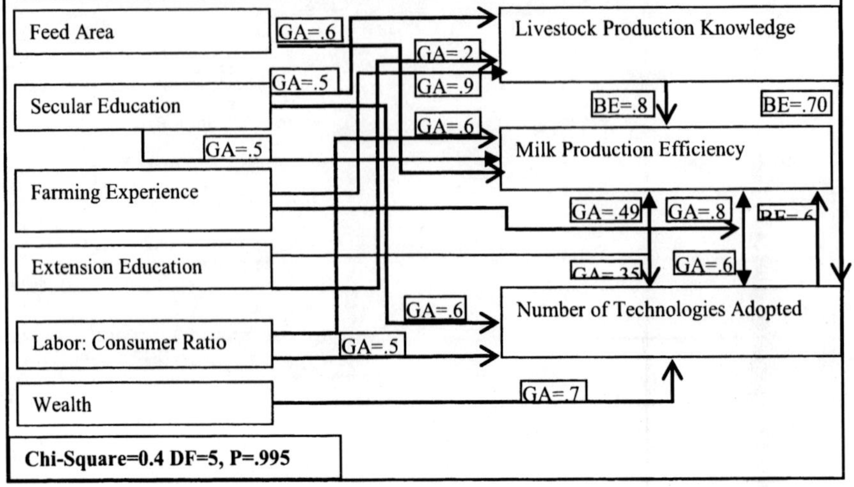

Fig. F3. Milk Production Efficiency & Socio-economic Variables (No. of Technologies Adopted=2, Ada)

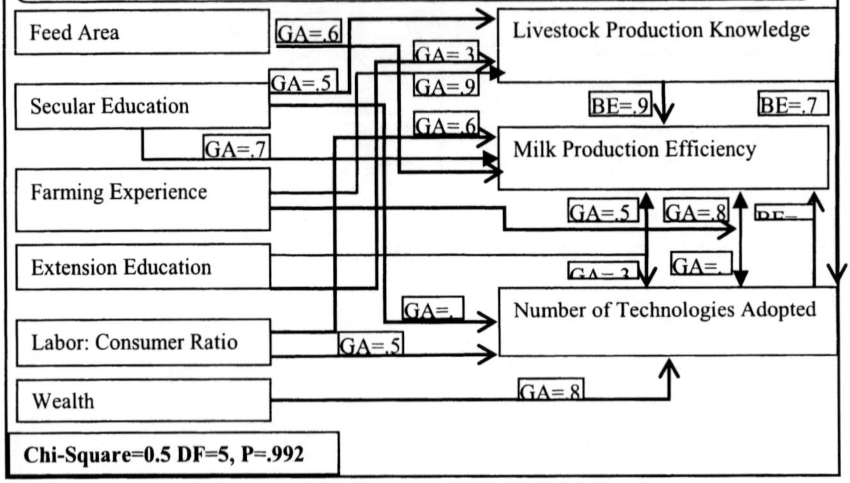

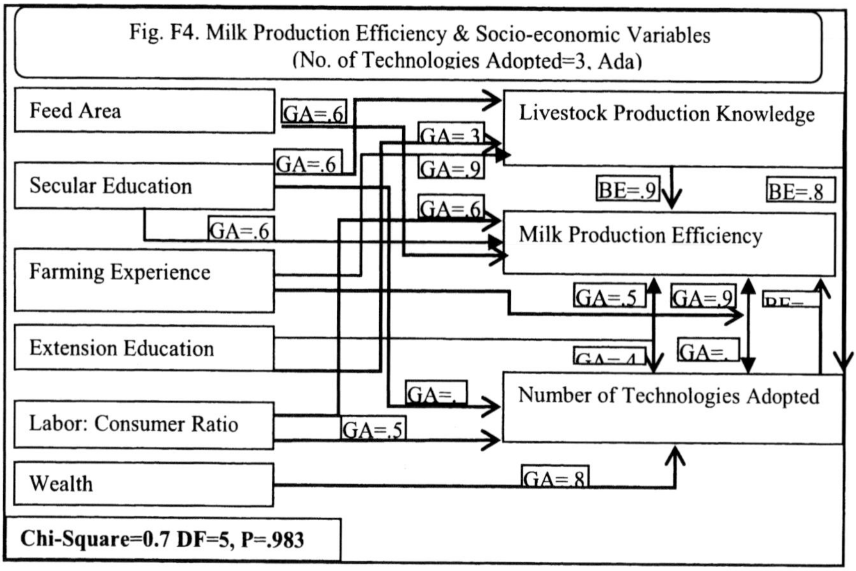
Fig. F4. Milk Production Efficiency & Socio-economic Variables (No. of Technologies Adopted=3, Ada)

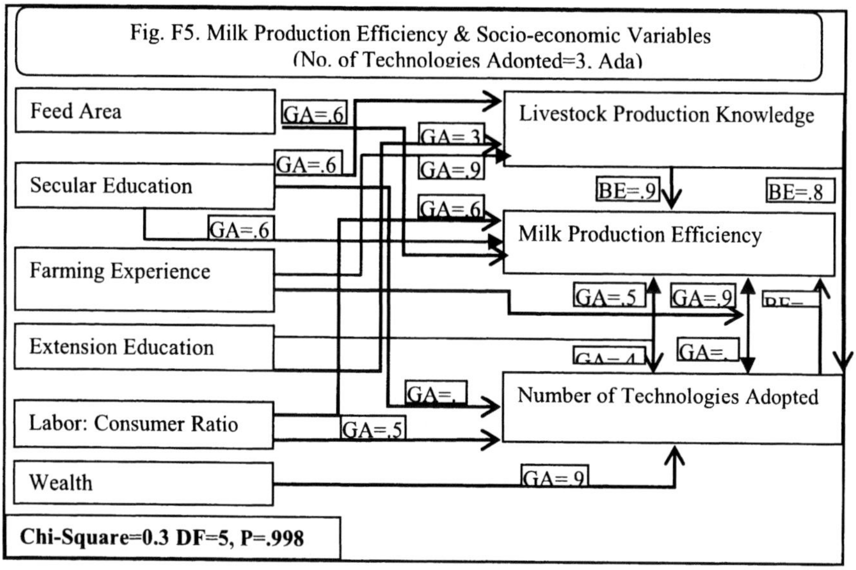
Fig. F5. Milk Production Efficiency & Socio-economic Variables (No. of Technologies Adopted=3, Ada)

Appendix 9g. LISREL Analysis of Selective Choice of Technologies and Crop Production Efficiency (Selale)

Fig. G1. Crop Production Efficiency & Socio-economic Variables (Non-Adopters, Selale

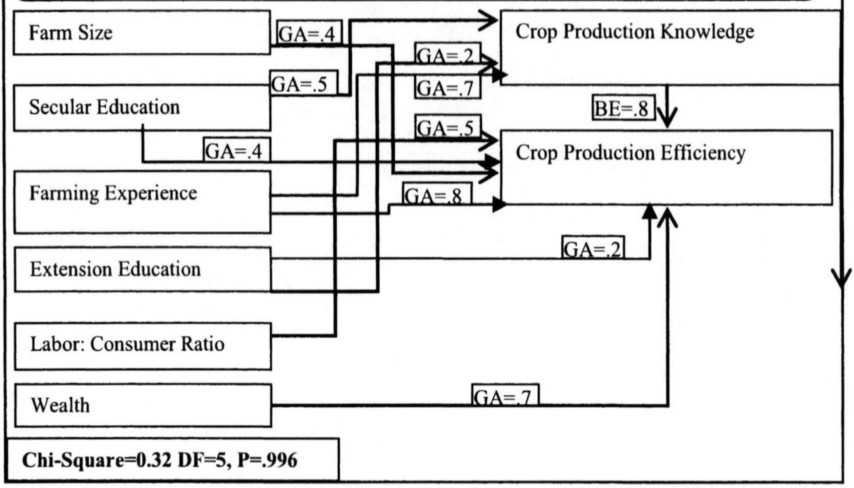

Fig. G2. Crop Production Efficiency & Socio-economic Variables (No. of Technologies Adopted=1, Selale)

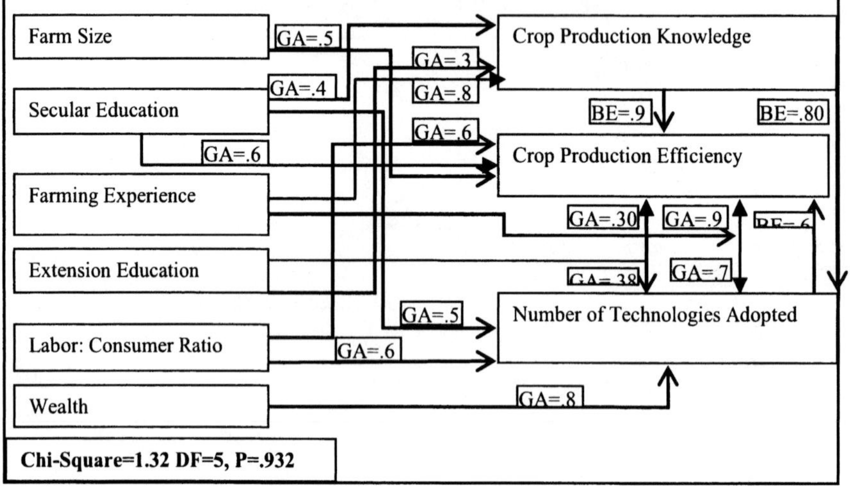

Fig. G3. Crop Production Efficiency & Socio-economic Variables
(No. of Technologies Adopted=2, Selale)

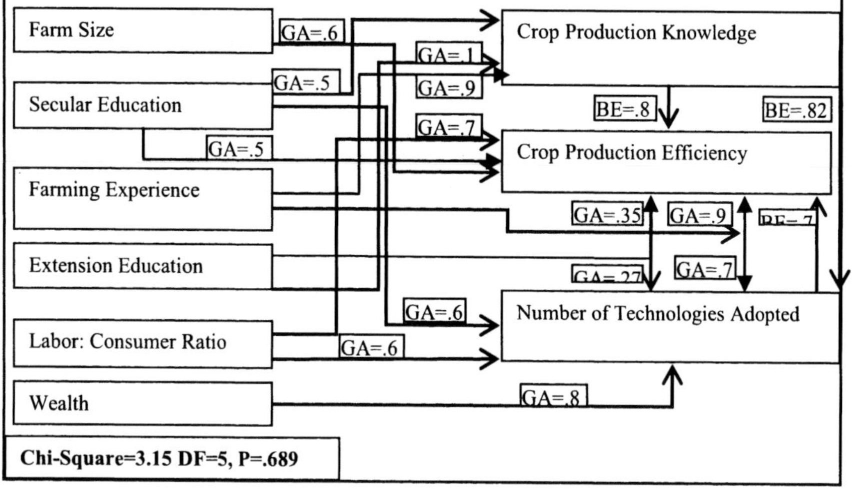

Fig. G4. Crop Production Efficiency & Socio-economic Variables
(No. of Technologies Adopted=3, Selale)

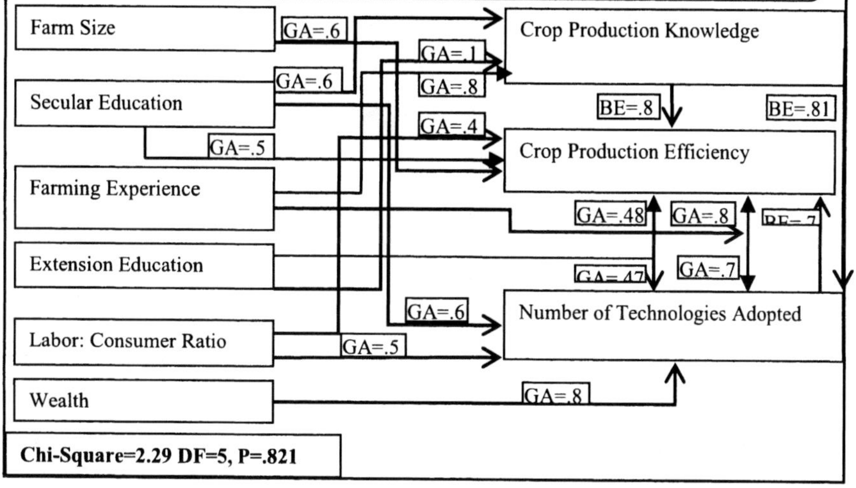

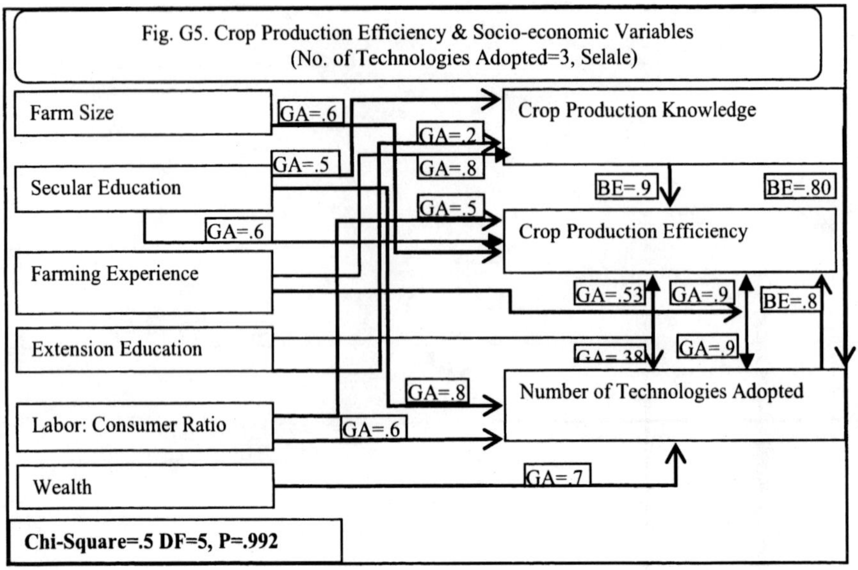

Appendix 9h. LISREL Analysis of Selective Choice of Technologies and Milk Production Efficiency (Selale)

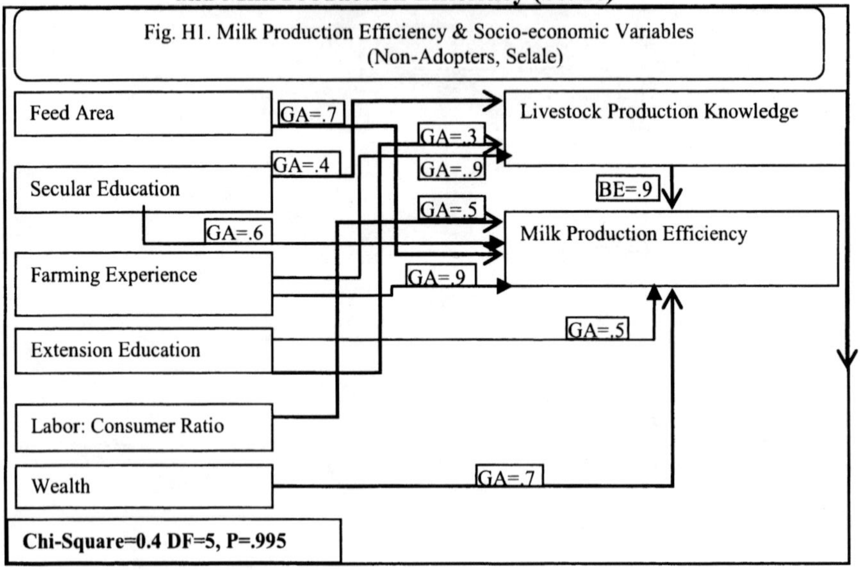

Fig. H2. Milk Production Efficiency & Socio-economic Variables
(No. of Technologies Adopted=1, Selale)

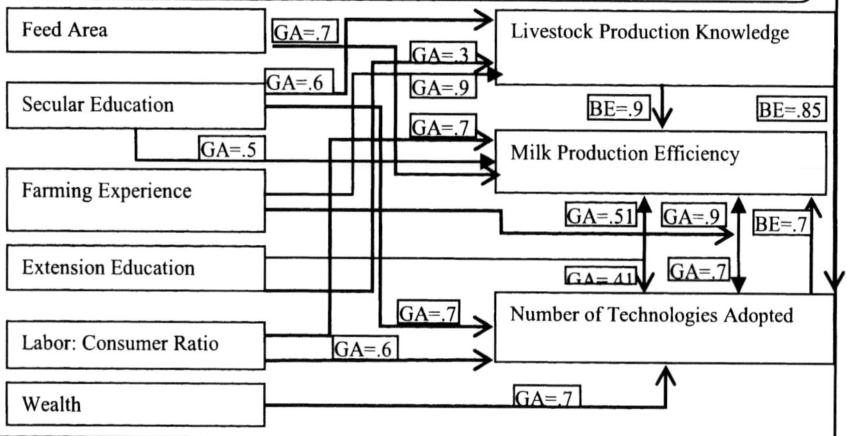

Chi-Square=1.4 DF=5, P=.924

Fig. H3. Milk Production Efficiency & Socio-economic Variables
(No. of Technologies Adopted=2, Selale)

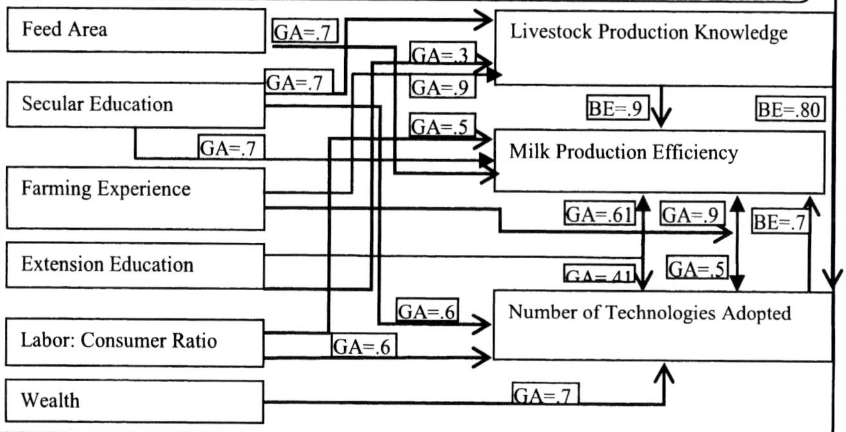

Chi-Square=3.4 DF=5, P=.683

Fig. H4. Milk Production Efficiency & Socio-economic Variables
(No. of Technologies Adopted=3, Selale)

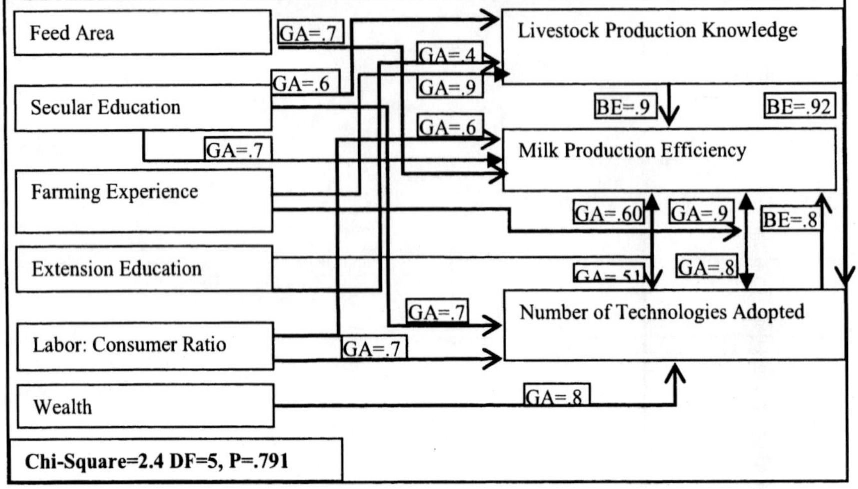

Fig. H5. Milk Production Efficiency & Socio-economic Variables
(No. of Technologies Adopted=3, Selale)

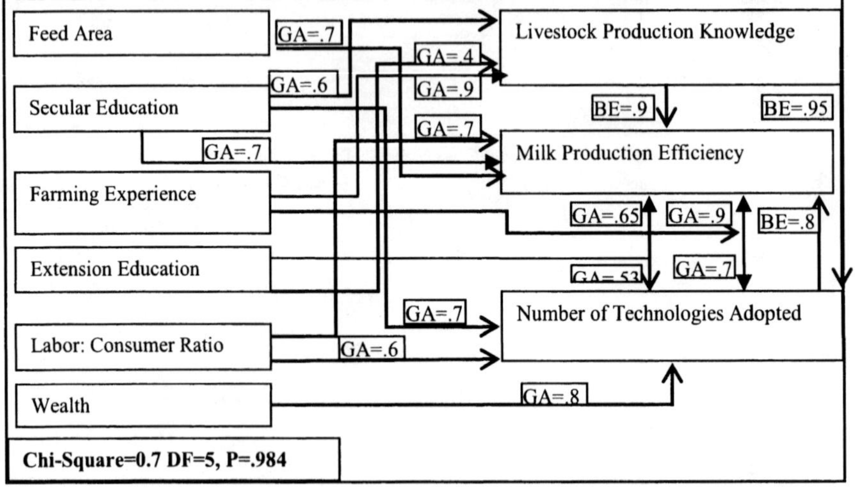

BIBLIOGRAPHY

Abbot, K.C. 1987. Agricultural Marketing Enterprises for the Developing World. Cambridge University Press.

Adelman, I., C.T. Morris. 1973. Economic Growth and Social Equity in Developing Countries. Stanford. Stanford University Press.

Ahmed, I. and V.W. Ruttan. 1988. Introduction. In I. Ahmed and V.W. Ruttan (eds) Generation and Diffusion of Agricultural Innovations: The Role of Institutional Factors. Grower Publishing Company Ltd. Aldershot:England. pp.1-18.

Ahmed, M., D. Aredo, H. Shiferaw. 1984. Agricultural Production and Rural Institutions in Post Revolutionary Ethiopia. August, Addis Ababa, Unpublished Report.

Ahmed, M., D. Aredo, H. Shiferaw. 1984. Land and Agriculture in Pre-Revolutionary Ethiopia. August, Addis Ababa, Unpublished Report.

Aigner, D., Lovell, C.A.K. and Schmidt, P. 1977. Formulation and Estimation of Stochastic Production Function Models. J. of Econometrics. 6:21-37.

Aklilu, B. 1974. Technological Change in Subsistence Agriculture: The Adoption and Diffusion of Fertilizer in Ethiopia's Minimum Package Areas. Ph.D. Dissertation, Boston University.

Aklilu, B. 1980. The Diffusion of Fertilizer in Ethiopia: Patterns, Determinants, and its Implications. The J. of Developing Areas. 14:387-399.

Alemaya College of Agriculture, 1975. Annual Research Report, Alemaya, Ethiopia.

Alemayehu, A. 1975. Identification of Opinion Leaders for the Adoption of Agricultural Innovations in two Northern Regions of Ethiopia. Ph.D. Dissertation, University of Maryland.

Amin, S. 1974. Accumulation on a World Scale: A Critique of the Theory of Underdevelopment. NY: Monthly Review Press.

Amin, S. 1976. Unequal Development: An Essay on the Social Formations of Peripheral Capitalism. NY: Monthly Review Press.

Anderson, J. R., J.L. Dillon, and J. B. Hardaker, 1977. Agricultural Decision Analysis. Iowa State University Press, Ames, Iowa.

Arnould, E.J. 1984. Marketing and Social Reproduction in Zinder, Niger Republic. In Netting, R. McC., R.R. Wilk, and E.J. Arnould. (ed). Households: Comparative and Historical Studies of the Domestic Group. Berkeley, University of California Press.

Arrow, K.J. 1971. Exposition of the Theory of Choice Under Uncertainty. In C.B. McGuire and R. Radner (eds). Decisions and Organization Amsterdam: North Holland. pp.19-55.

Assefa, S. 1983. An Analysis of Resource Allocation in a Household Resource Economy: Cases from Chilalo District, Ethiopia. Ethiopian J. of Development Research. 5-7:103-32.

Ayalew, S. 1986. Time use and labor supply. Paper presented at national workshop on food strategies for Ethiopia, Alemaya University of Agriculture, December.

Ayana, I. 1985. An Analysis of Factors Affecting the Adoption and Diffusion Patterns of Packages of Agricultural Technologies: A case Study in Two Extension Districts of Ethiopia. M.A. Thesis, Department of Economics, Addis Ababa University, June.

Banker, R.D., Charnes, A., and Cooper, W.W. 1983. Some Models for Estimating Technical and Returns to Scale Inefficiencies. Management Science.

Banker, R.D. 1984. Estimating Most Productive Scale Size Using Data Envelopment Analysis. European J. of Operational Research 17:35-44

Barlett, P.F. 1980. Agricultural Decision Making: Anthropological Contribution to Rural Development (ed). New York: Academic Press.

Barlett, P.F.1982. Agricultural Choice and Change. New Brunswick, NJ: Rutgers University press.

Barlett, P.F. 1980. Cost benefit Analysis: A test of Alternative Methodologies, P.F. Barlett (ed), Agricultural Decision Making: Anthropological Contribution to Rural Development, New York: Academic Press. pp.137-160.

Barnum, H.N., and L. Squire. 1979. A Model of an Agricultural Household. World Bank Staff Occasional Papers no. 27. Baltimore: John Hopkins University Press.

Barry, P. 1984. Risk Management in Agriculture (ed). Iowa State University Press, Ames, Iowa.

Barth, F. 1966. Ethnic Groups and Boundaries. Little, Brown, and Company, Boston.

Barth, F. 1967. Economic Sphere in Darfur. In R. Firth (ed). Themes in Economic Anthropology, London: Tavistock Publishing Ltd. pp.149-174.

Bates, R.H. 1981. Markets and States in Tropical Africa: The Political Basis of Agricultural Policies. Berkeley, University of California Press.

Becker, H. 1990. Labor Input Decision of Subsistence Farm Households in Southern Malawi. J. of Agricultural Economics. Vol.40, pp.162-171.

Belay, H.S. 1977. Problems, Practices and Strategies for Rural Development in Ethiopia. Debrezeit Agricultural Experiment Station. Addis Ababa University.

Belete, A. 1989. An Economic Analysis of Smallholder Mixed Farmers in Ethiopia. Ph.D. Dissertation, University of Armidale, Dept. of Agricultural Economics and Farm Management, Australia.

Belete, A. J.L. Dillon, and F.M. Anderson. 1991. Development of Agriculture in Ethiopia since the 1975 land reform. Agricultural Economics. 6:159-175.

Bell, C. 1972. The acquisition of agricultural technology: its determinants and effects. J. Dev. Studies. 9:123-59.

Bennet, J.W. and J.R. Bowen. 1988. Production and Autonomy: Anthropological Studies and Critique of Development (eds). Monograph in Economic Anthropology, No.5, University Press of America, NY.

Berry, S. 1984. Fathers Work for Their Sons: Accumulation, Mobility and Class Formation in an Extended Yoruba Community. Berkeley and Los Angeles: University of California Press.

Berry, S. 1988. Property Rights and Rural Resource management: The Case of Tree Crops in West Africa. In J.W. Bennet and J.R. Bowen (eds). Production and Autonomy: Anthropological Studies and Critique of Development. Monograph in Economic Anthropology, No.5, University Press of America, NY. pp.143-182.

Berry, S. 1980. Decision Making and Policy making in Rural Development. In P.F. Barlett (ed). Agricultural Decision Making: Anthropological Contribution to Rural Development, New York: Academic Press. pp.321-336.

Berry, S. 1986. Macro Policy Implications of Research on Rural Households and Farming Systems. In J.L. Moock (ed). 1986. Understanding Africa's Rural Households and Farming Systems. West View, Boulder. pp. 199-216.

Beyene, K. and l. J. Lambourne. 1985. The Status of Livestock Pasture and Forage Research and development in Ethiopia. IAR Proceedings.

Biggs, S.D. 1980. Informal Research and Development. Ceres, July/August 23-25.

Binswagner, H.P. and J. McIntire. 1987. Behavioral and material Determinants of Production Relations in Land Abundant Tropical Agriculture. Economic Development and Cultural Change. 36(1):

Binswagner, H.P. and M.R. Rosenzweig (eds). 1984. Contractual Arrangements, Employment, and Wages in Rural Labor Markets in Asia. New Haven, Conn.: Yale University Press.

Binswagner, H.P. and V.W. Ruttan. 1978. Induced Innovation: Technology, Institutions and Development. Baltimore: The John Hopkins University Press.

Binswagner, H.P. 1980. Altitudes towards Risk: Experimental measurement in Rural India. Am. J. of Agri. Econ. 62:395-407.

Binswagner, H.P. and M. R. Rosenzweig. 1986. Behavioral and material Determinants of Production Relations in Agriculture. The Journal of Development Studies. 22(3):503-539.

Bohannan, P. and L. Bohannan. 1968. Tiv Economy: Evanston, IL: Northwestern University.

Boserup, E. 1981. Population and Technical Change: A Case Study of Long Term Trends. Chicago: University of Chicago Press.

Boserup, E. 1965. The Conditions of Agricultural Growth: The Economics of Agrarian Change Under Population Pressure. London: Allen and Unwin.

Bransford, J.D. and N.S. McCarrel, 1983. A Sketch of a Cognitive Approach to Comprehension: Some thoughts about understanding what it means to comprehend. pp. 377-399. In P.N. Johnson-Laird and P.C. Watson (ed). 1983. Thinking: Readings in Cognitive Science. Cambridge University Press, New York.

Braun, D. P., and S. Plog. 1982. Evolution of Tribal Social networks: Theory and Prehistoric North American Evidence. American Antiquity 47:504-525

Bravo-Ureta, B.E. and L. Rieger. 1990. Alternative Production Frontier Methodologies and Dairy Farm Efficiency. J. of Agricultural Economics 41:215-226.

Bravo-Ureta, B.E. 1986. Technical Efficiency Measure for Dairy Farms Based on a Probabilistic Frontier Function Model. Canadian Journal of Agricultural Economics, 34:399-415.

Brown, L.A. 1981. Innovation Diffusion: A New Perspective. New York: Methuen.

CADU, 1968, 1971, 1972, 1973, 1974, 1978, 1982, 1984. Annual Reports. Assela.

Calkins, P.H. 1989. Towards a Unified Theory of Economic Development Theories. Paper Presented to the Canadian Agricultural Economics and Farm management Society Conference, Montreal, July.

Cancian, F.1980. Risk and Uncertainty in Agricultural Decision Making. In P.F. Barlett (ed), Agricultural Decision Making: Anthropological Contribution to Rural Development, New York: Academic Press, pp.161-176.

Cancian, F. 1979. The Innovator's Situation. Stanford: Stanford University Press.

Cary, J.W., and W.E. Holmes. 1982. Relationships among Farmers' Goals and Farm Adjustment Strategies: Some Empirics of a Multidimensional Approach. Aust. J. of Agr. Econ. 26:114-130.

Casely, D.J., and D.A. Lury. 1982. Monitoring and Evaluation of Agricultural Projects and Rural Development Practices. Baltimore: The John Hopkins University Press.

Chambers, R., and J. Jiggins. 1986. Agricultural Research for Resource-Poor farmers: A Parsimonious Paradigm. IDS, University of Susex, Discussion Paper 220.

Chambers, R.G. 1988. Applied Production Analysis: A Dual Approach. Cambridge: Cambridge University Press.

Chambers, R. 1983. Rural Development: Putting the Last First. Longman, Harlow, UK.

Charnes, A., W.S. Cooper, and E. Rhodes, 1978. Measuring the Efficiency of Decision Making Units. European J. of Operational Research. 2:429-44.

Charnes, A., and W.W. Cooper, 1980. Management Science Relations for Evaluation and Management Accountability. Journal of Enterprise Management, 2:143-162.

Chatman, E.A. 1986. Diffusion Theory: A Review and Test of a Conceptual Model in Information Diffusion. J. of the American Society for Information Science. 37(6):377-386.

Chenery, B. 1979. Structural Change and Development Policy. Oxford.

Chibnik, M. 1980. The Statistical Behavior Approach: The Choice Between Wage Labor and Cash cropping in Rural Belize. P.F. Barlett (ed), Agricultural Decision Making: Anthropological Contribution to Rural Development, New York: Academic Press. pp. 87-114. Chow, G.C. 1960. Tests of Equality Between Sets of Coefficients in Two Linear Regressions. Econometrica. 38:591-605.

Cohen, J.M. and Nils-Ivar Isaksson, 1988. Food production Strategy Debate in Revolutionary Ethiopia. World Development, Vol.10, no.3, pp.323-348.

Cohen, J.M. and D. Weintraub. 1975. Land and Peasants in Imperial Ethiopia: The social Background to a Revolution. Van Gorcum and Comp. B.V.-Assen., The Netherlands.

Cole, M. 1983. An Ethnographic Psychology of Cognition. In P.N. Johnson-Laird and P.C. Wason (eds). Thinking: Readings in Cognitive Science. Cambridge University Press, New York. pp. 468-482.

Coleman, G. 1983. The Analysis of memory Bias In Agricultural Labor Data Collection: A Case Study of Small Farmers in Nigeria. J. of Agricultural Economics. 34:79-86.

Collinson, M.P. 1982. Farming Systems Research in East Africa: The Experience of CIMMYT and Some National Agricultural Research Services, 1976-1981, MSU Rural Development Paper No.3, Michigan State University, East Lansing, Michigan.

Colmenares, H.J. 1970. Adoption of hybrid seeds and fertilizers among Colombian corn growers. Mexico city: Cimmyt.

Compton, J.L. 1980. Indigenous Folk Media and Rural Development. In D.W. Brokensha et al. (eds). Indigenous Knowledge Systems and Development. Lanham, Md.: University Press of America, pp.307-320.

Compton, J. Lin. 1989. Strategies and Methods for the Access, Integration and Utilization of Indigenous Knowledge in Agricultural and Rural Development. In D.M. Warren, L. Jan Slikkerveer, and S.O. Titilola. (eds) 1988. Indigenous Knowledge Systems: Implications for Agricultural and International Development. Studies in Technology and Social Change, No.11, Iowa State University, Ames, Iowa. pp.21-32.

Constable, M., 1983. Ethiopian Highland Reclamation Study: Summary, Ministry of Agriculture/FAO, Addis Ababa, Ethiopia.

Conti, A. 1979. Capitalist Organization of production through Non-capitalist Relations: Women's Role in a pilot Resettlement in Upper Volta. Review of African Political Economy 15/16:75-92.

Constable, M. 1985. Summary Report on Ethiopian Highland Reclamation Study. FAO/Ministry of Agriculture, Addis Ababa, Ethiopia.

CSO, 1989. Economic and Demographic Data, Addis Ababa, Ethiopia.

Cutie, J. 1976. Diffusion of hybrid corn technology: the case of El-Salvador. Mexico city: Cimmyt.

Dalton, G. 1967. Tribal and Peasant Economies: Readings in Economic Anthropology (ed). Garden City, N.Y.: Natural History Press.

Debreu, G.1951. The Coefficient of Resource Utilization. Econometrica. 19:273-92.

Debrezeit Agricultural Experiment Station, 1975, 1980, 1985. Annual Research Reports. Debrezeit, Ethiopia.

Dejene, A. 1990. Environment, Famine and Politics in Ethiopia: A View from the Village. Boulder, Colo., U.S.A.: Lynne Reinner.

Dewalt, B.R. and K.M. Dewalt. 1980. Stratification and Decision making in the Use of New Agricultural Technology. In P.F. Barlett (ed), Agricultural Decision Making: Anthropological Contribution to Rural Development, New York: Academic Press. pp.289-320.

Dillon, J. 1977. The Analysis of Response in Crop and Livestock Production, 2nd ed. Sydney: Pergmon Press Ltd.

Dillon, J.L. 1986. Institutional Possibilities to Minimize Risk at farm Level. paper Presented at Sixth development of Rainfed Agriculture Under Arid and semi-Arid Condition. World Bank, Washington, D.C., 6-10 January.

Dillon, J.L. and P. Scandizzo. 1978. Risk Attitudes of Subsistence Farmers in North East Brazil: A sampling Approach. American J. of Agr. Econ. 60:425-435.

Dillon, J. and J. Anderson. 1971. Allocative Efficiency, Traditional Agriculture and Risk. American J. of agricultural Economics. 60:425-435.

Dogramaci, A. and R. Fare. 1988. Application of Modern Production Theory: Efficiency and Productivity (ed). Kluwer Academic Publishers, Dordrecht.

Dommen, A.J. 1988. Innovation in African Agriculture. West View Press, Boulder, CO.

Douglas, M. and B. Isherwood. 1979. The World of Goods. New York: basic Books.

Dunn, E.S. Jr. 1971. Economic and Social development- A Process of Social Learning. Baltimore: The John Hopkins University Press.

Dunn, D.W. 1984. Applied Decision Analysis. McGraw-Hill Book Company. New York

Dunn, W.N. 1983. Conceptualizing Knowledge Use. In G.M. Beal, W. Dissayanake, and S. Konoshima (eds). Knowledge Generation, Exchange, and Utilization. Honolulu: University of Hawaii.

Durkheim, E. 1945. The Elementary Forms of the Religious Life. Glencoe.

Durrenberger, E.P. 1984. Chayanov, Peasants and Economic Anthropology (ed). Academic Press, Inc, NY.

Eisemon, T.O., and A. Nyamete. 1988. Schooling and Agricultural Productivity in Western Kenya. J. of East African Research and Development, 18:44-66.

Eisemon, T.O. 1988. Becoming a Modern Farmer: The Impact of Primary Schooling on Agricultural Thinking and Practice in Kenya and Burundi. In D.M. Warren, L. Jan Slikkerveer, and S.O. Titilola. (eds) 1988. Indigenous Knowledge Systems: Implications for Agricultural and International Development. Studies in Technology and Social Change, No.11, Iowa State University, Ames, Iowa. pp.41-67.

Ekanayake, S.A.B., and S.K. Jayasuriya. 1989. Change, Adjustment and the Role of Specific Experience: Evidence From Sri Lankan Rice Farming. Australian J. of Agr. Econ. 33(2):123-135.

Elder, J.W. 1968. Cultural and Social Factors in Agricultural Development. In Development and Change in Traditional Agriculture: Focus on South Asia. South Asian Series Occasional Paper No.7. East Lansing: Michigan State University, Asian Studies Centre.

Ellis, F. 1988. Peasant Economics: Farm Households and Agrarian Development, Cambridge: Cambridge University Press.

Engles, F. 1972. The origin of the Family, Private Property, and the State. Ed. E.B. Leacock. New York: International Publishers.

FAO, 1982, 1984. Mission Reports to The Ethiopian Highland Reclamation Study; Project Documents in collaboration with the Land Use and the Planning and Programming Department of Ministry of Agriculture, Addis Ababa, Ethiopia.

Fare, R., and Lovell, C.A.K. 1978. Measuring the Technical Efficiency of Production. J. of Economic Theory. 19(1):150-62.

Fare, R. and Grosskopf, S. 1983. Measuring Output Efficiency. European J. of Operational Research. 13:173-79.

Fare, R., S. Grosskopf, and C.A. K. Lovell. 1985. The Measurement of Efficiency of Production. Kluwer.Nijhoff Publishing, Dordrecht.

Farrel, M.J. 1957. The Measurement of Productive Efficiency. Journal of the Royal Statistical Society, 120 (III): 253-81.

Farrel, M.J., and F. Fieldhouse. 1962. Estimating Efficiency in Production Functions Under Increasing Returns to Scale. J of the Royal Statistical Society. A125:252-67.

Feder, G., R. E. Just, and D. Zilberman, 1982. Adoption of Agricultural Innovation in Developing Countries, Washington, DC: World Bank Staff Working Paper No.549.

Feder, G.R. 1981. Adoption of interrelated agricultural innovations: complementarity and impact of risk, scale, and credit. Amer. J. Agr. Econ. 64:94-101.

Feder, G.R., R. E. Just, and D. Zilberman, 1985. Adoption of agricultural innovations in developing countries: A survey. Econ. Dev. and Cult. Change. 33:255-98.

Fei, J.C., and G. Ranis. 1964. Development of the Labor Surplus Economy: Theory and Policy. Homewood, Ill.: Irwin.

Ferguson, C.E. 1971. The neoclassical Theory of Production and distribution. Cambridge: Cambridge University Press.

Finnida/Ministry of Agriculture. 1989. Base Line Survey: Selale Dairy Development Pilot Project. Unpublished Report.

Firth, R. 1969. Social Structure and Peasant Economy: The Influence of Social Structure upon Peasant Economies, pp.23-27. In C. Wharton (ed). Subsistence Agriculture and Economic Development. Chicago: Aldine.

Firth, R. 1966. Housekeeping Among Malay Peasants: London: Athlone, London School of Economics Monographs in Social Anthropology, No..7

Firth, R. 1939, 1965. Primitive Polynesian Economy. 2nd ed. Hamden, Conn: Shoe String Press.

Firth, R. 1952. Elements of Social organization. London: Watts and Co.

Fleisher, B., and L.J. Robison. 1985. Application of Decision theory and the Measurement of Attitudes Towards Risk in Farm management Research in Industrialized and Third World Settings. MSU International development paper No.6.

Forsund, F.R. and Hjalmarsson, L. 1974. On the Measurement of Productive Efficiency. Swedish J. of Economics. 76(2):141-54.

Foster, G.M. 1962. Traditional Cultures and the Impact of technical Change. NY.: Harper Brothers.

Foster, G.M. 1962. Traditional Cultures and the Impact of technical Change. NY.: Harper Brothers.

Frank, A.G. 1978. Dependent Accumulation and Underdevelopment. London: McMillan.

Frank, G. 1971. Capitalism and Under Development in Latin America. New York: Penguin.

Frank, A.G. 1967. Sociology of Underdevelopment and Underdevelopment of Sociology. Catalyst. 3:20-73.

Frisch, R. 1965. Theory of Production. Chicago: Rand McNally and Company.

Gafsi, S. 1976. Green Revolution: The Tunisian Experience. Mexico City, CIMMYT.

Galaty, J. 1988. Pastoral and Agro-Pastoral Migration in Tanzania: Factors of Economy, Ecology and Demography in Cultural Perspective. In J.W. Bennet and J.R. Bowen (eds). Production and Autonomy: Anthropological Studies and Critique of Development. Monograph in Economic Anthropology, No.5, University Press of America, NY. pp. 163-184.

Gartrell, J.W., and D.C. Gartrell. 1979. Status, Knowledge and Innovation. Rural Sociology. 44:73-94.

Geertz, C.1973. The Interpretation of Culture. New York: Basic Books.

Geertz, C. 1963. Agricultural Involution. Berkeley: University of California Press.

Geertz, C. 1983. Local Knowledge: Further Essays in Interpretative Anthropology. New York: Basic Books.

Getahun, A. 1978. Agricultural Systems in Ethiopia. Agricultural Systems. 3:281-293.

Getahun, A. 1980. Agro-Climates and Agricultural Systems in Ethiopia. Agricultural Systems. 5:39-50.

Ghai, D. L.D. Smith. 1987. Agricultural Prices, Policy and Equity in Sub-Saharan Africa. Lynne Rienner Publishers, Inc. Boulder, Colorado.

Ghose, A.K. 1985. Transforming Feudal Agriculture: Agrarian Change in Ethiopia Since 1974. The Journal of Peasant Studies. 22(1):127-49.

Gibbon, P., Yusuf Bangura, Arve Ofstad, 1992. Authoritarianism, Democracy and Adjustment: The Politics of Economic Reform in Africa (eds). The Scandinavian Institute of African Studies, Uppsala.

Gilbert, E.H., D.W. Norman, and F.Winch, 1980. Farming System Research: A Critical Appraisal. MSU Rural Development Paper, No.6, Michigan State University, East Lansing, Michigan.

Gill, G.J. 1974. Traditional Ethiopian Land tenure and new Inputs Ethiopian journal of Development Research, 1(1):81-101

Gladwin, C. 1980. A Theory of Real Life Choice: Applications to Agricultural Decisions. P.F. Barlett (ed), Agricultural Decision Making: Anthropological Contribution to Rural Development, New York: Academic Press. pp.45-84.

Gladwin, C. and R. Zabawa. 1987. Transformation of Full-Time Family Farms in the U.S.: Can they Survive?. in M Maclachlan (ed). Household Economics and Their Transformations. Lanham: University Press of America. pp.212-227.

Gladwin, C. 1979. Production Functions and Decision Models: Complementary Models. American Ethnologist. 6(4):653-674

Gladwin, C. and Murtaugh, 1980. The Attentive/Pre-attentive Distinction in Agricultural Decisions. In P.F. Barlett (ed), Agricultural Decision Making: Anthropological Contribution to Rural Development, New York: Academic Press. pp. 115-136.

Gladwin, C. 1979. Cognitive Strategies and Adoption Decisions: A Case Study of Non-Adoption of an Agronomic Recommendation. Econ. develop. and Cul. Change. 28:155-173.

Goldberger, A.S. and O.D. Duncan, 1973. Structural Equation Models in the Social Sciences (eds). New York: Seminar Press.

Goody, J. 1972. The evolution of the family. In Household and Family in Past time. In Peter Laslett and R. Wall (eds). Cambridge: Cambridge University Press. pp.103-24.

Granger, 1969. Investigating Causal Relationship by Econometric Models and Cross-Spectral Methods. Econometrica. 37:424-438.

Greene, W. H. 1991. LIMDEP Manual. Ergo Computing, Inc.

Greene, W.H. 1982. Maximum Likelihood Estimation of Stochastic Frontier Production Models. J. of Econometrics. 18:285-289.

Greene, W.H. 1980. Maximum Likelihood Estimation of Econometric Frontier Functions. J. of Econometrics. 13:27-56

Griffin, K. and R. Hay 1985. Problems of Agricultural Development in Socialist Ethiopia: An Overview and A Suggested Strategy. J. of Peasant Studies. 13(1):37-66.

Gryseels, and G. Assaminew, 1985. Links Between Livestock and Crop Production in the Ethiopian Highlands. ILCA Newsletter Vol.4, No.2, April.

Gryseels, G. 1988. Role of Livestock on Mixed Smallholder Farms in the Ethiopian Highlands. A Case Study from the Baso and Worena Near Debrebirhan. Ph.D. Thesis, Agricultural Economics, Wageningen University.

Gryseels, G. and K. de Boodt. 1986. Integration of Cross-bred cows (boran*Friesian) on Smallholder Farms in the Debrezeit Area of the Ethiopian Highlands. High Land Programme, ILCA, July.

Gryseels, G. F.M. Anderson, J.W. Durkin, and G. Asaminew. 1986. Draught Power and Smallholder Grain Production in the Ethiopian Highlands. ILCA Newsletter Vol.5, No.4, October.

Gryseels, G. and F.M. Anderson. 1983. Research on Farm and Livestock Productivity in Central Ethiopian Highlands: Initial Results, 1977-1983. ILCA Research Bulletin Report No.4.

Guba, E.G. 1968. Diffusion of Innovations. Educational Leadership, 25:292-295.

Guyer, J.I. 1986. Intra-Household Processes and Farming Systems Research: Perspective from Anthropology. In J.L. Moock (ed). 1986. Understanding Africa's Rural Households and Farming Systems. West View, Boulder. pp. 105-123.

Guyer, J. I. 1984. Anthropological Models of African Production. Man. 19(3):371-388.

Habtu, R. 1988. The Political Economy of Input Substitution Industrialization in Ethiopia: Capitalist and Socialist Variants. Unpublished Ph.D. Dissertation, Submitted to Faculty of Graduate Studies of the University of Massachusetts, Department of Economics.

Hagen, E.E. 1980. The Economics of Development. 3rd ed. Home Wood, Ill.: Irwin.

Hagen, E.E. 1962. On the Theory of Social Change- How Economic Growth Begins. Home Wood, Ill.: Dorsey Press.

Hammel, E.A. 1984. On the *** of Studying Household From and Function. In Netting, R. McC., R.R. Wilk, and E.J. Arnould. 1984 (ed). Households: Comparative and Historical Studies of the Domestic Group. Berkeley, University of California Press. pp.29-43.

Hardaker, J.B. and R.D. Ghodake. 1984. Using Measurement of Risk Attitudes in Modeling Farmer's Technology Choices. Economics Program, Progress report 60. January, ICRISAT, India.

Harper, W.M. and C. Eastman. 1980. An Evaluation of Goal Hierarchies for Small Farm Operators. American J. of Agri. Economics, 62(4):742-47.

Hazell, P. B.R., and R.D. Norton. 1986. Mathematical Programming For Economic Analysis in Agriculture. New York, Macmillan Publishing Company.

Harrel, F.E., 1982. The Logit Procedure. SAS (SUGI) Supplemental Library User's Guide, Version 5 Edition. NC: SAS Institute Inc., pp.269-293.

Harris B. 1972. Innovation adoption in Indian agriculture: the high yielding variety program. Modern Asian studies 6:

Hayami, Y. and V.W. Ruttan. 1985. Agricultural Development: An International Perspective. Rev. ed. Baltimore: The John Hopkins University Press.

Hayduk, L.A. 1987. Structural Equation Modeling with LISREL: Essentials and Advances. Baltimore: John Hopkins University Press.

Heath, A.F. 1976. Decision Making and Transactional Theory. In B. Kapferer (ed) Transaction and Meaning: Directions in the Anthropology of Exchange and Symbolic Behavior. Institute for the Study of Human Issues, Inc. pp.25-40.

Heathfield, D.F., and S. Wibe, 1987. An Introduction to Cost and Production Function. London: Macmillan Publishing Company.

Helleiner, G.K. 1975. Small Holder Decision Making: Tropical African evidence. In G. Reynolds (ed). Agriculture In Development Theory. New Haven, Conn.: Yale University Press.

Hendry, P. 1987. Research on Farming System Offers New Perspectives. Ceres (FAO), Vol.120, Vol.48. pp.13-15.

Henn, J. 1983. Feeding the Cities and Feeding the Peasants: What Role for Africa's Women Farmers? World development, 11,12.

Herath, H.M.G., B. Hardaker, and J.R. Anderson. 1982. Choices of Varieties by Sri Lanka Rice farmers: Comparing Alternative Decision Models. American J. Agri. Econ. 64:87-93.

Hicks, W.W., and R. Johnson, 1974. Population growth and the adoption of new technology in Taiwanese agriculture. Working paper in economics no. 1974-e6. Columbia: University of Missouri.

Hildebrand, P.E., 1986. Perspectives on Farming Systems Research and Extension (ed). Colorado: Lynne Reinner Publisher Inc.
Hirschman, A.O. 1958, 1968. The Strategy of Economic Development. New Have: Yale University Press.
Hirschman, A.O. 1984. A Dissenter's Confession: The Strategy of Economic development Revisited. In G.M. Meier and D. Seers. Pioneers in Development. Oxford University Press. pp.85-114.
Hoben, A. 1973. Land Reform Among the Amhara of Ethiopia. Chicago: Chicago University Press.
Homans, G.C. 1961. Social Behavior: Its Elementary Forms. New York: Harcourt Brace and World.
Hopkins, Terence, K., Immanuel Wallerstien, et, al. 1982. World-System Analysis: Theory and Methodology. Explorations in the world Economy. Publication of the Bernard Braudel Centre, Vol. I, Beverly Hills, Sage.
Hoselitz, B.F. 1960. Sociological Aspects of economic growth. 1960. NY.: The Free Press.
Huffnagle, H.P. 1961. Agriculture in Ethiopia. FAO, Rome. 484 pp.
Hyden, G. 1986. The Invisible Economy of Small Holder Agriculture in Africa. In J.L. Moock (ed). 1986. Understanding Africa's Rural Households and Farming Systems. West View, Boulder. pp.11-35.
Hyden, G. 1980. Beyond Ujaama in Tanzania: Underdevelopment and Uncaptured Peasantry. London: Heinemann.
Hyden, G. 1983. No Short Cuts To Progress: African Development Management in Perspective. Berkeley: University Of California Press.
IAR, 1976. Reports of Experiments in Animal production. Animal Production Department, No.1., July.
IAR, 1985, 1988. Annual Research report. Addis Ababa, Ethiopia.
IFPRI, 1992. Understanding How Resources are Allocated Within Households. IFPRI Policy Briefs no.8, IFPRI/The World Bank.
ILCA, 1981, 1985. Annual Research Report, Addis Ababa, Ethiopia.
ILCA, 1983. Old Technology Finds New Application. ILCA Newsletter Vol.2, No.4, October.
ILCA, 1981. Animal traction in Sub-Saharan Africa. ILCA Bulletin No.14, Addis Ababa, December.
ILCA. 1983. The single ox: An innovation for the small farmer. ILCA news letter, Vol.2, No.2, Addis Ababa, April.
Inkeles, A., and D. Smith. 1974. Becoming Modern. Individual Change in Six developing Countries. Cambridge, Mass.: Harvard University Press.
Isaac, B.L. 1989. Research in Economic Anthropology (ed), Vol. 13.
Jahnke, H.E. 1982. Livestock production Systems and Livestock Development in Tropical Africa. Kieler Wissenschaftsverlag Vank, Kiel, West Germany.
Jamison, D. and P. Moock. 1984. Farmer Education and Farm Efficiency in Nepal. World Development. 12:67-86.

Jamison, D.T. and L. Lau. 1982. Farmer Education and Farmer Efficiency. Baltimore: The John Hopkins University Press.

Janelid, I. 1979. Rural Development programmes and the Farm Household as a unit of Observation and Action. In Clio Presvelou and Saskia Spijkers-Zwart (eds), The Household, Women, and Agricultural Development. Proceeding of a symposium organized by the department of Home Economics, Agricultural University of Wageningen, The Netherlands, 18-20 January, H. Veenman and Zonen B.V.- Wageningen, pp.83-100.

Jodha, N.C., and A.C. Mascarenhas .1983. Adjustments to Climatic Variability in Self Provisioning Societies: Some Evidences From India and Tanzania, ICRISAT Economics Program Progress Report No.4, India.

Johnston, J. 1984. Econometric Methods. McGraw-Hill Book Company. New York.

Jondrow, J., Lovell, C.A.K., Materov, I.S. and Schmidt, P. 1982. On the Estimation of Technical Inefficiency in the Stochastic Production Function Model, J. of Econometrics. 19:233-238

Jones, J.R. and B.J. Wallace. 1986. Social Sciences and Farming Systems Research: Methodological Perspectives on Agricultural Development (eds). Westview Press, Inc.

Joreskog, K.G. 1981. Analysis of Covariance Structures. Scandinavian J. of Statistics. 8:65-92.

Joreskog, K.G. and D. Sorbom 1979. Advances in Factor Analysis and Structural Equation Models. Cambridge, MA: Abt Books.

Joreskog, K.G. and D. Sorbom. 1991. LISREL 7: A Guide to the Program and Applications. SPSS Inc., Chicago, Illinois.

Just, R.E., D.L. Heuth and A. Schmitz. 1982. Applied Welfare Economics and Public Policy. Prentice-Hall, Inc., Englewood, N.J.

Kalirajan, K.P. and R.T. Shand. 1985. Types of Education and Agricultural productivity: A Quantitative Analysis of Tamil Nadu Rice Farming. J. of Dev. Studies. 21:232-244.

Kataoka, S. 1963. A Stochastic Programming Model. Econometrica. 31:181-96.

Kebede, Y., G. Coffin, and J. Galaty, 1993. Adoption of Agricultural Technologies in Grain-Surplus Regions of Ethiopia. Working paper, Dept. of Agri. Economics, McGill University.

Kebede, Y. 1993. The Limits to Common Resource Management: The Bypassed Commons. Paper presented at the 20th Annual Meeting of the Canadian Anthropological Society, May 6-10,1993, Toronto, University of Toronto.

Kebede, Y. 1992. Experiential Knowledge and Strategic Household Decision Making: A Review and Empirical Evidence. Memo, Centre for Society Technology and Development, McGill University.

Kebede, Y., K. Gunjal, and G. Coffin 1990. Risk and Adoption of Agricultural Technologies in Ethiopia. Agricultural Economics, 4:27-43.

Kebede, Y. 1988. Economic Evaluation of Post-Drought Recovery Agricultural Project, The Case of Tegulet and Bulga District, Shoa Province, Ethiopia, M.Sc. Thesis, Department of Agricultural Economics, McGill University.

Kennedy, P. 1985. A Guide to Econometrics. Cambridge, Massachusetts: MIT Press.

Kiros, F.G. 1976. An Estimate of the Proportion of the Potential Work Year Allocated to Socio-Cultural Observances in Rural Ethiopia. Ethiopian Journal of Development Research. 2(2): 15-27.

Kiros, F.G. 1980. An Investigation into the Problem of Fragmentation of Agricultural land in Ethiopia in Post 1975 Period. Research report No.31. June, IDR, Addis Ababa University, Ethiopia.

Kiros, F.G. 1978. Technological Information and Adoption Patterns in Welmera Woreda. Addis Ababa University, February.

Kaminsky, M. 1979. A Comment. In A. Valdes, G.M. Scobie, and J.L. Dillon (eds). Economics and the Design of Small-Farmer Technology, Iowa State University Press, Ames, Iowa, pp.65-66.

Knight, F.H. 1921. Risk, Uncertainty and Profit. Boston: Houghton Muffin Co.

Kokeb, 1992. International Ethiopian Newspaper, February, Toronto, Canada.

Koopmans, T.C. 1957. Three Essays in the State of Economic Science. New York: McGraw Hill Book Company.

Koopmans, T.C. 1951. Activity Analysis of Production and Allocation (ed). Cowles Commission for Research in Economics, Monograph No.13, New York: John Wiley and Sons, Inc.

Kopp, R.J. and V.K. Smith. 1980. Frontier Production Function Estimates for Steam Electric Generation: A Comparative Analysis. Southern Economic Journal. 47:1049-59.

Korten, D.C. 1972. Planned Changes in a Traditional Society: Psychological problems of Modernization in Ethiopia. Praeger Publishers, NY.

Koutsoyiannis, A. 1981. Theory of Econometrics. 2nd. ed. Hong Kong: The Macmillan Press Ltd.

Kumbhakar, S.C. 1987. The Specification of technical and Allocative Inefficiency in Stochastic Production and Profit Frontiers. J. of Econometrics. 34:335-348.

Kunkel, J.H. 1970. Society and Economic Growth: A Behavioral Perspective of Social Change. NY: Oxford University Press.

Kusterer, K. 1989. Small-farmer Attitudes and Aspirations: AID program Evaluation Discussion paper No. 26., USAID, June.

Laclau, E. 1971. Feudalism and Capitalism in Latin America. New Left review, 67:19-38.

Larick, R. 1986. Age Grading and Ethnicity in the Style of Loikop (Samburu) Spears. World Archaeology. 18(2):268-282.

Laslett, P. 1984. The Family as a Knot of Household Interest. In R. netting ed. pp.353-382.

Lechtman, H. 1984. Andean Value Systems and the development of Prehistoric Metallurgy. Technology and Culture. 25:1-35.

Lee, L.F. 1983. A test for Distributional Assumptions for the Stochastic Frontier Functions. J. of Econometrics. 22:245-268.

Lele, Y. 1976. The Design of Rural Development: Lessons from Africa Baltimore: John Hopkins University Press.

Levine, D. 1974. Greater Ethiopia. Chicago: The University of Chicago Press.

Lewis, A. 1984. Development Economics in the 50's. In G.M. Meier and D. Seers. Pioneers in Development. Oxford University Press.

Lewis, A. 1984. The State of Development Theory. American Economic Review.74 (1):1-10.

Lewis, W.A. 1954. Economic Development with Unlimited Supplies of Labor. Manchester School of Economics and Social Studies Bulletin, London.

Leys, C. 1975. Under Development in Kenya. London: Heinemann. Educational Books.

Lipsky, G.A. 1962. Ethiopia- its People, its Society, its Culture. HRAF press, New Haven, Conn.

Lipton, M. 1968. The Theory of Optimizing Peasants. Journal of Development Studies, 4:327-351.

Lipton, M. 1977. Why Poor People Stay Poor? Cambridge: Harvard university Press.

Lirenso, A. 1986. Food Marketing and Pricing. Paper presented at national workshop on food strategies for Ethiopia, Alemaya University, December.

Lockheed, M.E., D.T. Jamison, and L.T. Lau. 1980. Farmer Education and Farmer Efficiency: a Survey. Econ. Dev. and Cult. Change, 29:37-76

Long, N. 1977. Introduction to the Sociology of Rural Development. London: Tavistock.

Longo, R.M. Juliano. 1990. Information Transfer and the Adoption of Agricultural innovations. J. of the American Society for Information Science. 41(1):1-9.

Lutz, M.A. and K. Lux, 1979. The Challenge of Humanistic Economics. Benjamin/Cummings, Menlo Park.

Maddala, G.S. 1977. Econometrics. New York: McGraw-Hill.

Maddala, G.S., 1983. Limited-Dependent and Qualitative Variables in Econometrics. Cambridge University Press, New York.

Malthus, T.R. 1933. Essay on Population. NY. Dutton

Marx, K. 1970. Contribution to the Critique of Political Economy. NY: International Publishers.

Mason, R. 1963. The Use of Information Sources by Influentials in the Adoption Process. Public Opinion Quarterly. 27:456-466.

Mason, R. and A.N. Halter. 1980. Risk Attitude and the Forced Discontinuance of Agricultural Practices. Rural Sociology. 45(3):435-447.

Mauss, M. 1966. The Gift. London: Cohen and West Ltd.

McCann, J., 1986. Oxfam Ox/Seed Project Evaluation. Boston University.

McClelland, D.C. 1961. The Achieving Society. NY.: D. Van Nostrand and Co.

McNamara, R.S. 1985. The Challenges for Sub-Saharan Africa. Sir John CrawFord Memorial Lecture, CGIAR, WB, Washington, D.C.

Meeusen, W., Broeck, J. van den. 1977. Efficiency Estimation from Cobb-Douglas Functions with Composed Error. International Economic Review. 81:435-444.

Meier, G. 1984. Leading Issues in Development Economics (ed). 4th edition.

Meier, G.M. 1976. Leading Issues in Economic Development. 3rd ed. Oxford University Press: New York.

Meillassoux, C. 1981. Maidens, Meals, and Money. Cambridge: Cambridge University Press.

Meillessoux, C. 1972. From Reproduction to Production: A Marxist Approach to Economic Anthropology. Economy and Society 1:93-105.

Mekonnen, G. 1983. Preliminary Evaluation of Holstein Breed and Their Half-Breeds for Milk Production. Ethiopian Journal of Agricultural Sciences. V(1):43-49.

Mellor, J.W. 1966. The Economics of Agricultural Development. Cornell University Press, Ithaca.

Mellor, J.W. 1969. In Wharton, C.R. Jr. Subsistence Agriculture and Economic Development. Aldine Publishing Company, Chicago. pp. 209-227.

Menger. K. 1950. Principles of Economics. Gelncoe, Illinois: Free Press.

Messing, S. 1962. The Abyssinian Markets. In P. Bohannan and Dalton (eds). Markets In Africa. Northwestern University Press. pp. 386-408.

Michie, B.H. 1986. Indigenous Technology and FSR: Agro-forestry in Indian Desert. In Jones, J.R., and B.J. Wallace (eds). Social Science and farming Systems research: Methodological Perspectives on Agricultural Development. West View Press, Boulder. pp.221-244.

Ministry of Agriculture, 1982. Data Book on Land Use and Agriculture in Ethiopia, Vol. I and II. Ethiopia, Addis Ababa.

Ministry of Agriculture. 1985, 1986. Annual Reports of Project Planning division, Addis Ababa, Ethiopia.

MOA, 1982, 1984, 1989, 1990. Annual Report, Addis Ababa, Ethiopia.

Molnar, J.J., and H.A Clonts, 1983. Transferring Food Production Technology to Developing Nations: Economic and Social Dimensions, (eds). Colorado: West View Press.

Moock, J.L. 1986. Understanding Africa's Rural Households and Farming Systems (ed). West View, Boulder.

Moock, P.R. 1981. Education and Technical Efficiency in Small Farm Production. Econ. Dev. and Cult. Change, 29(4):723-739.

Moore, S.F. 1963. Social Change. Englewood Cliffs, N.J.: Prentice Hall.

Morrison, Denten E., K. Krishna, E.M. Rogers and F.C. Fliegel. 1976. Stratification and Risk-Taking: A Further negative replication of Cancian's Theory. American Rural Sociology Review, 41:1083-89.

Moscardi, E.R. 1979. Methodology to Study Attitudes Towards Risk: The Puebela Project. In A. Valdes, G.M. Scobie and J.L. Dillon (eds) Economics and Design of Small Farmer Technology. Ames: Iowa State University press, pp.30-47.

Myint, H. 1975. Agriculture and Economic Development in the Open Economy. In L.G. Reynolds (ed), Agriculture in development Theory. Yale University Press, New haven.

Myrdal, G. 1968. The Asian Drama: An Inquiry in to the Poverty of Nations. NEw York: Pantheon

Nair, K. 1962. Blossoms in the Desert: The Human Factor in Indian development. NY: Praeger.

National Revolutionary Development Campaign and Central Planning Supreme Council (NRDC/CPSC), 1982. Evaluation on the Objectives and Future Prospects of the ten-year agricultural Development Plan 1984-1994. A Committee report. Addis Ababa (Amharic Version).

Netting, R. McC. Household Dynamics in a Nineteenth-Century Swiss Village. J. of Family History. 4:39-58.

Netting, R. McC. 1982. Some Home truths on household size and wealth. American Behavioral Scientist, 25:641-61.

Netting, R. McC., R.R. Wilk, and E.J. Arnould. 1984 (ed). Households: Comparative and Historical Studies of the Domestic Group. Berkeley, University of California Press.1984

Nixson, F. 1982. Import Substitution Industrialization. In M. Fransman (ed). Industry and Accumulation in Africa. London: Heinemann.

Norem, R.H., R. Yoder, and Y. Martin. 1988. Indigenous Agricultural Knowledge and Gender Issues in Third World Agricultural Development. Prepared for the Joint Meetings of the Society of Social Studies of Science and the European Association of Science and Technology.

Norman, D.W., M. Newman, and I. Quedraogo. 1981. Farm and Village Production Systems in the Semi-Arid of West Africa. ICRISAT.

Norman, D.W. 1978., Farming System Research to Improve the Livelihood of Small Farmers. American Journal of Agricultural Economics, 60:813-818.

Norman, D.W. 1980. Rationalizing Mixed Cropping Under Indigenous Conditions: The Examples of Northern Nigeria. J. of Development Studies. 11(1):3-21.

Nurkse, N. 1959. Patterns of Trade and Development. Stockholm. Almquist and Wicksell.

O'Mara, G., 1983. The Micro-Economics of Techniques Adoption by Small Holding Mexican Farmers. In R.D. Norton and L. Solis (editors), The Book of Chac: Programming Studies for Mexican Agriculture. The John Hopkins University Press, Baltimore, pp.250-289.

Ortiz, S. 1980. Forecasts, Decisions, and Farmer's response to Uncertain Environments. P.F. Barlett (ed), Agricultural Decision Making: Anthropological Contribution to Rural Development, New York: Academic Press. pp.177-202.

Ortiz, S. 1983. What is decision analysis about? The Problem of Formal Representation. In S. Ortiz (ed). Economic Anthropology: Topics and Theories. Monograph in Economic Anthropology, No. 1., NY.: University Press of America. pp.249-300.

Pankhurst, R. 1966. State and Land in Ethiopia (ed). Addis Ababa, Oxford University press.

Pankhurst, R. 1967. Ethiopian Royal Chronicles. Addis Ababa: Oxford University Press.

Pankhurst, R. 1961. An Introduction to Economic History of Ethiopia. London: Lalibela House.

Parkin, D. 1972. Palms, Wine and Witnesses. San Francisco: Chandler.

Parsons, T., and E.A. Shils, 1951. Towards a General Theory of Action (eds). Cambridge, Mass.: Harvard University Press.

Perrin, R.K., D.L. Winkelman, E.R. Moscardi and J.R. Anderson, 1976. From agronomic data for farmer recommendation: An Economics training manual. Information bulletin no.27. Mexico: Cimmyt.

Peters, P.E. 1986. Household Management in Botswana: Cattle, Crops and Wage Labor. In J.L. Moock (ed). Understanding Africa's Rural Households and Farming Systems. West View, Boulder. pp. 133-145.

Peterson, W. and Y. Hayami, 1977. Technical Change in Agriculture, In L.R. Martin (ed) A Survey of Agricultural Economics Literature, Minneapolis: University of Minnesota Press, Vol. II, 1977, pp.497-540.

Piaget, J. 1972. Psychology of Intelligence. Littelfiled, Adams and Co. Totowa, NJ.

Pindyck, R.S. and D.L. Rubinfeld. 1981. Econometric Models and Econometric Forecasts. Second Edition., New York: McGraw Hill.

PMAC, 1977. Basic Documents of the Ethiopian revolution. Addis Ababa: Berhanna Selam Press.

PMAC, 1976. A Proclamation to Provide for the Rural Land Use Payment and Agricultural Activities Income tax. Negarit Gazeta, Proclamation no.77.

PMAC, 1975. Public Ownership of Rural Lands. PMAC Proclamation no.31, Addis Ababa, Ethiopia.

PMAC. Proclamation to provide for the Organization and Consolidation of Peasant Associations. Negarit Gazeta, Proclamation No. 71.

Polyani, K. 1944. The great Transformation. New York: Rinehart and Co., Inc.

Polyani, K., C.M. Arensberg, and H.W. Pearson. 1957. Trade and Market in the Early Empires. Glencoe, Ill: Free Press.

Prebish, R. 1964. Towards a New Trade Policy for Development. UNCTAD, United Nations, Geneva.

Putterman, L. 1985. Extrinsic Versus Intrinsic Problems of Agricultural Cooperation: Anti-Incentivism in Tanzania and China. J. of Development Studies. 21(2):175-204

Rahm, M.R., and Huffman, W.E., 1984. Adoption of Reduced Tillage: the Role of Human Capital and other Variables. Amer. J. Agr. Econ. 66:405-413.

Redfield, R. 1956. Peasant Society and Culture: An Anthropological Approach to Civilization. The University of Chicago Press.

Rey, R.P. 1975. The Lineage Mode of Production: Critique of Anthropology. 3:27-29.

Rhamato, D. 1984. Agrarian Reform in Ethiopia. Trenton: The Red Sea Press.

Richards, P. 1985. Indigenous Agricultural Revolution. Hutchinson, London.

Ricket, James. 1991. Economic Development in Ethiopia: Agriculture, the market and the Sate, OECD, Paris.

Roe, T., 1983. Risks and Information: Farm Level Impediments to Transforming Traditional Agriculture. In J.J. Molnar and A.A. Clonts (editors), Transferring Food Production Technologies to Developing Nations. West View Press, Colorado, pp.85-105.

Rogers, M. 1971. Communication of Innovations. 2nd ed. NY: Free Press

Rogers, E.M. and F.F. Shoemaker. 1971. Communications of Innovations. New York: Free Press.

Rogers, E.M. 1983. Diffusion of Innovations. 3rd. ed. New York: The Free Press, Collier, Macmillan.

Rostow, W.W. 1960. The Stage of Economic Growth: A Non-Communist Manifesto. London: Cambridge University Press.

Roumasset, J. 1979. Introduction and States of the Arts. In J. Roumasset, J.M. Boussard and I. Singh (eds). Risk, Uncertainty and Agricultural Development. Agricultural Development Council, NY. pp.1-22.

Roumasset, J.A. 1976. Rice and risk: Decision making among low income farmers. New York: North-Holland publishing company,1976.

Roumasset, J.A. 1979. Unimportance of risk for technology design and agricultural development policy." In A. Valdes, G.M. Scobie, and J.L. Dillon (eds). Economics and the Design of Small-Farmer Technology, Iowa State University Press, Ames, Iowa, pp.48-67.

Roy, A.D. 1952. Safety-First and the Holding of Assets. Econometrica. 20(3):431-449.

Rubenson, S. 1984. Proceedings of the Seventh International Conference of Ethiopian Studies (ed). University of Lund, 26-29 August, 1982. Berlings, Arlov, Sweden, 1984.

Rubenstein, H. 1975. The Family as a Non-group: Domestic Organization in an Eastern Caribbean Village. In proceedings of the Congress of the Canadian Ethnology Society II, pp.311-23. Ottawa: National Museum of Canada.

Ruttan, V. 1977. The green revolution: seven generalizations. Int. De. Rev. 19:16-23.

Ruttan, V.W. 1978. Induced Institutional Change. In H.P. Binswagner and V.W. Ruttan (eds), Induced Innovation: Technology, Institutions and Development. Baltimore: The John Hopkins University Press.

Rymon, D., and U. Or. 1991. Accelerating Technology Transfer by Means of ATTA. J. of Sustainable Agriculture. 2(1):103-118.

Sahlins, M. D. 1976. Culture and Practical Reason. Chicago University Press, Chicago..

Sahlins, M. 1972. Stone Age Economics. Chicago: Aldine.

Saith, A. 1985. The Distributional Dimensions of Revolutionary Transition: Ethiopia. J. of Development Studies. 22(1):150-79.

Sarris, W.E. 1984. Causal Modeling in Non-Experimental Research: An Introduction to the LISREL Approach. Amsterdam: Sociometrics Research Foundation.

Schejtman, A. 1988. The Peasant Economy: Internal Logic, Articulation and Persistence. In Wilber, C.K. 1988. (ed). The Political Economy of Development and Underdevelopment. 4th ed. Random House. pp.364-392.

Schluter, M.G.G. and T.D. Mount. 1976. Some Management Objectives of the Peasant Farmer: An Analysis of Risk Aversion in the Choice of Cropping pattern, Surat District, India. J. of Development Studies. 12:246-261.

Schmidt, P. 1985-86. Frontier Production Functions. Econometric Reviews. 4:289-328.

Schultz, T.P. 1981. Economics of Population. Reading, Mass.: Addison-Wesley.

Schultz, T.W. 1980. The Economics of Being Poor: Noble Lecture. Journal of Political Economy, Vol.88, pp.639-651.

Schultz, T. 1975. The value of the ability to deal with disequilibrium. J. Econ. Liter. 13:827-826.

Schultz, T.W. 1953. The Economic Organization of Agriculture. NY: McGraw-Hill.

Schultz, T.W. 1964. Transformation of Traditional Agriculture. Yale University Press. New Haven.

Schultz, T.W. 1978. (ed). Distortions of Agricultural Incentives. Indiana University Press. Bloomington.

Scott, J.C. 1975. The moral Economy of the Peasant: Rebellion and Subsistence in Southeast Asia. New Haven, CT: Yale University Press.

Seddon, D. Aspects of Kinship and Family Structure among the Ulad Stut of Zaio Rural Commune, Nador Province, Morocco. In Mediterranean Family Structure, ed. J. G. Peristiany, pp.173-94. Cambridge: Cambridge University Press.

Sen, A. K. 1984. Poverty and Famines: An Essay on Entitlement and Deprivation. Oxford University Press, New York.

Sengupta, J.K. 1987. Stochastic Optimization and Economic Models. Boston, MA. D. Reidel.

Sengupta, J.K. 1989. Efficiency Analysis by Production Frontiers: The Non-Parametric Approach. Kluwer Academic Publishers. Dordrecht: The Netherlands.

Sexton, T.R. 1986. The Methodology of Data Envelopment Analysis, In(ed). R.H. Silkman. Measuring Technical Efficiency: An Assessment of Data Envelopment Analysis, San Francisco, CA: Jossey-Bass.

Shahabudin, Q., and S. Mestelman. 1985/86. Uncertainty and Disaster Avoidance Behavior in Peasant farming: Evidence from Bangladesh. J. of development Studies. 22:740-752.

Shakya, P.B., and J.C. Flinn, 1985. Adoption of Modern Varieties and Fertilizer Use on Rice in the Eastern Terrai of Nepal. J. of Agr. Econ.36:409-419.

Shaner, W.W., P.F. Philipp., and W.R. Schmehl 1982. Farming Systems Research and Development: Guidelines for Developing Countries. Colorado: West View Press.

Short, E.C. 1973. Knowledge Production and Utilization in Curriculum. A Special case of the General Phenomenon. Review of Educational Research. 73:237-301.

Silkman, R.H. 1986. Measuring Technical Efficiency: An Assessment of Data Envelopment Analysis (ed), San Francisco, CA: Jossey-Bass.

Sisay, S. 1980. Agricultural Systems in Ethiopia: A Review Article. Agricultural Systems. 5(1):29-30.

Sisaye, S. and E. Stommes 1978.. Agricultural Development in Ethiopia: Government Budgeting and development Assistance in the Pre and Post 1975 Periods. J. of Development Studies, pp.156-185.

Smith, P. 1990. The Use of Performance Indicators in the Public Sector. J. of the Royal Statistical Society. A153:53-72.

Smith, et al., 1984. Introduction. In J. Smith, I. Wallerstein and H. Evers (eds). Households and the World Economy. Beverly Hills: Sage Publications. pp.7-13.

Smith, A. 1776, The Wealth of Nations (ed) E. Cannan. New York: The Modern Library, 1937.

Smith, C.A. 1983. Regional Analysis in World-System Perspective: A Critique of Structural Theories of Uneven Development. In S. Ortiz (ed). Economic Anthropology: Topics and Theories. Lanham, MD: University Press of America. pp.

Solow, R.M. 1957. Technical Change and Aggregate Production Function. Review of Economics and Statistics. 39:3(312-20.

Somer, L. 1954. Exposition of a New Theory of the Measurement of Risk. Econometrica. 22:23-36.

Sperling, L. 1986. The Labor Organization of Samburu Pastoralism Unpublished Dissertation, Department of Anthropology, McGill University.

Spijkers-Zwart, S.I. 1979. The Household and Householding; Some Conceptual Considerations. In Ed. by Clio Presvelou and Saskia Spijkers-Zwart, "The Household, Women, and Agricultural Development. Proceeding of a symposium organized by the department of Home Economics, Agricultural University of Wageningen, The Netherlands, 18-20 January, 1979. Veenman and Zonen B.V.-Wageningen, 1980. pp.69-74.

Stevens, R.D. and C.L. Jabara. 1988. Agricultural development Principles: Economic Theory and Empirical Evidence. The John Hopkins University Press.

Stevenson, R.E. 1980. Likelihood Functions for Generalized Stochastic Frontier Estimation. J. of Econometrics. 2:367-374.

Steward, J. S. 1955. The Theory of Culture Change. Urbana: University of Illinois Press.

Streeten, P.P. 1981. The Rise and Fall of Development Economics; In Essays in Trespassing: Economics to Politics and beyond, New York: Cambridge University Press.

Streeten, P. 1982. A Look at the Outward-Looking Strategies for development. World Economy. Vol. 5, No.2.

Tagoe,, C. 1983. Ethiopian Highlands Reclamation Study: A Tentative Review of Agriculture in the Highlands. Ministry of Agriculture and FAO, Addis Ababa, Ethiopia.

Taylor, T.G., E. Drummond, and A.T. Gomes, 1986. Agricultural Credit Programs and Production Efficiency: An Analysis of Traditional Farming in Southeastern Minas Gerais, Brazil. Amer. J. of Agri. Econ. 68(1):110-119.

Taylor, T.G. and Shonkwiler, S.J. 1986. Alternative Stochastic Specification of the Frontier Production Function in the Analysis of Agricultural Credit Programs and Technical Efficiency. J. of Development Economics. 21:149-160.

Tecle, T., 1975. Application of Probit Analysis to Adoption of New Agricultural Practices. Ethiopian J. of Dev. Research, 2:501-522.

Tecle, T. 1975. An Economic Evaluation of Agricultural package Programmes in Ethiopia, Ph.D. Dissertation, Cornell University.

Teka, T. and N. Tenassie. 1984. Rural Poverty Alleviation: The Case of Ethiopia. FAO, Rome.

Theil, H. 1971. Principles of Econometrics. New York: Wiley.

Thrupp, L.A. 1988. Legitimizing Local Knowledge: Scientized Packages or Empowerment of Third World People. In D.M. Warren, L. Jan Slikkerveer, and S.O. Titilola. (eds) 1988. Indigenous Knowledge Systems: Implications for Agricultural and International Development. Studies in Technology and Social Change, No.11, Iowa State University, Ames, Iowa. pp.138-153.

Tittarelli, F. 1990. Smallholder Dairy Marketing Pattern in Central Ethiopian Highland, Selale Awaraja: A Case Study. Centro per lo Sviluppo Agricolo e Rurale, Assisi, Italy.

Toborn, J. 1978. Farmers Use of fertilizers: A Survey in Two Extension Areas. Addis Ababa, March.

Todaro, M.P. 1989. Economic development in the Third World. New York: Longman. (4th ed).

Tversky, A. 1972. Elimination by Aspects: A Theory of Choice. Psychological Review. 79(4):281-291.

Ullah, A. 1988. Nonparametric Estimation and Hypothesis Testing in Econometric Models. Empec, 13:223-249.

Uphoff, N.T., and W.F. Ilchman. 1972. The Political Economy of Development. Berkeley and Los Angeles: University of California Press.

Valdes, A. G.M. Scobie, and J.L. Dillon. 1979. Economics and Design of Small Farmer Technology (eds). Ames: Iowa State University Press.

Verner, C. and P.M. Gubbels. 1967. The Adoption or Rejection of Innovations by Dairy Farm operators in the Lower Fraser Valley. The University of British Columbia. The Agricultural Economics Research Council of Canada. Canada.

Vierich, H. 1986. Agricultural Production, Social Status and Intra-Compound Relationships. In J.L. Moock (ed). 1986. Understanding Africa's Rural Households and Farming Systems. West View, Boulder. pp. 155-164.

Vinod, H.D. and A. Ullah. 1986. Flexible Production Function Estimation by Nonparametric Kernel Estimators. Paper Presented at a Canadian Econometric Study group meeting at Montreal, Canada.

Waktola, A. 1980. Assessment of the diffusion and adoption of agricultural technologies in Chilalo, Ethiopia. Eth. J. Agri. Sci. 2:51-68.

Waktola, A. 1975. Assessment of the Development of Diffusion and Adoption of Packages of Agricultural Innovations in Chilalo, Ethiopia. Ph.D. Dissertation, The Ohio State University.

Waldman, D.M. 1984. Properties of Technical Efficiency Estimators in the Stochastic Frontier Model. J. of Econometrics. 25:353-364.

Wallerstein, I. 1974. The Modern World System. NY.: Academic Press.

Wallerstien, I. 1979. The Capitalist World-Economy. Cambridge.: Cambridge University press.

Ward, R.D. and A. Zahalka 1983. Evaluation of Haitian Agricultural Development With the Use of Principal Component. J. Agr. Econ. 35:243-55.

Warren, D.M., L. Jan Slikkerveer, and S.O. Titilola. 1988. Indigenous Knowledge Systems: Implications for Agricultural and International Development. Studies in Technology and Social Change, No.11, Iowa State University, Ames, Iowa.

Warren, D.M. 1988. Linking Scientific and Indigenous Agricultural System. In J.L. Compton (ed). The Transformation of International Agricultural Research and Development. Boulder: West View Press.

Webb, P. 1988. Intra-household Decision Making and Resource Control: The Impact of New Rice Production Technology and Commercialization in a West African Setting. IFPRI, Washington, June 1988.

Weber, M. 1947. The Theory of Social and Economic Organization. NY: Free Press.

Weins, T. Peasant Risk Aversion and Allocative Behavior: A Quadratic programming Experiment. American Journal of Agricultural Economics. 58:629-635.

Wharton, C.R. Jr. 1969. Subsistence Agriculture and Economic Development. Aldine Publishing Company, Chicago.

Wharton, C.R. Jr. 1971. Risk, Uncertainty and the subsistence farmer: Technological Innovation and Resistance to Change in the Context of Survival. In G. Dalton (ed). Studies in Economic Anthropology. Washington, DC: American Anthropology Association, Anthropological Studies, No.7, pp.151-178

Wibe, S. 1984. Engineering Production Functions: A Survey. Economica, 51:401-411.

Wiessner, 1982. Beyond Willow Smoke and Dogs' Tails: A comment on Binford's Analysis of Hunterer-Gatherer Settlement Systems. American Antiquity, 47:171-178.

Wilber, C.K. 1988. The Political Economy of Development and Underdevelopment (ed). 4th ed. Random House.

Wilk, R.R., and R. McC Netting. 1984. Households: Changing Forms and Functions. In Netting, R. McC., R.R. Wilk, and E.J. Arnould. (ed). Households: Comparative and Historical Studies of the Domestic Group. Berkeley, University of California Press. pp.1-28.

Wilk, R. R. 1984. Households in Process: Agricultural Change and Domestic Transformation among the Kekchi Maya of Belize. In Netting, R. McC., R.R. Wilk, and E.J. Arnould. 1984 (ed). Households: Comparative and Historical Studies of the Domestic Group. Berkeley, University of California Press. pp. 217-244,

Wilk, R.R. 1989 (ed).The Household Economy: Reconsidering the Domestic Mode of Production. Westview Press, Boulder.1989.

Wilk, R.R. 1989. Decision Making and Resource Flows within the household: Beyond the black Box. R.R. Wilk. pp.23-54.

Wilk, R. 1991. The Household in Anthropology: Panacea or Problem? Reviews in Anthropology, 20:1-12.

Wolde-Mariam, M. 1986. Food security problems and prospects in Ethiopia. Paper presented at national workshop on food strategies for Ethiopia, Alemaya University, December 1986.

Wolde-Mariam, M. 1984. Rural Vulnerability to Famine in Ethiopia, 1958-77. Vikas Publishing House ltd.

Wolf, C.J. 1955. Institutions and Economic Development. American Economic Review. 45:867-83.

Wolgin, J. 1975. Resource Allocation and Risk: a Case Study of Small Holder Agriculture in Kenya. Am. J. Agri. Econ. 57:622-30.

World Development Report. 1981, 1982, 1983, 1986, 1988. Oxford University Press. Published for IBRD.

World Bank, 1981, 1982, 1984. Country Reports for Kenya, Sudan and Tanzania.

World Bank. 1983. Ethiopia: Recent Economic Developments and Future Prospects. Volume I.

World Bank, 1983, 1984, 1986, 1988. Annual report.

Wubineh, M. and Y. Abate. 1988. Ethiopia: Transition and Development in the Horn of Africa. Boulder, Col.: Westview.

About the Author

Yohannes K.G. Mariam (AKA Yohannes Kebede or Yohannes Kebede Gebremariam) is an environmental and energy economist, and social scientist. He received his B.Sc. degree in agricultural economics from Addis Ababa University (now Alemaya University, Ethiopia), his M.Sc. degree in agricultural economics and his Ph.D. degree from a multidisciplinary program involving the departments of agricultural economics, economics, Anthropology and applied cognitive psychology from McGill University, Montreal, Canada. He worked as environmental economist and social scientist for Environment Canada from 1994-1999, as a senior economist for Washington Utilities and Transportation Commission from 1999-2007, and as a senior economist for Government of the District of Columbia since 2007. He is also an adjunct professor economics and management with University of Phoenix and Florida Keys Community College.